Process Modelling and Simulation in Chemical, Biochemical and Environmental Engineering

Process Modelling and Simulation in Chemical, Biochemical and Environmental Engineering

Ashok Kumar Verma

CRC Press
Taylor & Francis Group
Boca Raton London New York

CRC Press is an imprint of the
Taylor & Francis Group, an **informa** business

CRC Press
Taylor & Francis Group
6000 Broken Sound Parkway NW, Suite 300
Boca Raton, FL 33487-2742

© 2015 by Taylor & Francis Group, LLC
CRC Press is an imprint of Taylor & Francis Group, an Informa business

No claim to original U.S. Government works

Printed on acid-free paper
Version Date: 20141002

International Standard Book Number-13: 978-1-4822-0592-3 (Hardback)

Library of Congress Cataloging-in-Publication Data

Verma, Ashok Kumar.
  Process modeling and simulation in chemical, biochemical, and environmental engineering / author, Ashok Kumar Verma.
       pages cm
  Includes bibliographical references and index.
  ISBN 978-1-4822-0592-3 (hardback)
  1. Chemical processes--Computer simulation. 2. Chemical engineering--Mathematics. I. Title.

TP155.7.V47 2014
660'.28--dc23   2014034065

Visit the Taylor & Francis Web site at
http://www.taylorandfrancis.com

and the CRC Press Web site at
http://www.crcpress.com

Dedicated to my wife, Geeta Verma,

and sons, Anurag Srivastava and Anupam Srivastava.

Contents

Preface

The quest for understanding various phenomena is human nature, and the development of science and engineering is based on it. Observation, searching for a trend in it, is followed by invention. Theoretical knowledge helps in predicting possible phenomena without carrying out an experiment. Chemical engineers are involved in the safe and economic design and operation of process plants and in increasing the productivity of existing process plants.

Today, simulation is an accepted way of studying various process industries. It may be carried out to study the behaviour of unit operations or a plant without fabricating any equipment, or to study the entire plant without any real loss or damage to the real equipment or environment. However, simulation requires a process model, which is a mathematical representation of a simplified picture of the process. Simulation studies are carried out by experimenting with the model. Thus, modelling and simulation both exist together.

The engineering of process systems was initially qualitative in nature. At present, it is founded on engineering sciences, computer-aided tools and the system-theoretical methodologies. The use of simulation has helped in developing an integrated approach to process design. The effects of environmental issues, choice of the controller, process safety and so forth on the economic design of the process plant can now be considered.

Earlier modelling efforts focused on developing analytical models. Various mathematical techniques were applied to obtain analytical solutions of various chemical engineering problems. Later, various numerical techniques were also applied to solve more rigorous problems. The books on modelling during those days covered these topics comprehensively. At present, many of these techniques such as estimation of the inverse of a matrix, solution of non-linear equations, numerical integration and so on are available with many types of software. The focus on modelling is now to understand the process in more detail and represent it in mathematical form. This book is written with this aim. It attempts to explain the simplification of a complicated process at various levels with the help of a 'model sketch' and to proceed in a systematic manner.

The models have been divided into four categories. The simple models are those which are based on simple laws such as Fick's law. The next category of models consists of generalised equations such as equations of motion. These deterministic models do not have any random phenomena and consider all variables as continuous variables. The next category covers discrete-event models and stochastic models which consider at least one variable as a discrete variable. The models based on population balance have also been covered with it. The boundary between these models is not sharp; however, they certainly reflect increasing levels of difficulty and require different approaches.

The dynamic models have no separate treatment as many good books on dynamic modelling and process control are already available. Computational fluid dynamics has been purposefully left out due to the different nature of the modelling. It is not possible to cover everything related to modelling and simulation in this volume. The field of modelling and simulation is rapidly advancing. However, I hope that the concepts discussed here will be useful in future also.

Most of the software have their own syntax and semantics. Therefore, choosing a software was not an easy task. Finally, it was decided that the examples will be solved in MATLAB®. The code is short, and more attention can be given to the representation of process in mathematical form. The code also runs on open-source SCILAB, with minor modifications if no specific toolbox was used.

The vector notation of laws of conservation has been avoided because it is not intended to illustrate the derivation of the analytical solution. The Nomenclature does not provide unit as empirical correlations are not used in this book. The model equations are independent of units.

Chapter 1 introduces the complex nature of chemical processes and an introduction to the terms 'modelling' and 'simulation'. The advantages of simulation and the role of modelling in simulation have been studied.

Chapter 2 describes steps in simulation, various types of models, the approach to modelling, types of model equations, solution strategies, sources of equations and some of the common assumptions that are made to simplify the complexity of the process.

The few simple models that are still useful have been described in Chapter 3. The equation of state, Newton's law, Fourier's law, Ficks' law, Henry's law, Arrhenius law and adsorption isotherms, film model were discussed and their application was illustrated in few simple problems. In Chapter 4, the models based on laws of conservation of momentum, mass and heat transfer have been described. These models are applicable to systems involving a single phase only. Chapters 5 and 6 cover the models based on laws of conservation in multiphase systems without and with chemical reactions. The examples include the stationary dispersed phase (packed beds) and moving dispersed phase (fluidised beds and bubble columns).

Models based on population balance, discrete-event models and stochastic models have been covered in Chapter 7. These models generally involve random variables to describe discrete or even continuous variables. The simulation strategy is different in the sense that no analytical or numerical solution is sought as in the case of deterministic models.

The artificial neural network (ANN)–based models follow a black-box approach. They relate the data through a network of nodes and have a different nature. Feedforward-type ANNs and hybrid-type ANNs, which combine the advantage of the first principle model, have been described in Chapter 8.

A model has to be validated for either experimental observations or benchmark problems. After validation, the model is acceptable for simulation purposes. The procedures for model validation and parametric sensitivity analysis have been presented in Chapter 9.

To illustrate the model development methodology, four case studies have been presented in Chapter 10: an analytical model, a numerical model, a stochastic model and an ANN-based model. The detailed reasoning while making assumptions, the development of model equations and simulation methodology are discussed. The second example of a numerical model, the breaking of a model into several sub-models during model development, was discussed. The complete MATLAB code is given in Appendix B.

The last chapter, Chapter 11, discusses the various approaches used for simulation of a large plant. The methodology needed to break down a large problem into sub-problems, the handling of recycle streams and other topics are discussed. A brief discussion on the nature of the problem related to batch process scheduling is included.

The techniques for solving model equations are outside of the scope of the book. Hence, various topics on solving differential equations, optimisation, genetic algorithms and so on have not been included.

It was not possible to include every good model in this book. Only those models were chosen which I could access and for which I found a fit at a right place in the manuscript.

I hope that this book will be helpful to senior undergraduate and postgraduate students and fulfil the expectations of its readers.

Ashok Kumar Verma

MATLAB® is a registered trademark of The MathWorks, Inc. For product information, please contact:

The MathWorks, Inc., 3 Apple Hill Drive, Natick, MA 01760, USA,
Tel: 508-647-7000; Fax: 508-647-7001;
E-mail: info@mathworks.com; Web: www.mathworks.com

Acknowledgements

I sincerely thank Professor R. Kumar of the Indian Institute of Sciences, Bangalore, for introducing the subject when I was a student there.

I acknowledge Professor D. P. Rao of the Indian Institute of Technology, Kanpur, with whom the stochastic model presented in Chapter 10 was developed during my PhD programme.

I also acknowledge Ankit, Department of Chemical Engineering & Technology, IIT (BHU), Varanasi, who helped in keeping my laptop and desktop in working condition and in searching old volumes of journals. My son, Anupam Srivastava, was also always helpful in such cases by searching solutions and helping me, even at odd hours.

It is a pleasure to acknowledge my wife Geeta's efforts in always encouraging me to complete the manuscript by ensuring that I get maximum time for this work.

I sincerely thank the entire team of editors, especially Dr. Gagandeep Singh, Kate Gallo and Christian Munoz, who were always helpful in bringing the book in this form. Thanks also to Viswanath Prasanna for a wonderful job with the typesetting.

Ashok Kumar Verma
Indian Institute of Technology (Banaras Hindu University), Varanasi

Author

Ashok Kumar Verma is a professor in the Department of Chemical Engineering and Technology at the Indian Institute of Technology (Banaras Hindu University) Varanasi. He holds a BSc degree from Allahabad University, a BE degree in chemical engineering from University of Roorkee (now Indian Institute of Technology, Roorkee), an ME degree in chemical engineering from the Indian Institute of Sciences, Bangalore, and a PhD in chemical engineering from the Indian Institute of Technology, Kanpur.

Dr Verma joined the Institute of Technology, Banaras Hindu University, Varanasi [now known as Indian Institute of Technology (BHU) Varanasi] in 1984. He carried out postdoctoral research at the Department of Chemical Engineering, University of Illinois, Chicago. His research experience includes modelling liquid fluidised beds and heat transfer in slurry bubble columns. He has taught courses in modelling and simulation, process engineering, computer-aided design, artificial intelligence in chemical engineering, transfer processes and process control. His current research areas are chemical microprocessing and modelling of microreactors.

Dr Verma has authored or co-authored about 10 research papers that have appeared in a range of chemical engineering journals. He also authored or co-authored numerous papers in national and international proceedings. He has made presentations at several conferences and seminars and presented invited lectures at several workshops organised at other universities, institutions and industries.

Dr Verma and his family reside in Varanasi, India. He can be reached at akverma.che@itbhu.ac.in or vermaakg@rediffmail.com.

Nomenclature

a	parameter in equation of state
a_i	interfacial area per unit volume
$a\Delta t$	probability of bubble interacting with fluid
a_c	area of catalyst surface per unit volume
a_{ij}	parameter in equation of state
A	chemical species, cross-sectional area, constant in an equation, area of heat or mass transfer
A_1	constant
Ar	Archimedes number, Ar, is given by $\mathrm{Ar} = \dfrac{\rho_g \left(\rho_p - \rho_g\right) g d_p^3}{\mu_g^2}$
A_s	area of the interface or transfer surface
b_i	difference in gas total stoichiometric coefficients due to chemical reaction
b_j	bias of the jth neuron
b, b_{ij}	parameter in equation of state
B	chemical species, constant in an equation
Bi	Biot number based on k_r
c	decay constant
$c\,(d_i, d_j)$	coalescence rate, product of the collision frequency and coalescence efficiency for dispersed-phase units of sizes d_i and d_j
C_1, C_2	constant obtained while solving a differential equation
C_A	concentration of chemical species A
C_{A0}	initial concentration of chemical species A
C_{Ag}, C_{Al}	concentrations of A in bulk of gas and liquid phases, respectively
C_{Al0}, C_{Bl0}	concentrations of A and B, respectively, in the liquid phase at the inlet
C_{AM}, C_{BM}	concentrations of species A and B in the membrane
C_D	drag coefficient
C_p	specific heat
C_s	concentration of solids in slurry at any axial position z, mass-solid/slurry volume

$\bar{C}_A, \bar{C}_B$	average concentrations of A and B in the catalyst surface, respectively
$\bar{C}_s$	average solid concentration at the bottom
$C_{i0}^{aq}, C_{i0}^{org}$	initial concentrations of ith species in aqueous and organic layers, respectively
C_s^0	local solid concentration at the bottom
C_A^*	equilibrium concentration of A at the interface
C_{As}^s	concentration of A at the surface of the solid
C_s^s	concentration of the crystals in the suspension, concentration at the surface of the biofilm at any axial position
d	shortest distance between particle and wall
d_{av}	average bubble diameter
d_b	bubble diameter
d_e	eddy size
d_o	hole diameter in sparger
d_p	particle diameter
d_r	inter-particle distance in the fluidised bed
D	column diameter
D_{AB}	diffusivity or diffusion coefficient
D_{Ag}	diffusion coefficient of A in the gas phase
D_{AK}	Knudsen diffusion coefficients
D_c	critical radius of the particles (i.e. minimum size of crystal which is visible)
D_E	area of zone of sparger plate containing orifice/perimeter of the plate
D_{eff}	effective diffusivity of A in ash layer
D_p	particle diameter
E	activation energy of a reaction, energy dissipation rate per unit mass, turbulence energy spectrum
Eo	Evötös number
E_l	liquid dispersion coefficient
E_s	solid dispersion coefficient
E_y	induced electric field (i.e. electrokinetic potential between the plates)
ΔE	change in energy
f	fraction of the catalyst surface exposed to the gas, activation function, friction factor, fugacity

$f(D)$	number probability density function
$f(x)$	probability distribution function
F_b	buoyancy force
F_B	difference of buoyancy and gravity forces
f_{bp}	volume fraction of the biofilm
F_{gi}	gravitational force
$f(x, D, t)$	number probability density function by particle vector
F_{pi}	pressure force due to hydrostatic head
$F(x)$	function of x
F_D, F_d	drag force acting on the particle
Fr	Froude number
$F_{l,lg}$	interaction forces between the gas and liquid phases
g	acceleration due to gravity
g_{ANN}	function in the form of an artificial neural network
G_1	growth rate of the slowest growing face
$G(D, t)$	growth rate of dispersed system
$\bar{G}_i$	global sensitivity measure
h	heat transfer coefficient
$h(t, z)$	local instantaneous heat transfer coefficient
$h_{0,ij}, h_{f,ij}$	initial and critical film thicknesses, respectively, of the film between nearby bubbles
h_1, h_2	heat transfer coefficients on one side of the heat transfer surface
h_{cgi}, h_{rgi}	heat lost to the ambient air from the top cover by convection and radiation, respectively
h_{cp}	convective heat transfer coefficients between the absorber plate and inner glass plate
h_f, h_p	fluid convective and particle convective heat transfer coefficients, respectively
h_j	heat transfer coefficient between the air and jth particle
$h_{out, k}$	enthalpy of kth inlet stream
$h(D, D')$	collision frequency (i.e. the frequency of collisions between the dispersed-phase units of sizes D and D')
h_r	radiative heat transfer coefficient
h_{rgi}, h_{rp}	radiative heat transfer coefficients between the inner and outer glass plates, and the absorber plate and inner glass plate, respectively
H_i	Henry's constant

H_L	minimum bed height, at which the bubble coalescence is complete and stable slug spacing is achieved
i	chemical species, electrical current
I	improvement coefficient
I_c	conduction current
$I_{j,b}$	improvement coefficient due to the jth sweep of particles in the presence of background fluid mixing
$I_{j,F}(I_{j,b}, t)$	improvement coefficient at time t due to the jth sweep and represents a quantitative measure of a 3D flow field around the particle
I_s	streaming current
J	Jacobian
j	mass diffusion flux
J_A	molar diffusion flux
k	individual mass transfer coefficient, constant
$k(v, v_k)$	size distribution when a particle of size v breaks to form particles of size v_k
k'	kinetic rate expression
k_0	constant
k_{2n}	rate constant for secondary nucleation
k_a	axial thermal diffusivity
k_f, k_g, k_1, k_2	thermal conductivity of fluid
K_g, K_1	overall mass transfer coefficients
k_{gc}	growth rate constant (i.e. the mass transfer coefficient)
k_L	mass transfer coefficient
k_m	thermal conductivity of the medium
k_r	radial thermal diffusivity
KE	kinetic energy
K_a, K_d	rates of adsorption and desorption, respectively
K_{Agl}	mass transfer coefficient at the gas–liquid interface
K_{av}	average mass transfer coefficient
l_{bij}	mean distance between bubbles of sizes d_i and d_j
l'_b	bubble turbulent path length
L	length or height of the column or bed, length of the boundary layer, length parameter, length of pore, pipe, plate and so on
L_1, L_2	width of the solid
m	number of chemical species, mass of the particle or constant

m_1, m_2, m_3, m_4	slope of the lines in a two-film model
M_5	model parameter
M_b	constant
$\bar{M}_i$	global sensitivity measure
M_w	molecular weight of the crystals
n	number of inlet streams, Richardson and Zaki's exponent, number of dispersed-phase units per unit volume, wave number (i.e. number of cycles per unit length), mass flux
n_A	mass flux relative to spatial coordinates
n^0	number density in dilute dispersions
n_B	wave number at which $E(n) = 0$
n_h	number of holes in sparger
n_i	stoichiometry coefficient for ith component
$n_k(x,v_k,t)$	number of bubbles
N	number of chemical species, number of holes in the distributor, ratio of column diameter to particle diameter
N_A, N_B	mass transfer rates of A and B species, respectively; molar flux
N_{Ax}	mass transfer flux of A
N_H	number of nodes in hidden layers
N_I	number of inputs to the artificial neural network
N_i	molar flux relative to fixed spatial coordinates
N_T	number of training data sets
Nu	Nusset number
$\widetilde{N}$	normally distributed random numbers
P	pressure
P_c	critical pressure of gas
P_{ek}	probability that a bubble changes its velocity from v_k to v_e during time interval Δt
Pe_p	Peclet number
Pe_{p1}	Peclet number
Pe_{p1}	Peclet number of particle, $\left[= U_g^c D / E_s \right]$
Pe_{p2}	Peclet number of particle, $\left[= U_g D / E_s \right]$
p_i	partial vapour phase
Pr	Prandtl number
P_T	total pressure

$P_c(d_i, d_j)$	coalescence efficiency $P_c(d_i, d_j)$ for dispersed-phase units of sizes d_i and d_j
p_{Ag}	partial pressure of A in gas phase
$P_b(d_i, d_j)$	dispersed-phase breakup efficiency
ΔP_{mf}	pressure drop in the bed at minimum fluidisation
$(p_B)_{ln}$	logarithmic mean of p_{B1} and p_{B2}
q	heat transfer flux
$\dot{q}$	heat transfer rate
Q	volumetric flow rate, $m^3 \cdot s^{-1}$, solar radiation
q	heat flux, WK^{-1}
q_0	adsorption capacity of the adsorbent
q_{edge}	edge losses from solar collector
q_{top}	heat loss from the top
r	radial position in a cylindrical or spherical coordinate system, measured from the axis or centre, respectively
r_i, r_o	inner and outer radii of a tube
$-r_A$	rate of reaction of A
r_a	radius of ash layer in fluid–solid reaction
$r_{H_2,WS}$	reaction rate due to water-shift reaction
$r_{H_2,FT}$	reaction rate due to Fischer–Tropsch synthesis
$r_{H_2,MF}$	reaction rate due to methanol formation
R	universal gas constant, residual of an equation, electrical resistance
$R_{2n}(t)$	rate of secondary nucleation
Re	Reynold's number
Re_g	Reynolds number of gas $[= D\rho_l U_g / \mu_l]$
Re_p	Reynolds number of particle $[= d_p \rho_l U_{t\infty} / \mu_l]$
R_{gi}	heat transfer resistance between the inner and outer glass plates
R_{go}	heat transfer resistance between the top glass cover and ambient condition
R_{gp}	heat transfer resistance between the absorber plate and inner glass plate
R_i	relative growth rate of the ith face of the crystal
R_{ij}	radius of the film between two adjacent bubbles which may coalesce
R_{ov}	overall resistance
S	concentration of substrate

Sc	Schmidt number
S_i	first-order sensitivity index
S_p	surface per unit volume of spherical particle
St	Stantan number
S_{Ti}	total sensitivity index
$s_{\phi i}^y$	absolute sensitivity
$S_{\phi i}^y$	normalised sensitivity
$S_{i,...,j}$	sensitivity indices
$s_{\phi i}^T$	first-order sensitivity coefficient
t	time, time elapsed since the arrival of the jth particle
t_c	time of contact
$\overline{t_c}$	average time of contact
t_e	time of staying at the transfer surface by the eddy in surface renewal models
t_s	time taken by the particle to move distance y in still fluid
$t_{j,m}$	time at which the maximum value due to the jth particle takes place
T	temperature
$T_a, T_p, T_{gi}, T_{go}, T_s$	temperatures of ambient conditions, absorber plate, inside glass cover, outside glass cover and sky, respectively
T_b	temperature in bulk
T_c	critical temperature of gas
T_D	dew point
T_f, T_s	temperatures in the bulk fluid and at the surface, respectively
T_j	time at which the jth particle arrives at the transfer surface
T_0	initial temperature
T_p	plate temperature
$T_{p(i,j)}$	temperature of particle, i and j are horizontal and vertical positions in the column
T_s	surface temperature, K
T_W	wall temperature
T_s^s	temperature at the surface of the solid
u_c	combined uncertainty
u_{ci}	critical velocity to break the bubble
u_i	velocity of ith species
u_{te}	turbulent velocity

$u_{w,i}$	velocity of the wake behind the leading bubble
U	overall heat transfer coefficient, superficial gas velocity
U_b	bubble velocity
$U_{b\infty}$	terminal velocity of an isolated bubble
$\bar{U}_{bs}$	average bubble velocity
U_c	parameter in sedimentation–dispersion model corresponding to settling velocity
U'_c	settling velocity in bubble column
U_f	eddy velocity
U_g	superficial gas velocity
U_g^c	critical velocity to suspend the solids
U'_g	settling velocity in the bubble column
U_{gc}	the critical velocity (i.e. the minimum gas velocity to suspend the solids)
U_{ms}	minimum slugging velocity
U_{ov}	overall heat transfer coefficient
U_s	slip velocity of the solid
U_{sl}	superficial velocity of the slurry
$U_{t,\infty}$	terminal velocity of a single particle
$\bar{u}$	average velocity of all the species
$u(v_k)$	number of daughter dispersed-phase units of volume v_k
v_i	volume of the ith class of bubbles
v_k	bubble velocity with a position between x and $x+\Delta x$ at time t
v_s	specific volume
v_p	particle velocity
v_z	local fluid velocity in the axial direction
v	velocity profile, velocity scale of the micro-eddies
$\overline{v_z}$	average axial velocity
V	volume, voltage
V_C	volume of the immobilised enzyme
V_f	final volume of the bubble
V_i	conditional variance of the expectation (i.e. mean value)
$V_{lf},\ V_s$	velocity
V_p	slip velocity
$V_{w,i}$	volume affected by the wake of a bubble belonging to the ith class

$\widetilde{V_p}$	random axial velocity
w_{ji}	weight associated with the interconnection between the jth input neuron and the ith neuron in the present layer
W_A	mass transfer rate of A
We	Weber number
x	direction in a Cartesian coordinate, distance from transfer surface
x_A, x_B	compositions of species A and B, respectively
x_{Ag}	mole fraction in equilibrium with y_A
x_i, x_j	mole fraction of ith or jth species in a liquid stream
X	concentration of biomass, biomass concentration
X_A	conversion of A
X_i	conversion of A in the input stream
X_m	moisture content on dry basis
X_o	conversion of A in the output stream
X_W	total mass of the biomass
y	output of a model, variable in an equation, ratio of the final volume to the expansion-stage volume of a bubble during its formation
y_{Al}	mole fraction in equilibrium with x_A
y_i	mole fraction of ith species in vapour or in a particular stream, output of the jth neuron
y_{iANN}	data predicted by the artificial neural network model
Y	yield coefficient
z	coordinate in the axial direction or in the direction of one-dimensional flow
z_i	mole fraction of ith species in the feed
$z_{in,jk}$	mole fraction of kth species in the jth stream
Z	compressibility factor, dimensionless axial position

Greek Symbols

α	ratio of the rate of evaporation of the moisture to the rate of change of moisture, thermal diffusivity, adaptive learning rate, constant
β	fixed bias error, coefficient of thermal expansion

$\beta(D, D')$	probability that a dispersed-phase unit of size D will be produced if a unit of size D and D' breaks
δ	half of the distance between two infinite parallel plates, thickness of a laminar sub-layer in turbulent flow
δ_h	thickness of thermal film
δ_m	diffusional film thickness
δ_{ϕ_i}	deviation in the input parameter
δ_y^{acp}	variation of the model output
δ_c	thickness of the diffusional boundary layer
δ_{ij}	binary interaction coefficient for a pair of species i and j in equation of state
$\varepsilon_{go}, \varepsilon_{gi}, \varepsilon_p$	emissivities of the top and lower glass covers and absorber plate, respectively
ε	fraction of solid in the slurry, holdup, porosity of the membrane support
ε_b	emissivity of bulk and particle
$\varepsilon_{org}, \varepsilon_{aq}$	holdup of organic and aqueous phases, respectively
ε_μ	random error
ε_{mf}	porosity at minimum fluidisation velocity
ϕ_i	ith model parameters
ϕ_{AA}, ϕ_{BA}	swelling parameters
ϕ_j	volume fraction of solids in phase j
ϕ	catalyst deactivation factor
ϕ_p	shape factor of the crystals
Γ_{ij}	function of the ratio of the mean distance between bubbles of sizes d_i and d_j and the bubble turbulent path length
η	apparent viscosity
η_l	length scale of the micro-eddies
η_e	effectiveness factor
λ	mean
λ_l	latent heat of vapourisation
λ_n	eigenvalues of appropriate equations
$\lambda(D, D')$	collision efficiency (i.e. the fraction of the collisions that result in successful coalescence)
θ	average contact time for micro-eddies, average contact time in surface renewal model
$\bar{\theta}$	average resistance time of the particle at the surface
$\bar{\Sigma}_i^2$	global sensitivity measure

μ	viscosity, momentum in the learning rate, mean
μ_{max}	maximum specific growth rate
μ_a	apparent viscosity
ν	kinematic viscosity
ρ	density
ρ_B	density of the bulk solid
ρ_{bf}	density of the dry biofilm
ρ_{bp}	density of the particle with biofilm
ρ_{b0}	density of the fully dried material
ρ_c	density of the crystals
σ	standard deviation
$\sigma^2(T)$	second-order sensitivity coefficient
τ	time taken to drain the film between bubbles, tortuosity of the membrane support, time interval between the successive arrival of the particle
τ_{max}	time at which the maximum value occurs
τ_{yx}	shear stress
τ_w	shear stress at the wall
τ_{ij}	time required to drain the liquid from the film between the bubbles
ζ	thermo-gradient coefficient, standard deviation
$\omega(d_i, d_j)$	collision frequency
Ψ_{AM}, Ψ_{AS}	partition coefficients
ξ_A	effectiveness factor
ξ_g, ξ_s	fractions of gas and solid, respectively, in gas–liquid and liquid–solid dispersed systems

---

Superscript

*	value at the interface

Subscript

0	initial value (i.e. at $t = 0$)
a	property in axial direction, or property of air

gl	adjacent to, at or from gas–liquid interface
gs	adjacent to, at or from gas–liquid interface
g	gas, or property in gas
in	input stream
l	liquid, or property in liquid
ls	adjacent to, at or from liquid–solid interface
max	maximum value or related to
mf	minimum fluidisation
out	output stream
r	property in the radial direction
R	radius of the column or sphere
s	solid
W	wall

1

Introduction to Modelling and Simulation

Today, simulation is an accepted way of studying various processes that are practised in the chemical and biochemical industries, including processes to control air and water pollution. These processes may be a single-unit operation or involve an entire plant. Simulation may be carried out to study the behaviour of these unit operations or the plant as such without fabricating any equipment or the entire plant. Therefore, there is no real loss or damage to the real world (i.e. life, equipment or even the environment). However, simulation studies can be used to predict such losses.

The engineering of process systems, now called 'process systems engineering', was initially qualitative in nature. With time, it has evolved and is now founded on engineering sciences, computer-aided tools and system-theoretical methodologies (Stephanopoulos and Reklaitis 2011). Computers have helped in solving complex problems involving systems of non-linear equations, optimisation problems, partial differential equations etc. This has helped in the development of an integrated approach to process design. The effect of environmental issues, choice of the controller, process safety and the like on the economic design of a process plant can now be considered. It has been made possible only by the use of simulation.

Simulation is the representation of the behaviour or characteristics of a real system by using a system of mathematical equations, which may be either linear or non-linear algebraic equations, in the form of inequalities and ordinary or partial differential equations. Often, the equations are difficult to solve analytically and hence are solved numerically using a computer. Therefore, 'simulation' is frequently referred to as 'numerical simulation' or 'computer simulation'.

1.1 Chemical Processes

The first step in simulation is to represent the process in terms of mathematical equations; however, the physical process may be so complex that it would be impossible to formulate it in terms of mathematical equations. The processes used in chemical industries are complex in nature. Classroom problems are simplified representations of these complex processes. They are repeated so many times that the students accept the descriptions as the final versions of the representations of the processes by the set of equations. For example, Poiseuille's equation, taught in lower-level classes, seems to be the only equation to describe the flow behaviour in a pipe until one realises that not all the fluids are Newtonian. Thus, the flow is not always laminar; it may also be 'turbulent'. The tubes may be short, for which 'end effects' have to be considered.

The physical representation of a process requires understanding all those phenomena which characterise it. These phenomena should be expressed in terms of mathematical equations using known laws of physics. A detailed description of a process will result in a large set of mathematical equations. This enhances the degree of difficulty in solving these equations.

As aforementioned, the chemical processes are complex in nature. Even a single-unit operation might be too complicated to describe. Let us understand why it is so through the following examples.

1.1.1 Unit Process: Fixed Bed

Fixed beds find many applications in chemical, biochemical and environmental engineering. These are used as reactors, absorbers, distillation columns and many more. Let us consider a fixed bed in which a fluid enters from the bottom of the column through a distributor (Figure 1.1). The pressure drop due to a single particle can be determined from the drag force

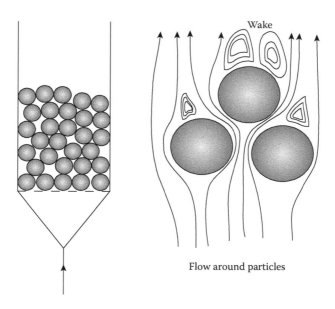

Wake

Flow around particles

FIGURE 1.1
Fixed bed.

obtained by knowing the flow field around a single particle. One such relationship is Stokes' law, which describes the drag force on a sphere at a very low Reynolds number. However, it is still not clear as to how to use this information to determine the pressure drop across a bed of particles. Let us examine the complexities present in a fixed bed.

1. Not all fixed beds contain the same types of particles. Various types of packing are used in industrial applications.

2. All particles are not of the same shape and size. Variation in shape and size is observed.

3. The particles (or packing) may not have the same orientation. The inter-particle distance may be different.

4. The fluid entering the column flows through the interstitial space, which is not the same around all the particles. Therefore, the drag force for each particle will be different.

5. The fluid velocity near the distributor will not be uniform. The fluid enters the bed through many holes and then spreads.

6. In the case of shallow beds, the fluid may not spread due to insufficient height of the beds (i.e. the flow is not fully developed). The behaviour of shallow beds is different from that of typical fixed beds.

7. The fluid velocity in fixed beds exhibits radial variation. The fluid velocity is minimum adjacent to the wall and is maximum at the centre.

8. The flow regime may be laminar, transition or turbulent. All fixed beds may exhibit these regimes at various conditions.

9. The change from laminar to transition and from transition to turbulent flow regime occurs be at different Reynolds numbers depending on the shape of the particle.

10. In the case of a reaction, the flow rate may change in the axial direction due to a change in the volume.

11. The volume of a gas depends on temperature and pressure. Thus, varying the temperature results in a change in the gas flow rate along the column.

12. If the packing height is high, the pressure of the gases decreases with increasing bed height. It also results in an axial variation of the flow rate.

13. Since the properties of the gas depend on temperature and pressure, the flow regimes may change at elevated temperatures and pressures. As a result, fixed beds operating at elevated temperatures and pressures behave differently from those operating at ambient conditions.

There may be many more factors which have not been mentioned here. The complexities mentioned above have been tackled by researchers either by using average properties (e.g. porosity) or by specifying certain properties (e.g. packing factors). Computational fluid dynamics (CFD) simulation may be carried out to consider many of the complexities, but even for a small variation in the definition of the problem, the CFD simulation must be carried out again. Parameters such as 'superficial velocity' have been used in various correlations. Superficial velocity is the fluid velocity considering that the column does not have any solids in it. Thus, superficial fluid velocity is hypothetical but is being used without much discussion; this is because of the need to find a way to describe the hydrodynamic behaviour of the fixed bed. Due to the wide range of operating conditions and geometrical parameters, various models and correlations for flow in fixed beds are reported in the literature. It is not an easy task to propose a universal equation to define the pressure drop in fixed beds.

Let us now look at an example of an entire plant and examine its complexity.

1.1.2 Sulphuric Acid Plant

Sulphuric acid is manufactured industrially using a source of sulphur as a raw material. A schematic flow diagram of the process is given in Figure 1.2. In a typical sulphuric acid plant, sulphur is burnt in excess of air to produce sulphur dioxide, which is converted to sulphur trioxide in a four-stage fixed-bed reactor. After each stage in the fixed bed, the gases are passed through heat exchangers to cool to a desired temperature and then sent to the next stage. The outlet gases from the fourth stage of the fixed bed are cooled and sent to the absorption column, where the sulphur trioxide

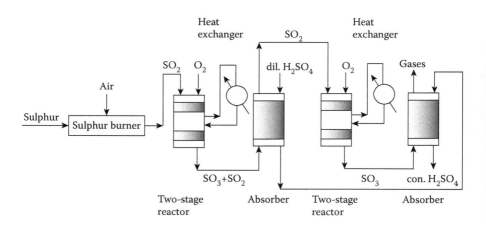

FIGURE 1.2
Sulphuric acid plant.

gets absorbed into 98% sulphuric acid. This absorption column is also a fixed bed. The recent trend is to use two absorption columns (Figure 1.2). The gases leaving the second stage of the fixed-bed reactor are sent to the first absorber. The unabsorbed gases leaving the absorber are sent to the third stage of the reactor. The gases from the fourth stage are sent to the second absorber. This process is called the double conversion double absorption (DCDA) process.

The other units present in the sulphuric acid plant are storage tanks, sulphur burners, heat exchangers, waste heat boilers, pumps, and compressors. Several different types of equipment and several similar kinds of equipment, but of different sizes, are used in a process plant, as is seen in the case of a sulphuric acid plant. A mathematical description of the process is thus much more complicated than that of a single unit.

Several fixed-bed reactors and absorbers are thus used in this process. If a set of equations is used to describe the behaviour of a certain fixed bed, then a similar set of equations can be used to describe all other fixed beds. However, the operating conditions and the reactions in each fixed bed may be different, resulting in different behaviours. Each fixed bed can be considered a sub-model of a model for the entire plant.

Sub-models cannot always be solved independently. The four-stage reactor, absorbers and a network of heat exchangers are arranged in such a manner that this portion of the plant involves a recycle stream which is unknown. The recycle stream is chosen to minimise the cost of production. Increasing the recycle stream reduces the conversion and hence increases the size of the absorbers. An increased bed height results in an increased outlet temperature from the fixed-bed reactor, but at the same time it reduces equilibrium conversion.

1.1.3 Complex Nature of Chemical Processes

Most chemical processes are complex in nature as they exhibit non-linear and non-equilibrium behaviour. This complexity is due to simultaneous and often coupled momentum-, heat- and mass-transfer phenomena and kinetic processes taking place at different scales. The bubbles and drops have dimensions ranging from 10^{-3} to 10^{-4} m. The catalyst particles are much smaller, ranging from 10^{-9} to 10^{-6} m (Charpentier 2009). Various geometrical dimensions of the reactor may be of the order of a few metres. At a different scale, the physical phenomena governing a process are different. This multiscale nature of chemical processes makes them even more complex. Charpentier (2010) proposed a triplet 'molecular processes–product–process engineering (3PE)' integrated multiscale approach for chemical product design and manufacturing.

The demands of a product vary with time. The process industry should be able to handle this problem. Fluctuations in product demand and supply of the raw materials have forced process engineers to address this problem.

Supply chain management and time scheduling of multiple products are some of the problems that have been addressed recently, resulting in strategies meant to increase profits and handle uncertainties. Today, with the help of simulation studies, chemical engineers are able to tackle more complex problems.

1.2 What Is Simulation?

A mathematical model to capture the behaviour of a process plant may be formulated. It may involve several algebraic and ordinary and partial differential equations. It is not always possible to obtain an analytical solution to a mathematical model. However, it is possible to obtain a numerical solution using a suitable numerical technique. Can solving a set of equations be called 'simulation'?

Simulation is the process of analysing a whole process or a part of it, using the set of equations describing it. The purpose of the analysis may include optimising operating conditions, analysing the effect of input variation on the properties of the output stream or troubleshooting a problem in the process. Simulation studies thus involve several parameters and studying the effects of these variables on the solution (i.e. response of the process). Korn (2007) states, 'Simulation is experimentation with the models'.

Problems can be divided into two types. When the inputs to the process are known and the output variables are to be determined, the problem is called a 'rating problem'. The problem can be solved by solving the model equations directly. For example, for a given configuration of a heat exchanger, the geometry of the heat exchanger, as well as the flow rate and physical properties of the process fluid and heating–cooling fluid, is known. The requirement is to determine the output temperature of the process fluid or the flow rate of the cooling–heating liquid. The other type of problem is called a 'design problem'. The outputs of the system are the desired values, and one must determine the input variables which provide the desired output. This class of problem can be solved by inverting the model equations (Charpentier 2010). The solution methodology consists of treating the problem as an 'optimisation problem'. In the case of design of a heat exchanger, geometrical parameters – such as tube length, tube diameter, number of tubes, shell diameter, baffle spacing and flow rate of the cooling–heating fluid – are to be determined if the flow rate and input and output temperatures of the process fluid are known. A number of solutions can be obtained; however, the desired solution is that which is most economical. The solution of the model equation is obtained by using optimisation techniques.

1.2.1 Types of Simulation

Simulations can be classified into various types based on the type of model used. Models may be steady-state or dynamic, deterministic or stochastic, or continuous or discrete events. The formulation of the problem, the types of variables and equations, and the solution methodology in each simulation type may be different.

1.2.1.1 Steady-State Simulation

Many processes work in continuous mode. The variables and parameters remain constant and do not change with time. When the laws of conservation of mass, energy or momentum are applied, the accumulation term remains zero. The latter represents the variation of mass, energy or momentum with time. The differential equations thus obtained do not have any temporal derivative. The model equations are simple since 'time' is not present in these equations. However, the steady-state simulations are not capable of predicting the dynamic behaviour of the process.

1.2.1.2 Dynamic Simulation

A process industry consists of processes which operate under batch or continuous mode. Batch processes are of a dynamic nature. Reactions are carried out in batch reactors to achieve the desired selectivity in a cost-effective manner. Many separation processes such as adsorbers and regenerators also operate in batch mode. Start-up and shutdown of a unit operating continuously also constitute a dynamic process. The design of process control systems requires knowledge of the dynamic response of the process to disturbances. These aspects cannot be studied by steady-state simulations. A simulation study that considers time as a parameter is called 'dynamic simulation'. Dynamic simulation studies can be used to obtain optimum operating conditions for a process. The time needed for the start-up and shutdown of a process can be minimised. A safe, feasible and economical procedure for process start-up and shutdown can be developed using dynamic simulation.

Dynamic simulation–based systems can be used to train plant personnel. Plant operators become familiar with the behaviour of unit operations and control systems. After training, they quickly respond to the changing behaviour of the system and take appropriate action. These systems can be used to verify the responses of the operators. The learning process can be carried out even in the absence of an instructor.

1.2.1.3 Stochastic Simulation (Monte Carlo Simulation)

Several models do not consider the random nature of the phenomena and do not involve random variables. Instead, they use the average properties. For example, transfer processes in fluidised beds may be modelled without considering the particle velocity. Such models are called 'deterministic models'. If a simulation model involves random variables, it is called a 'stochastic model'. Simulation studies using such a model are known as 'stochastic simulation'. This allows the use of more basic laws of physics. Simulation based on deterministic models directly provides average properties, and the results of stochastic simulation are averaged to get them. This type of simulation is also known as 'Monte Carlo simulation'.

1.2.1.4 Discrete-Event Simulation

Many variables change continuously. The conduction in a slab at constant wall temperature is a continuous process. The conduction may be a steady-state or even unsteady-state process. There are several events which take place suddenly at a particular time. For example, the arrival of particles on the wall of a fluidised bed is of a discrete nature. Addition of reactants into a fed-batch reactor may take place only at fixed times. This process is also discrete in nature. Discrete-event simulation usually involves random variables. The mathematical description involves discrete variables.

1.2.1.5 Molecular Simulation

To understand the processes at the molecular level and their relationship with the macroscale processes, the model must consider the effect of the molecule on the process. Molecular dynamic simulations that consider the velocity of each molecule, the charges on a single molecule and the forces between all molecules have been carried out to predict many physical properties such as coefficient of thermal expansion, surface tension, viscosity, vapour–liquid equilibria and liquid–liquid equilibria (Gupta 2003). Such simulations can help in studying diffusion at the nanoscale, or the development of new drugs, enzymes and biocatalysts. However, molecular simulations are time consuming, requiring extensive computational power. At present, supercomputers are used to carry out molecular simulation.

Note that the classification discussed here is very broad. Simulation to handle a large system of equations in cases of increased degrees of complexity is continuously being attempted. The methodologies to solve the equations are also being refined day by day. A few decades ago, solving a momentum balance equation in the case of a steady state and three dimensions was considered extremely difficult, even numerically. Today, it is a reality and being simulated using dedicated CFD software such as FLUENT, STAR-CD and COSMOL Multiphysics.

1.2.2 Applications of Simulation in Chemical Engineering

Simulation is an important tool for chemical and process engineers. It is used at various steps, including plant design, plant operation, troubleshooting, shutdown and start-up operations. Simulation is based on a model or a set of sub-models. Many processes are so complex that the traditional method of changing a variable at a time and studying its effect on other parameters experimentally is not possible. However, it is possible using simulation studies. Some of the wide areas of application in which simulation plays an important role are discussed in this section.

1.2.2.1 Process Synthesis

Before commissioning a new process plant, a techno-economic feasibility study is carried out. If only laboratory data are available, a pilot plant is required to generate the data for the development and validation of a model. Later, the model is used to simulate the process plant of any size. If the technology is already developed (i.e. the required simulation model is available), then simulation studies are carried out to design a plant for the required capacity. Even in such cases, there are several alternatives which depend upon the size of the plant. For example, the production of citric acid at low capacities is economical when it is produced using citrus fruit. For production at large capacities, however, the fermentation route is economical.

At the stage of techno-economic evaluation, the simulation may involve simple models. Rigorous models are used during detailed design of the equipment. Simulation studies have made it possible to consider process and control instrumentation during the early stages of design.

Let us consider the case of a plant that is profitable but running at low efficiency. It must increase the efficiency of the process. If a choice of one process alternative out of many is to be made, then a good understanding of the process behaviour is required. Simulation studies can be carried out to assess various process alternatives, which reduces the risk involved in choosing an alternate process.

1.2.2.2 Equipment Design

A simulation model may be used to design a single piece of equipment or an entire plant. The design of individual equipment involves determination of its optimum size, which can be achieved by repeated simulation of the equipment. Computer programmes for the design of equipment are procured or developed. Simulation helps in deciding on the design variables. For example, while designing a shell and tube heat exchanger, the number of passes and the length of the heat exchanger are not known, but the heat transfer area depends upon these parameters. An economical number of shell and tube passes can be determined using a simulation model.

1.2.2.3 Retrofitting

Quite often, process modifications should be made by introducing new equipment, such as a new heat exchanger to save energy, or changing existing equipment to another type. To properly evaluate the performance of the modified process, some experiments may be required. Such experiments cannot be performed as they require not only shutting down of a running plant but also fabricating new equipment, both of which involve cost and time. Simulation studies can be carried out without disturbing the operation process. The process can be analysed for conditions before making any process modification or change in the operating or resource policy.

The development of computer-aided design methodologies has made simulation studies easy. Since the design is considered a repeated simulation, several parallel runs can be made at different locations simultaneously to save time. Supercomputers use parallel computing. Cloud computing may possibly be used in future.

Sometimes, it is argued that simulation cannot mimic the real behaviour of a system. Even in such cases, however, it gives an idea of the process behaviour and increases our confidence before implementing a process change. Even if simulation discards any proposed alternative or finds an alternative to be infeasible, it is still a meaningful finding.

1.2.2.4 Process Design

For the simulation of an entire process, several flowsheeting programmes, such as ASPEN PLUS, HYSIS and UNISIM, are available. A few open-source flowsheeting programmes, such as COCO and DWSIM, are also available.

If a simulation model is reliable, then it results in savings with regard to time and design cost. Simulation can be used to explore new alternatives without experimenting with the equipment or process. Thus, simulation can also be considered an experiment and can eliminate the studies performed with costly pilot plants.

A simulation can be used to study the effect of many variables within a short period of time compared to what is required for an actual experiment. It helps the process designer to analyse his or her concepts and test the feasibility of any possible modification of the process. Based on the simulation results, changes in the process are suggested and can be implemented in the real system with more confidence.

Today, saving time in process design is of utmost importance in the light of the tough competition and rapid technological advancements.

1.2.2.5 Process Operation

When a plant is in operation, it might seem that the only objective is to keep it running in the same condition for as long as possible. However, the main objective is to run the plant at optimum conditions (i.e. to yield maximum profits). Fluctuations in the availability and price of raw materials, the price of the product, demand for the product affect the aforementioned 'optimum conditions'.

Simulation can be used to rapidly estimate the optimum conditions of plant operation and help the management to take appropriate action.

1.2.2.6 Process Control

Quality control of a product is essential for various reasons. Properties such as colour, shape and uniformity are desirable to enhance its consumer appeal. The yield in a reactor can be increased by controlling temperature, pressure and other operating conditions. A drop in yield will increase the load on the separator and decrease the production. When uncontrolled, a distillation column may not produce the distillate.

The quality control of a product in a process is achieved by using several instruments. These include measuring instruments, controllers, valves and other devices actuated by the input from the controller. Process control requires a good understanding of the dynamics of a process. The dynamic behaviour of the measuring elements and final control element is also essential in controller design. Due to the complex nature of the processes and the disturbances, dynamic behaviour cannot always be expressed analytically. Expressing the dynamic behaviour in the Laplace domain and frequency domain has yielded valuable methods. Still, the performance of the controller in a process is studied only in the time domain.

Advanced control techniques, such as model reference adaptive control or model predictive control, require a model for the process. The controller design for feedforward control requires a process model.

1.2.2.7 Process Safety

A simulation can be used to study various aspects related to hazards and ways to handle them. The spread of a fire or toxic chemicals in an environment, the effect of vapour leakage from a pressurised tank on the surroundings, spillages from tanks etc. can be studied through simulation. Processes to reduce environmental pollution can be developed, designed and evaluated.

1.2.2.8 Personnel Training

Simulation may be used to train industry personnel to teach both trainees and practitioners the basic skills in systems analysis, statistical analysis and decision making. Process simulators for a given industry simulate the process and display the outcome in real time and a real environment. The output of the simulator is rendered to the display devices used in an actual industry. It helps personnel to better understand the process and prepares them to act quickly in case of any abnormal situations.

1.3 Modelling

The success of a simulation rests on the model being used. Modelling thus becomes a prerequisite for simulation. High-level modelling languages have made model building quite easy. Such a software would have many built-in sub-models. One has to choose the appropriate sub-model out of many options available and connect them properly. Syntax and semantics for writing code for user-defined sub-models are available. Object-oriented programming and graphical user interface have made model building fast and easy. However, it still needs experience, imagination, good understanding of the process and model validation.

1.3.1 What Is a Model?

To explain the working of a machine or a process to young students, physical models are used. Students are encouraged to make models. A model may be either a static model or a dynamic model, and it is a visual representation of the actual process. Static models mimic a visual picture of a machine or structure. Static models are also used to explain the internal construction of equipment. The static model of a boiler is shown in Figure 1.3.

FIGURE 1.3
Static model of a boiler.

The internal construction of a boiler cannot be seen until the boiler is dismantled, but this is time consuming and costly. The use of a small-scale static model with a sectional cut at the desired location is a better option. One can observe the external as well as internal arrangement of various parts of the boiler. It is an interesting and time-saving way of understanding the construction of any equipment. Dynamic models, on the contrary, demonstrate the working of a device.

Thus, the purpose of a model is to mimic the performance of a physical phenomenon. As scientists and engineers require a quantitative description of a process, a concept mimicking the process is required. It may also be said that idealisation or approximation of physiochemical processes is called 'modelling' (Rice and Do 1995). However, the model should not lose the basic characteristics of the process. The concept should be simple enough so that it can be expressed in terms of mathematical equations. Thus, a representation or description of a physical phenomenon by a system of mathematical equations is also called a model.

When the idealised physicochemical phenomena are still at the conceptual stage, a model may be termed a 'conceptual model'. By this time, all the approximations and assumptions are clear. It helps us to choose the appropriate laws of physics or chemical kinetics and other laws out of several that are known to us. However, the equations are yet to be written.

The model's equations are, as mentioned, based on laws of physics, and they describe the mechanisms of the process. For example, material balance, heat balance and momentum balance equations are based on the laws of conservation of mass, energy and momentum, respectively. Few of the equations used in a model may be of an empirical nature. Sometimes, the system of equations is referred to as a 'mathematical model'. The set of equations, if solved numerically, may be called a 'numerical model'. The model equations may also have few differential equations. In such cases, the assumptions help us to specify the required set of boundary conditions.

A model must be able to predict the behaviour of a process. After solving the model equations, various states of a process can be estimated. The solution need not predict every aspect of the process. Many models describing few limited aspects may be assembled together. Frequently, separate models describing the kinetics, heat transfer and mass transfer are used. The assembly of model equations usually consists of not only algebraic and transcendental equations but also differential equations. Thus, many times, model equations for all these individual aspects are solved simultaneously. For example, the model for a drying operation involves heat and mass transfer equations which are solved simultaneously.

It has been pointed out several times that representation of a system by a model does not capture reality. It is confined to a certain perspective of reality which is considered relevant in the context in which the model is supposed to be used (Klatt and Marquardt 2009).

1.3.2 Role of Modelling in Simulation

Simulation is based on the set of equations representing a process. This set of equations is called a model. A good model is necessary for better understanding of the process. The model equations require a good understanding of the characteristics of the process or equipment involved. The rheology of fluids and the behaviour of bubbles, drops, slurries, emulsions, foams, gels etc. should be well described by appropriate equations. Morphology, particle size distribution, particle shape and porosity, and pore size distribution may also require attention.

1.3.3 Limitations of Models

Many variables have an insignificant effect on the performance of a process. Mathematical equations to describe a process without considering such variables may predict the behaviour of that process satisfactorily. Variables affecting the process significantly should always be included in the process model. In other words, the process may be more sensitive to some variables as compared to others. This information is useful while designing a control scheme.

Although simulation is a powerful tool to study the performance of an entire process or a part of it, it does not mean that the information obtained by simulation studies is accurate and complete (Ramirez 1997). The predictions of the process rely upon the accuracy of the model used. A model in itself is an idealistic mathematical representation of a process, and hence in principle it can never describe the process' exact behaviour. Since a model is acceptable only after its validation based on the experimental data, the accuracy of the model also depends on the accuracy of the experimental data. Often, when experimental data obtained at the laboratory scale are used to validate a model developed for large-scale equipment, the results are far from satisfactory because of the attempt to apply a model developed on one scale to another scale. For example, extending an equation that describes the flow in a tube to describe the flow in a microchannel may not be appropriate. The depth of the microchannel is of the order of a few microns. The interactions between the wall and fluid molecules are significant under such conditions.

Various models to describe a part of the process may be available in the literature. One might think that any one of the models may be chosen since all are correct and are being used regularly by various researchers. For example, a model to describe heat or mass transfer may be based on 'surface renewal' theory, 'boundary layer concept' or 'film theory'. One has to make a proper choice among these. Different models are obtained by having different assumptions (i.e. considering different variables as insignificant variables). Therefore, these models are applicable under different situations. A proper choice is required so that simulated results have the least error for the range of operation. For these reasons, the process of developing a simulation is time consuming. It consists of time spent collecting data for validation of the model, developing models and integrating small models to obtain a complex model. But once a proper model is found acceptable for the purpose of simulation, it can be used with confidence for similar applications. It saves time and cost of simulation.

The modelling activity thus involves the simplification or idealisation of a complex process. However, after the simplification, the model should be able to capture the basic features of interest. The boundary layer theory, the film model for transfer processes and the residence time distribution theory are some of the simple models. These simple models have been successfully applied in many cases. Whenever these simple models fail to predict the characteristic features, a more detailed model is developed. Making a model unnecessarily complex always increases the computational effort and may show only marginal improvement.

1.4 Summary

Chemical engineering processes are complex. The static and dynamic behaviour of the process may be predicted using simulation studies. Simulation helps in various chemical engineering activities (process synthesis, process design, process control, process monitoring, process safety etc.).

A process model is required to carry out simulation studies. The models are mathematical representations of the process, and they are developed by writing mathematical equations and solving them followed by their validation. The scope of the model should always be kept in mind. Simulation studies cannot provide the real behaviour of a process due to the limitations of models.

References

Charpentier, J.C. 2009. Perspective on multiscale methodology for product design and engineering. *Comput. Chem. Eng.* 33: 936–946.

Charpentier, J.C. 2010. Among the trends for a modern chemical engineering, the third paradigm: The time and length multiscale approach as an efficient tool for process intensification and product design and engineering. *Chem. Eng. Res. Design.* 88: 248–254.

Gupta, S., and Olson, J.D. 2003. Industrial needs in physical properties. *Ind. Eng. Chem. Res.* 42: 6359–6374.

Klatt, K.U., and Marquardt, W. 2009. Perspectives for process systems engineering—Personal views from academia and industry. *Comput. Chem. Eng.* 33: 536–550.

Korn, G.A. 2007. *Advanced Dynamic-State Simulation.* Hoboken, NJ: John Wiley.

Ramirez, W.F. 1997. *Computational Methods for Process Simulation*, 2nd ed. Oxford: Butterworth-Heinemann.

Rice, R.G., and Do, D.D. 1995. *Applied Mathematics and Modeling for Chemical Engineers.* New York: John Wiley.

Stephanopoulos, G., and Reklaitis, G.V. 2011. Process systems engineering: From Solvay to modern bio- and nanotechnology. A history of development, successes and prospects for the future. *Chem. Eng. Sci.* 66: 4272–4306.

2

An Overview of Modelling and Simulation

Various activities related to chemical engineering require using simulation as a tool to develop the technology to produce a new product or increase the productivity of an existing plant. A good understanding of chemical engineering fundamentals and of process models is a prerequisite to conducting simulations. Familiarity with the software being used for the simulation is an added advantage. The development of web-based simulation environments and the use of the Internet for communicating and transmitting data and other relevant information in various formats have made it possible for simulations to be performed at remote locations also. For example, troubleshooting at the site of industry may be carried out at a different place where the simulator and the simulating personnel are present.

Earlier modelling efforts focussed on the development of models that could be solved analytically. Therefore, textbooks of that time treated the modelling activity as the application of mathematical techniques to obtain analytical solutions. The use of various transforms, perturbation techniques, complex numbers, solutions in terms of series, matrix algebra solved ordinary and partial differential equations representing various processes. Application of these techniques to solve various chemical engineering problems was presented. Some of these books are given at the end of this chapter (see Davis 2001; Ramirez 1997; Rice and Do 1995).

The development of numerical techniques encouraged modellers to solve more rigorous problems which earlier could not be solved analytically. However, the analytical solutions still serve as benchmark solutions to test the effectiveness of the new numerical methods under development. The literature on modelling focussed on the use of various matrix methods to solve simultaneous linear equations, optimisation techniques to solve non-linear equations and solutions for ordinary and partial differential equations (Davis 2001; Ramirez 1997; Rice and Do 1995). The emphasis had been on solving model equations, and this is called an 'equation-solving approach' (Shacham 1982).

At present, various methods are available to solve not only ordinary but also partial differential equations coupled with algebraic equations. Commercial programmes for computational fluid dynamics (CFD) (FLUENT, COSMOL-MULTIPHYSICS, STAR-CD etc.) and flowsheeting programmes for process simulation (ASPEN PLUS, HYSIS, CHEMCAD, UNISIM, PROSIM, COCO, DWSIM etc.) are available. These software should be treated as tools only and are helpful for developing technology for process industries.

The modeller can now focus on the representation of a process in mathematical equations. The problem of solving these equations involves expressing them using the syntax and semantics of the chosen software.

2.1 Strategy for Simulation

The simulation studies are carried out following a systematic approach. The steps followed to carry out simulation studies involve 'common-sense steps' (Ramirez 1997), such as choosing an appropriate model for the given problem and then experimenting with the model. Various steps in a typical simulation normally involve the activities that are discussed in this section. These also have been discussed in the literature (Chung 2004) and are shown in Figure 2.1.

2.1.1 Problem Definition

The first step in carrying out any simulation study is to understand the problem. The goal or aim of the simulation studies should be clearly understood. For example, let us take a reaction such as the production of aluminium sulphate by the reaction of alumina and sulphuric acid. From a chemist's point of view, the problem is to obtain the reaction temperature, the amount of the reactants' kinetic data and the yield. From a chemical engineer's point of view, the type of the reactor, controlling of the reaction temperature, feeding of the reactants to the reactor are important. An engineer's goal is to determine the operating conditions of the reactor that are safe and economical. Therefore, the reaction should be visualised at an industrial scale, and the problems associated at this scale should be clear.

2.1.2 Understanding the Process

The second step of a simulation study is to understand the physical picture and identify the physical phenomenon and chemical transformation taking place in the process. At this stage, one must understand the process details

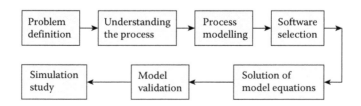

FIGURE 2.1
The steps in a simulation study.

and identify the fundamental mechanisms of the process. The depth of knowledge required depends upon the scope of the problem. In the example given in Section 2.1.1, it is realised that alumina is in the form of powder and hence has to be added slowly to the sulphuric acid. A chemical engineer may arrive at the conclusion that the reaction is very fast, and hence the kinetic data are not required as the reaction is mass transfer controlling. Therefore, the aim might be to determine the rate of addition of the alumina powder so that the temperature of the reaction remains constant.

The number of phases, the method of contact of these phases, the rate of reactions at the phase boundaries and the devices to provide contact between the phases are all important. It may be important to identify the parameters which affect the performance of the process, as well as possible insignificant parameters. These will later help in making the assumptions.

2.1.3 Process Modelling

The process can now be modelled. Based on the information available in the literature and the knowledge of the person involved in the modelling, the physical picture of the process has to be transformed into mathematical equations. A physical picture of the model can be represented in the form of a sketch, which is a better way of expressing the idea compared to words. The model sketch at this stage may be just the main mechanisms of the process. Dobre and Marcano (2007) described the various steps of the process of modelling. This first step in the modelling was called the 'model by words'. The equations can now be written with the help of the model sketch. At this stage, the sketch of the model never needs to resemble the sketch of the process. For example, scraped surface heat exchangers are used for heating or cooling of very viscous fluids. They consist of concentric cylinders, with the inner rotating cylinder having scrapers and an outer jacketed stationary cylinder. A model based on the surface renewal concept has deficiency; hence, other models such as three-dimensional (3D) lubrication theory were developed (Fitt et al. 2007). The renewal of the fluid element at the surface is quite different from the actual scraping of the surface; using a model based on the surface renewal concept is the result of forgetting that there is no fluid layer which is actually replaced. Also, the boundary layer is only fictitious; it is only a concept. The difference between fictitious and real things should always be clearly understood. This point will be clarified with examples in Chapters 4 to 6.

The sketch of the model should show a simplified process in terms of a sufficient number of well-understood phenomena. If any of the phenomena are not well understood, they should be investigated before developing the model. The model sketch helps us in understanding the interrelations of various phenomena, which may also be called 'sub-models'. These sub-models may be interrelated by parallel or in-series mechanisms or a combination of both.

The model equations are then written with the help of the model sketch consisting of well-understood sub-models, which are relatively simple, small and reliable. The model sketch also helps modellers in writing the model equations and indicates to some extent whether the model equations are to be solved sequentially or simultaneously. Expressing the equation in the form of an information flow diagram is helpful in obtaining a stable solution strategy to solve model equations. It also explains interrelationships between various parameters. The role of the model sketch in explaining the models will be discussed in Chapters 4 to 6.

2.1.4 Software Selection: Factors Affecting the Selection

There are many programming languages and specific-purpose languages which may be used for simulation purposes. The following factors affect the selection of suitable software.

2.1.4.1 Availability

In the 1980s and 1990s, computer programmes were written in FORTRAN and C. Later versions of FORTRAN are quite powerful and have syntax similar to that of C. These languages were available to many investigators and were the obvious choice at that time. These languages may still be used if supported by library functions.

Currently, many high-level languages such as MATLAB®, MATHMATICA, MATHCAD, gPROM, STELLA, ACCEND and GAMS are available. Some of these are very specific. It is also possible to use a combination of software packages.

2.1.4.2 Cost

The cost of the software is also an important factor. Commercial packages may be purchased and used. Another option is to use open-source languages such as SCILAB, ACCEND and LIBRA Office CALC.

2.1.4.3 Trained Personnel

Any software has its own syntax and semantics. The user has to learn the software before using it, which is time consuming. The user sometimes prefers to continue using the software which he or she is familiar with, ignoring the advantages of a new software or programming language. Thus, it is very important to start with appropriate software which can be used for a long period.

2.1.4.4 Suitability

The software chosen should be suitable for the purpose. It must have features which reduce the work while writing the computer programme for solving equations. At the same time, it must be flexible enough. Some desirable properties include easy syntax and semantics, provisions to use previously developed modules using FORTRAN and C and write user-defined modules, and capabilities to communicate with a wide range of input and output devices and file formats.

Many analytical solutions of simple models involve functions such as Bessel's function and an error function. Function evaluations involve the summation of a series. Almost all simulation languages, software and spreadsheets have implemented these functions. Other functions which are not available can also be easily incorporated by writing small user-defined function sub-programmes.

The simulation languages should have features to solve systems of algebraic and differential equations, inequalities etc. The results are presented in various graphical and tabular forms. A general-purpose high-level programming language, such as C++, FORTRAN and Visual C, requires an extensive time-consuming programming effort which may be reduced if the libraries for various functions and procedures are available. Other programming languages (MATLAB, OCTAVE, SCILAB, gPROM etc.) have simple commands to solve various types of equations and optimisation problems and to plot various types of graphs, and various input–output functions to access databases stored in files. The programming effort is less, and hence the time required to develop a model is much less. Such high-level languages have features that are commonly required. However, there is loss of flexibility.

Programming languages can be classified into two types: procedure oriented and object oriented. Earlier programming languages were procedure based. These involved a main programme using various functions and subroutines. The functions returned values while the subroutines were used to perform various tasks. The functions and subroutines were user defined and were usable components of the software. High-level languages such as MATLAB and SCILAB are primarily procedure based. A vast amount of code developed in FORTRAN, and C is procedure based. A deviation from the procedure-based programming will require rewriting the code. The computer programmes using the procedure-oriented languages are presented in the form of system flowcharts. The system flowcharts represent the manner in which the definition of input–output, the presence of loops (FOR...NEXT, WHILE...DO etc.), the calling of subroutines and reference to the database are handled. The system flowchart is thus a graphical representation of the important inputs, outputs and data flow among the key points in the system. For example, let us examine the system flow diagram for the estimation of dew point (Figure 2.2). The computer programme starts by reading the number of components, M; total pressure, P_T; initial guess for the dew point, T_D; constant, S; feed composition, z_i; $i = 1 \ldots M$ and tolerance EPC to specify the desired accuracy. Since at dew point the composition of the vapour does not change, y_i is taken as equal to z_i. The initial guess for the liquid compositions, x_i, is made. To avoid entering any of the endless loops due to convergence problems, two counters (Counter 1 and Counter 2) are initiated to check the two loops in the programme. The loops start checking the values of Counter 1 and Counter 2.

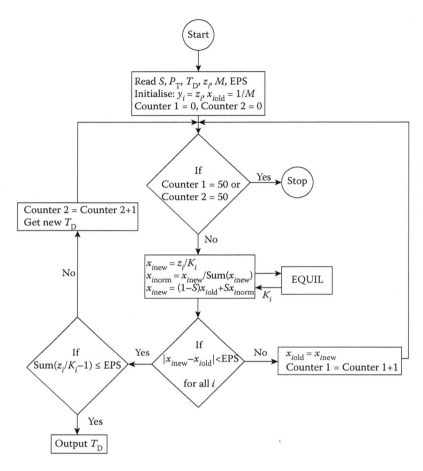

FIGURE 2.2
Flow diagram for dew point calculation.

If the convergence is not achieved, then further execution of the programme is stopped; otherwise, the subroutine EQUIL is called. This subroutine determines the composition of the liquid in equilibrium with that of vapour at a particular temperature. The values of x_i are renewed. The first loop (on the right of the flow diagram) is repeated after checking the values of the tolerance (EPS) until the difference between the old and new values of x_i (for all i) is less than the tolerance. After successful execution of the first loop, the second loop (on the left of the flow diagram) is executed. It checks the condition for the existence of dew point. If the value of T_D is not equal to the required value, then a new value of T_D is guessed by an appropriate method such as the bisection method. When the guess value of T_D is equal to the desired value, then the second loop is executed. The value of T_D is accepted as output at the required port (printer, file, temporary storage etc.).

It has been more than a decade since object-oriented programming languages (OOPLs) have become more acceptable than procedure-based programming (Stephanopoulos et al. 1987). In the OOPL approach, an object (often defined as a class) is the reusable software component. The 'class' contains various methods similar to subroutines in procedure-based languages. It also has properties similar to variables in procedure-based languages. The unique features of the object are that various similar classes can be created with a single line code called an instance of the class. Once an instance is created, more functionality can be added to it, including other available classes. Thus, a class can inherit the features of more than one class. It not only reduces the programming efforts but also is easier to reuse. To use a class, one should know the function and properties of the class, which is accessible from outside. C++, Visual BASIC and JAVA are OOPLs. For modelling of chemical engineering problems, ASCEND was developed using object-oriented programming. Such languages have predefined classes such as TANK and HEAT EXCHANGER. Now MATLAB has provisions to create classes now and supports OOPLs.

The structure of the computer programme is shown by the class inheritance. A programme usually has a main class consisting of various other classes (called 'children'). The class inheritance for the dew point calculation is presented in Figure 2.3. The main class is 'DEWPOINT' which consists

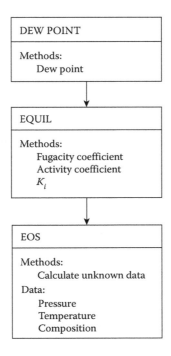

FIGURE 2.3
Class heritance for dew point estimation.

of the classes 'EQUIL' and 'EOS'. The EOS uses equation of state and holds data (temperature, pressure and composition of the mixture) and methods to calculate the unknown from the remaining data. The EQUIL uses this class to calculate fugacity, activity coefficient and hence K_i. For simplicity, it is assumed that a cubic equation of state such as Peng–Robinson is predicting liquid properties also. Otherwise, methods such as UNIFAQ, UNIQUAK and ASOG might be required.

2.1.5 Solution of Model Equations

The model is coded using the syntax of the chosen software. For example, to solve a set of simultaneous equations, the coefficients may be entered as a matrix. The representation of the matrix may depend upon the software used. When two or more types of software are used for simulation purposes, the transportability of the data among them should be taken care of. The equations are then solved. There may be more than one way to solve a set of equations. For example, a differential equation may be solved using an ordinary differential equation (ODE) solver, or using symbolic mathematics to express it in terms of Laplace transform and solve it as an algebraic equation. The user-defined modules should be debugged at various stages to avoid any programming error. Dedicated software has various error-checking methods and informs the user.

2.1.6 Model Validation

Solving the model does not mean that the model has been developed successfully. A model must mimic the performance of the process or system. Therefore, it is important to estimate various parameters using the model and compare them with experimental data. Getting a satisfactory agreement between the experimental data and model prediction is called 'model validation'. A model is ready to use for simulation studies only after successful validation. Few models have adjustable parameters or terms which may or may not be experimentally measurable, but their presence in the model is essential for predicting the correct trend of the experimental data. These parameters can be obtained by fitting the experimental data with the model equations. This process is known as 'parameter identification'.

2.1.7 Simulation Study

The model is used several times with an aim to study the effect of operating parameters on the performance of the process. The optimum operating conditions can also be determined if desired. The results are presented in various formats. These may be in the form of tables or graphs. Many graphical plots are topic specific, such as a triangular plot to represent liquid–liquid equilibrium data. Various high-level languages can plot the graphs in many formats and print them. Few software are capable of producing even animated movies to show time-varying phenomena. If these facilities are not available, then the results should be processed after simulation. Graphs may be plotted using spreadsheets or programmes such as ORIGIN.

2.2 Approaches for Model Development

The development of a model for complex chemical engineering processes involves the omission of several details of the process. It, however, results in a reduced scope of the model. Even though a simple model sometimes does not represent a large system, it can still be used as a sub-model for a part of the process or system. A model may consist of one or more sub-models which may be interconnected in series or in parallel. A model representing a large process in mathematical form can be developed in any of the three ways shown in Figure 2.4.

A simple model that is already available may be chosen as a base model. Rice and Do (1995) considered the hierarchy of models in which a simple model was considered at the lowest level. It was like a black box. The model at the highest level is capable of describing the process completely (i.e. transport processes such as pressure drop, flow behaviour, temperature and concentration profiles, thermodynamic properties etc.). All these models form a family of models. The scope and depth of the models in this family of models determine their complexity. The model equations of the base model (the model at the lower level in the hierarchy of models) may be modified by adding more terms. These terms correspond to phenomena that were already ignored while developing the base model. In other words, one or more restrictions as a consequence of assumptions in the

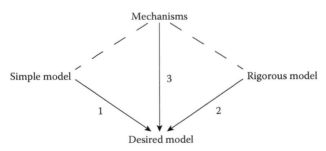

– – Dotted lines indicate that the model is already developed and available

FIGURE 2.4
Three approaches for model development.

original model are relaxed (dropped or modified), making the new model wider in scope. Such models are often termed as 'generalised models'. For example, the surface renewal model may be modified for a new type of age distribution. This means that the restriction of uniformly distributed age is relaxed. Analytical solutions for many simple models are available. The parameters in the analytical solutions are represented in mathematical functions. The constants in the functions are determined using curve-fitting methods.

In the second approach, an available generalised model is modified by ignoring some of the terms or adding a few simple terms corresponding to significant factors not considered earlier. In solving processes involving transfer processes, the Navier–Stokes equation may be considered the generalised model. A few terms may be dropped depending upon the type of flow. For example, while using the energy balance to determine temperature profile, heat generation due to various reasons or phenomena may be added. These differential equations may be solved for various boundary conditions. CFD software can solve the Navier–Stokes equation for various cases, and its use can be considered for this approach. When some of the terms in the generalised model are ignored, this is called a 'model reduction'. This method is frequently used while modelling systems of many reactions (e.g. reactions of petroleum fractions or biochemical reactions). The insignificant parameters are decided by the use of principal component analysis (PCA) or surface response technique.

Another way to develop a model may be to start from scratch (i.e. using the basic laws of physics to explain the mechanism of the process). This approach involves understanding the complexity of the model and then deciding the factors to be ignored. This approach requires more effort than the other two approaches, but during the development of the model the understanding of the process increases, which makes the modification of model equations much easier than in the other approaches. The models are expressed in terms of the variables which are available or measurable. The steps in the modelling using this approach have been discussed by Rice and Do (1995). In brief, they involve drawing a sketch of the system to be modelled, identifying various parameters, writing a differential equation using the laws of conservation and appropriate boundary conditions followed by solving the model equations. The most important suggestion is to consider various approximations for the model equations and the boundary conditions so that an acceptable solution is obtained.

In all three approaches, the model results must be validated before carrying out any kind of simulation study. In case of failure, the model equations are modified to reformulate the problem. Developing a model and solving model equations should be considered at the same time. The experience in these activities helps in the development of more complex models.

2.3 Types of Models

Based on the types of simplification, the models may be classified in the same manner as the simulation. Familiarity with classification of the models provides us with a few key terms, which will help readers understand the complexity of the model. However, these do not really help in the development of a model but only provide an idea of the model equations that are involved. They also help in choosing an appropriate methodology for solving the model equations. The discussion on the classification is therefore presented from the point of view of understanding the nature of the processes usually encountered. A classification of various types of models is presented in Table 2.1.

TABLE 2.1

Types of Models and Their Key Characteristics

Types of Models	Key Characteristics
Deterministic models	No random variable is considered.
Lumped parameter models	Ordinary differential equations consisting of usually either a temporal derivative or any one spatial derivative without representing true spatial distribution within the system.
Distributed parameter models	Partial differential equations consisting of spatial derivatives. Dynamic models consist of temporal derivatives also. ODE in case of 1D steady-state model.
Steady-state, or static, models	No temporal derivative.
Unsteady-state, or dynamic, models	Temporal derivatives are present.
Stochastic models	Random variables are considered in the form of either probability distribution functions or pseudo-random numbers.
Population balance models	Treatment of variation of numbers of individual entities in disperse phase systems based on 'birth and death' processes.
Agent-based models	Each part of the model, called an 'agent', is capable of taking independent decisions.
Discrete-event models	The states, even time, are considered discrete events.
Artificial neural network–based models	Input and output variables are related through a network of nodes and interconnections. It is a black-box approach.
Fuzzy models	Use of fuzzy variables.

2.3.1 Deterministic Models

In several cases, the model equations are such that the solution may be expressed as a simple mathematical function, an algebraic equation or a differential equation. The equations can be solved analytically or numerically. In the case of analytical solutions, the outputs can be expressed by explicit or implicit functions. All such models are known as 'deterministic models'. As the name suggests, the model predictions are well defined and can be determined exactly. These models do not consider any random variation of any of the variables. Also, the model predictions do not have any uncertainty.

2.3.2 Lumped Parameter Models

The application of several laws results in differential equations. The derivatives may be spatial or temporal. Spatial derivatives exist within the system or subsystem. Many times, the models do not consider any phenomena within the system. Therefore, these models do not consist of spatial derivatives. The changes taking place within the system are combined (or 'lumped') in a single variable. Let us consider an example of a completely stirred tank reactor (CSTR), in which the reaction A→B is being carried out (Figure 2.5). The steady-state material balance for species A can be written as

$$\begin{bmatrix} \text{Rate of A} \\ \text{entering} \end{bmatrix} + \begin{bmatrix} \text{Rate of A} \\ \text{leaving} \end{bmatrix} = \begin{bmatrix} \text{Rate of conversion} \\ \text{of A into B} \end{bmatrix} \tag{2.1}$$

The terms on the left-hand side of Equation 2.1 are the flow rates. The right-hand side of the equation corresponds to the rate of A which is converted into B due to the chemical reaction taking place in the reactor. It can also be written

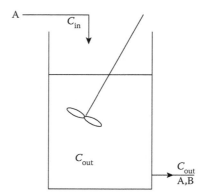

As soon as A enters into CSTR, it is completely mixed

FIGURE 2.5
Completely stirred tank reactor.

as the 'rate of disappearance of A', which emphasises that while writing the material balance for A, we are concerned with the reactions involving the species A. This term, when written in terms of rate of reaction, may involve other species also depending upon the type of the reaction. The effect of the spatial distribution of the concentration of A on its rate of disappearance may be avoided by making an assumption that involves the rate of reaction on the size of the reactor but does not take into account the spatial distribution of A within the reactor. The following assumption serves exactly this purpose:

> *All the fluid elements in the reactor have same chance (probability) of leaving the reactor irrespective of the time they have spent in the reactor.*

At first, it looks quite unreasonable that a fluid element can reach the exit immediately after it enters the reactor. To some extent, this is true also. If the reactor is vigorously mixed (i.e. the fluid element has large velocity so that the time to travel from entrance to exit is quite small in comparison to the average time spent in the reactor by all such fluid elements), under such conditions, the present assumption may be an acceptable approximation of the situation. The assumption is described by a single term such as 'well mixed' or 'completely stirred'. It may be shown that the above assumption implies that the time spent by a fluid element in the reactor may be specified by the exponential probability distribution. Though the model for CSTR based on the above assumption involves a probability function, it does not require any spatial distribution of any of the variables or spatial derivative. In this sense, it is a 'lumped parameter model'. This simple model does not require any 'stochastic simulation'. As a consequence of the above assumption, all spatial distributions of the concentration are lumped into one variable 'rate of disappearance of A', which is estimated as the concentration of A at the exit.

The 'lumped parameter models' generally involve only one or two independent variables. These may involve time or time and any one spatial derivative. If only one variable is involved, then ODEs are obtained. If two variables are involved, then partial derivatives are involved. However, the spatial derivative does not truly represent the spatial distribution within the system. If the temporal derivative is involved, the model is a dynamic model; otherwise, the model is a steady-state model. The lumped parameter models are also deterministic in nature. This family of models has been used in a wide range of applications in chemical engineering, biochemical engineering, environmental sciences etc., and even has been used to handle dynamic processes and process control (Luyben 1996).

2.3.3 Distributed Parameter Models

'Distributed parameter models' require a description of the process behaviour due to the spatial distribution of parameters within the system. The parameters might be the concentration of various chemical species, temperature, pressure etc. The model equations depend on the extent of the knowledge or information about the spatial distribution of these relevant variables or parameters that are available. The spatial derivatives describe the spatial variation of the parameters within the system. If all three spatial derivatives are present, then the model equations can be solved only numerically. Commercial CFD software makes use of equations of change and equations of continuity and is capable of solving these equations. The absence of any one or more spatial derivatives reduces the computational effort. Accordingly, these models are known as one-dimensional (1D), two-dimensional (2D) or three-dimensional (3D) models if the number of spatial derivatives is 1, 2 or 3, respectively.

A proper choice of a coordinate system may reduce the number of spatial derivatives. For example, in case of a laminar flow in a circular pipe, the fluid velocity is axial. The radial and angular velocities are absent. If a cylindrical coordinate system is used, the model involves only one spatial derivative. However, if the Cartesian or spherical coordinate systems are used to describe the flow, then not only do the model equations use three spatial derivatives but also the number of equations is more than one.

Let us consider an example of a plug flow reactor (PFR) with the same reaction as in the case of CSTR (Figure 2.6). Equation 2.1 describes the material balance of A in this case also. In the case of PFR, it is assumed that fluid flow through the reactor is such that no fluid element overtakes or mixes with any other element ahead or behind. Therefore, the reaction of each fluid can be treated without interference by other fluid elements. Since the concentration of the A varies within the reactor, the rate of conversion of A into B also depends upon the position of the fluid element. Due to an orderly flow, the position of the fluid element and the time spent in the reactor are directly related with each other by the following equation:

$$t = C_{A0} \int_{X_i}^{X_o} \frac{dX_A}{-r_A} \tag{2.2}$$

Since the exit concentration, corresponding to the conversion while leaving the reactor, X_o, can be expressed in terms of the total time, t, knowledge of concentration variation in the PFR is not used. A simple assumption of orderly flow has idealised the flow rate in such a way that the reaction rate within the reactor follows a well-defined variation, which on integration gives the conversion as a function of the composition of the reactant entering

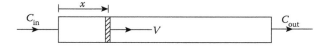

All fluid element at a given axial distance move with same velocity in a PFR

FIGURE 2.6
Plug flow reactor.

the flow rate and the length of the reactor. For an isothermal reaction in a PFR, the model seems to be a lumped parameter model. However, the model can also be called a distributed parameter model in some sense since the concentration at any length of PFR can also be determined using Equation 2.2, which is the solution for a 1D steady-state model.

2.3.4 Steady-State Models

The temporal derivatives are derivatives with respect to time. Models which do not contain temporal derivatives are called 'steady-state models'. These models cannot be used to predict the dynamic behaviour of the process under consideration. Due to the absence of temporal derivatives, the solution of steady-state models can be obtained using much simpler techniques in comparison to the models used to describe the dynamic behaviour of a process. The values of parameters used in the models are also time-averaged values.

2.3.5 Dynamic Models

The presence of temporal derivatives in the model is essential while studying the dynamic behaviour of the process. These models are called 'dynamic models' or 'unsteady-state models'. The dynamic models can give the steady-state behaviour of the process because an unsteady-state process, if stable, always approaches steady state as time approaches infinity. A steady-state model is easier to solve than an unsteady-state model due to the absence of the temporal derivative. Dynamic systems are always required to solve the problems related to process control, plant shutdown or plant start-up operations (Korn 2007; Luyben 1996).

2.3.6 Stochastic Models

The random numbers are characterised by their moment. The first moment is known as 'average' or 'expectation'. A model may consider only average properties, thus avoiding the complexity of the process due to the randomness of the properties.

Many models consider random behaviour of the parameters, although the outputs of the models are average properties. These models also are probabilistic in nature. Various moments of the random variables may be determined analytically or by Monte Carlo simulation (Rubinstein and Kroese 2008). The random variables follow a probability distribution function. The choice of the distribution is determined from the assumptions involved. Some of the stochastic models make use of Markov chains, a name given to the processes in which the present states are functions of only just previous states (Ramkrishna 2000). Another class of models dealing with random numbers is the population balance models, which are discussed in Section 2.3.7.

2.3.7 Population Balance Models

In multiphase systems – which are encountered in chemical engineering, biochemical and biomedical engineering, pharmaceutical applications, aerosol formation, particulate systems, etc. – the dispersed phases are present in the forms of particles, droplets or bubbles. The particles can agglomerate or break, thus changing the number of identities of the dispersed phases present. The bubbles and droplets break and coalesce. The breakage and coalescence or agglomeration are analogous to the death and birth of live species. Therefore, these processes are sometimes also called 'birth and death processes'. Each entity of the dispersed phase is called an individual, and the number of the entire entity is called a population. Due to breakup and coalescence or agglomeration, the sizes of the individuals of the dispersed phase change. The number and size of dispersed phases change with time. An appropriate consideration of these provides us with information about the 'particle-size distribution', 'bubble-size distribution' or 'droplet-size distribution'. The approach is equally valid for the crystallisation process, which also involves the nucleation and growth of crystals (Ramkrishna 2000). The population balance models require consideration of few discrete variables. The population balance models can also be classified under the category of stochastic and discrete-event models.

2.3.8 Agent-Based Models

The problems related to the supply chain are quite complex in the way in which the decisions are taken by various agents involved in the supply chain. Such problems can be modelled by a method called 'agent-based modelling'. It consists of modelling various agents and their interrelationships. Agents in the model are software entities that are autonomous, reactive, proactive and capable of interacting with other agents. This type of models is limited to situations where strong human behaviour is involved (Dam et al. 2007). The model is treated as discrete events.

2.3.9 Discrete-Event Models

In most cases, all the variables are considered continuous variables. There are many situations in which the variables are not continuous but discrete. Even time may be considered a discrete variable. This is true when considering the molecular level. Stochastic models and agent-based models are based on discrete events. The molecular dynamic simulation, used for estimation of thermodynamic properties, is also a discrete-event simulation. The definition of states as discrete events must be clear. It may be achieved by making use of Petri nets or many other techniques (Wainer 2009). Modelling of discrete events requires abstraction of data, for which object-oriented programming is better suited than the procedural languages. Many situations involve discrete events as well as continuous events. Such processes are also modelled as discrete-event models. Dessouky and Roberts (1997) have described such events as 'combined events'. The basic requirements of programming languages and their development to handle continuous simulation were discussed.

2.3.10 Artificial Neural Network–Based Models

In several cases, it is required only to establish a relationship between the inputs and output variables. The exact relationship is almost impossible to establish within the time available. Under such circumstances, it may be desirable to look for a black-box approach. Models based on artificial neural networks (ANNs) fall in this category. ANNs usually consist of interconnected nodes arranged in layers. The information from one layer is used to pass values to the next layer, as shown in Figure 2.7. The first layer is the input layer. Each input enters at one mode. From each node of this layer, the processed data is sent to each node in the next layer. In Figure 2.7, it has been shown by the connecting arrows. The last layer is called the 'output layer'. Each node of this layer gives the output. The layers between the input and output layers are called 'hidden layers'. While the number of input layers and output layers is equal to the model inputs and outputs, respectively, the number of hidden layers and number of nodes in each hidden layer are not known. They are fixed by using optimisation techniques known as training. At the most, three hidden layers are sufficient to relate inputs and outputs, provided that a relationship exists. The number of nodes in each hidden layer is determined by training. Although ANN-based models do not increase any understanding of the phenomena, an algorithm for estimation of model

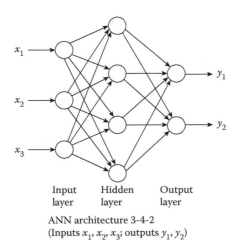

Input layer Hidden layer Output layer

ANN architecture 3-4-2
(Inputs x_1, x_2, x_3; outputs y_1, y_2)

FIGURE 2.7
Structure of artificial neural network.

outputs from the known inputs is obtained if a relationship exists. The use of ANN has the advantage that as the amount of data grows, the relationship between the input and output becomes refined.

ANN-based models may be used as sub-models in a large model. ANN has also been used to implement 'artificial intelligence' or 'expert systems'. Its ability to solve narrowly defined, specific industrial problems has attracted many (Stephanopoulos and Han 1996). It is frequently used as a software sensor, also known as an 'electronic nose', and also in process control applications such as controller and fault diagnosis system. (Hussain 1999). ANN-based models are also used for very complex systems to correlate various parameters and then predict the outcomes from the knowledge of the information available. The only disadvantage is that no interpretation can be given confidently.

2.3.11 Fuzzy Models

Fuzzy models take inputs which are not well quantified. Qualitative parameters, such as 'large', 'very large' and 'small', are used as inputs. Such variables are called 'fuzzy variables'. These are quantified with membership functions which are different from probabilities. The outputs of a model may also be fuzzy variable. These models are based on 'fuzzy' mathematics or 'fuzzy logic' involving fuzzy variables. Qualitative terms used as inputs and outputs are used almost daily in our life. Instead of mentioning the temperature of water, it is usually said that the water is 'warm', 'cold', 'very cold' and 'chilled'.

A properly developed fuzzy system can model non-linear functions of arbitrary complexity. It can make use of 'expert knowledge' and can be blended with conventional mathematics. Various modelling software has toolboxes which can handle fuzzy systems.

The calculations are made by converting the qualitative states into numbers. This step is called 'defuzzyfication'. After calculations, the results are again converted into qualitative terms that are easily understood by the user. This step is called 'fuzzification'. Fuzzy models are frequently used in expert systems (Stephanopoulos and Han 1996) and in process control applications (Zhang and Liu 2006).

Many other terms quite commonly used are not listed above. Many simple models consist of equations which can be solved analytically. These models are called 'analytical models'. These models have attracted modellers for a long time and are still attractive since they can be interpreted easily. The behaviour of the process in unknown conditions can be predicted with full confidence. The analytical models, for this reason, are used as benchmarks to test the effectiveness of new algorithms. While the models solved numerically may face the problem of convergence, the analytical solutions are free from it. At the same time, they almost always require less computational effort. The analytical solutions can be defined in the form of a 'function', 'subroutine' or 'class' (in OOPLs) and used as a sub-model. The numerical solutions of the sub-models cannot be integrated confidently into other sub-models or models.

The models involving many variables in algebraic or differential equations cannot be solved analytically and have to be solved numerically. Such models are called 'numerical models'. The results are often presented in the form of several graphs (depending upon the complexity of the problem). The development of numerical models requires extensive validation using not only experimental data but also benchmark problems. This is essential due to several factors ('round-off' errors, wrong initial guesses leading to undesirable minima or maxima, the choice of a small 'step size' etc.). Sometimes, the numerical scheme does not converge at all.

From the point of view of the nature of the models, the major differences are between the deterministic, discrete-event models and the stochastic models. The deterministic models consist of equations in continuous variables. All the discrete-event models handle discrete states (time or any other variable). Stochastic models take care of both random variables and discrete events. The simulation methodology for deterministic models differs from that of discrete-event models and stochastic models. The difference between static models and dynamic models is based on the presence of time and their capability to handle dynamic behaviour. The difference between analytical models and numerical models is based on the way the solutions are obtained. The ANN-based models are algorithm based and require no prior knowledge about the interrelationship and dependence of variables on each other. The fuzzy models are based on the use of fuzzy logic and can be used when crisp knowledge from the experts is not available.

It should also be clear that not all the available software can handle all types of models. There is dedicated software for a particular type of model. For example, CFD software can handle problems related to only flow of the fluid and cannot be used for the development of any other type of model. It is quite common to use one or more types of software for the simulation of large plants. Computer languages, such as MATLAB, have a number of toolboxes to handle various types of models; still, they cannot handle CFD problems. FEMLAB was developed to fill this gap. The general-purpose language requires coding of the model equations using the syntax and semantics of the software. The model equations are obtained by the user. The dedicated software such as CFD software generates an appropriate set of models of equations and hence has limited scope. At the same time, it also becomes necessary that output of such software may be rendered in various formats, and inputs from various devices should be acceptable. These facts should be kept in mind while choosing software for modelling and simulation work.

2.4 Types of Equations in a Model and Solution Strategy

The discussion in this chapter gives an idea of the types of equations in a model. The equations may be categorised into the types discussed in this section.

2.4.1 Algebraic Equations

These equations are easy to use to understand the interrelations between various parameters. If all equations are linear, then the model is called a linear model. The set of linear algebraic equations can be solved easily by any method based on matrix inversion (Davis 2001; Ramirez 1997). Almost all software today, including spreadsheets (e.g. Microsoft Excel and Openoffice's CALC), has functions to determine the inverse of a matrix. Software such as MATLAB, SCILAB, MATHEMATICA and MATHCAD has very simple syntax to determine the inverse of a matrix. For example, let us examine the command used in MATLAB. If **A** is the matrix whose inverse is to be determined, then the following small code determines the inverse of **A**.

```
inv(A)
```

If any one or more of the equations is non-linear, then the model is called a non-linear model. Not all, but only a few, of the non-linear equations can be solved analytically. These analytical solutions are useful for the sub-models or are used as benchmarks for evaluating the numerical methods used.

The non-linear models can also be solved numerically either by converting the model into the linear model by linearisation (Shacham and Mah 1978) or by using methods for non-linear optimisation (Ramirez 1997).

A vast literature, including various textbooks on the methods for non-linear optimisation, is available. The optimisation problems can be solved by software such as GAMS and MATLAB with optimisation toolbox. MATLAB also has an 'fsolve' function to solve non-linear equations. MATHCAD has a 'find' function to solve systems of non-linear equations. Spreadsheets, such as Excel, also have a Solver add-in to solve optimisation problems. This approach requires formulation of the problem as an optimisation problem. For example, the non-linear equation $F(x) = 0$ may be written as $F(x) = R$, where R is the residual. Now, let us consider $|R|$ as an objective function which is to be minimised. In this manner, a non-linear equation can be converted as an optimisation problem.

A set of non-linear equations may be expressed by

$$F_i(\mathbf{x}) = 0; \text{ for } i = 1 \text{ to } N \text{ and } \mathbf{x} = x_1, x_2, \ldots, x_N \tag{2.3}$$

The equation may be written as $F_i(\mathbf{x}) = R_i$, $i = 1$ to N. There are many residuals, and an appropriate objective function combining all of them may be defined. Since the optimisation methods use search methods, it is possible to specify the range of variable in which the solution exists. The problems associated with the optimisation approach are that (1) the obtained optima may be local optima and (2) the solution does not converge or stick at particular values. For obtaining global optima, techniques such as a genetic algorithm, simulated annealing and the like are employed. The problem of convergence may be overcome by using an efficient optimisation method, or several times by changing the initial guess. Unfortunately, there is no single optimisation technique which can be used efficiently for all problems. An appropriate optimisation technique has to be specified. It comes with the experience in identifying the problem and knowledge of the optimisation method available with the software. Software used for optimisation problems has various options available which may be chosen by a simple statement. For example, GAMS uses a statement such as

```
Solve mymodel using nlp minimizing cost.
```

In this example, *mymodel* is the name given for a set of equations. The word *minlp* tells the solver to use non-linear programming. The 'cost' is the objective function which is to be minimised.

The initial guess is also one of the important data required for obtaining the good and quick estimates of the optimum solution. Fortunately, the problems related to chemical engineering are real, and good initial guesses can be made in most cases.

Another approach which has been used in a lot of software and by many researchers is to linearise the non-linear equations. The set of non-linear equations may be written as

$$F_i(\mathbf{x}) = -\sum_j J_{ij}(\mathbf{x})\Delta x_j \quad \text{for } i = 1 \text{ to } N \tag{2.4}$$

where $J_{ij}(\mathbf{x}) = \dfrac{\partial F_i(\mathbf{x})}{\partial x}$.

The equation is again iterative. The right-hand side is completely known, and hence Equation 2.4 is linear. An initial guess for the vector $\mathbf{x}$ is required, and the estimation of the terms of the Jacobian matrix uses the values of the function estimated using the old values of x. The iteration is carried out until the right-hand side becomes less than the tolerance.

The derivatives may be estimated analytically in many cases; otherwise, they are estimated numerically. This approach requires only function estimations and the inversion of a matrix.

2.4.2 Differential Equations

Differential equations are obtained after applying laws of conservation of mass, energy or momentum. Population balance models with 'birth and death' processes also produce differential equations. If only one type of derivative is involved, then the ODEs are obtained. ODE is generally expressed as

$$y' = \frac{dy}{dt} = f(x,y,t) \tag{2.5}$$

It is true only for the unsteady-state (temporal derivative) lumped parameter model (no spatial derivative) or steady-state (no temporal derivative) 1D distributed parameter model (only one spatial derivative). In all other cases, partial differential equations (PDEs) are obtained. These equations involve derivatives with respect to more than one parameter, for example:

$$\frac{\partial y}{\partial t} = f\left(x, y, \frac{\partial y}{\partial x}, t\right) \tag{2.6}$$

Sometimes, PDEs are converted into simultaneous ODEs or algebraic equations. In cases where the phenomena vary with time, the spatial variables are discretised and 'time' is considered a continuous variable. The PDEs are thus converted into a large system of coupled ODEs with derivatives with respect to time. It happens in the case of parabolic and hyperbolic equations obtained in unsteady-state models. This method has been termed a 'method-of-lines'.

When the time is constant, such as in steady-state models, the resulting equations after discretisation of the spatial coordinates result in algebraic equations which can be solved by methods used for solving algebraic equations. Such a situation occurs in elliptic equations.

Solving the differential equations depends upon the boundary conditions, which are obtained by applying the laws of physics. In several cases, the solutions may be obtained analytically. The analytical solutions simplify a complicated model if sub-models having these solutions are used. These solutions are also used as benchmarks while developing codes for solving differential equations.

Complex problems often require numerical methods to solve the differential equations. The most widely used method is the Runge–Kutta (RK) method (Davis 2001; Ramirez 1997). Various algorithms based on this approach have been developed to solve differential equations of various types, often called 'non-stiff' and 'stiff' problems. The RK method is an initial value problem (i.e. it requires values specified at some point). The boundary value problems can also be solved using the RK method. In such cases, the initial values are guessed so that they provide the desired final value.

Various ODE solvers are available in software such as MATLAB, SCILAB, MATHCAD and R (Soetaert et al. 2010). These solvers are easy to use. For example, solving an ordinary differential using MATLAB requires definition of the ODE in a function and then using the following one-line command:

```
[t,y] = ode45(@myfun,[tinitial tfinal],[yinitial tinitial]).
```

A file named myfun.m defines the differential equation. The above command specifies the initial values of y and the initial and final values of t.

Higher order differential equations are written in terms of many first-order differential equations, which are then solved using the RK method. Guidelines to use an appropriate type of solver are documented and should be read.

Though the RK method has been widely used, another approach known as the predictor–corrector method is becoming more popular now. The RK method requires information about the solution at a single point $x = x_n$. Then the method proceeds to obtain the value of y at the next point $x = x_{n+1}$. Therefore, this method is a one-step method. Multistep methods make use of information about the solution at more than one point. Since information on some of these points may not be available, two equations are solved. The first equation, called the 'predictor formula', is used to get an approximate value of the solution. This solution is used in the second formula, called the 'corrector formula', to obtain a refined solution. The predictor formula is explicit (open type), and the corrector formula is implicit (closed type). An Adams–Bashforth fourth-order formula may be used as a predictor formula, with an Adams–Moulton fourth-order formula being used as a corrector formula (Rice and Do 1995). The current trend is to use the predictor–corrector formula in place of the well-known RK method.

2.4.3 Differential–Algebraic Equations

Many models consist of differential equations as well as algebraic equations. These types of mixed equations are called as differential-algebraic equations (DAEs). Such equations are generally obtained in several unsteady-state unit operations involving unsteady-state material and energy balance equations, equilibrium relationships, consistency equations etc. Simultaneous solutions of mixed types of equations cannot be solved easily.

DAEs are solved generally by first converting into ODEs, followed by using ODE solvers (Soetaert et al. 2010). The algebraic equations are differentiated and, after appropriate substitutions of variables, DAEs are converted into a system of ODEs. The minimum number of differentiations to convert DAEs into ODEs is called an 'index', which plays an important role in the solution of DAEs. The transformation of DAEs into ODEs depends upon the type of equations.

Once the DAE has been converted into ODE, software such as MATLAB, SCILAB, MATHCAD, R etc. also can be used to solve DAE.

The degree of difficulty in solving the model equations, in increasing order, is algebraic equations < ODE < PDE < DAE.

2.5 Sources of Equations

During conceptualisation of a simple picture of the complex process, a search for the source of the equations to be used for the modelling and simulation is also kept in mind. Though this step precedes solving model equations, it is being discussed after understanding the types of equations and the difficulty associated with the methodology in solving the model equations. The source of the equations depends on the field of application and the knowledge of the modeller. The source of equations for the development of models are conservation equations for mass, energy, momentum and other quantities such as electric charge, equilibrium laws at the interface(s), constitutive laws, kinetic expressions, boundary conditions and optimisation criteria which are discussed in the literature (Rodrigues and Minceva 2005). The source of equations can be classified into the categories discussed throughout the remainder of this section.

2.5.1 Empirical Equations

The empirical equations are usually obtained by fitting the experimental data in terms of important parameters with some simple functions. Most of the equations allow us to estimate the parameters explicitly. However, it is not necessary, and implicit equations are also used. For example, the Von Karman equation (McCabe et al. 1993) for friction factor is an implicit equation. The friction factor, f, appears on both sides of the equation. It cannot be estimated without using a numerical technique involving an iterative procedure.

$$\frac{1}{\sqrt{f/2}} = 2.5\ln\left(\mathrm{Re}\sqrt{f/8}\right) + 1.75 \tag{2.7}$$

Earlier experimental works correlated the experimental data using empirical correlations. Power law, exponential and polynomials were quite commonly used. Equations for f in the laminar and turbulent regimes, and various formulae for the predictions of diffusivity, vapour pressure, and correlations for mass and heat transfer in various geometries fall under this category. Many equations are purely empirical in nature and do not have any sound theoretical basis. Though these empirical correlations are easy to use, care must be taken not to extend the use of these equations beyond the range of application. These equations are not exact; hence, the error in the predicted values should be kept in mind. Use of empirical equations for simple cases does not help much in enhancing understanding of the phenomena. However, these may be used in simple sub-models or as initial guesses in rigorous models.

2.5.2 Equations Based on Theoretical Concepts

Many equations are based on simple laws. For example, ideal gas law relates the pressure, temperature and volume of a gas. It is true at low pressures and high temperatures and can be used for engineering purposes. When an equation of a state of order more than three is required, the ideal gas law still is used to obtain the initial guess for iterative calculations. The equations based on simple laws can be used in simple models and sub-models more confidently than pure empirical equations.

Most of the thermodynamic methods for the estimation of various properties are of this type. These include various equations of states and equations to describe vapour pressure and enthalpies.

Another approach used for estimation of thermodynamic properties is the group contribution methods. In this approach, it is assumed that the property of a substance depends upon the type and number of groups that are present in a compound and the composition of the mixture that has these compounds. The methods, such as UNIFAC, UNIQUACK and ASOG, have been developed for the estimation of liquid fugacity, which is essential for the estimation of vapour–liquid equilibria. Group contribution methods can be used to estimate the properties such as critical properties, heat capacity and Gibb's free energy (Poling et al. 2001). Based on this approach, the experimental data are fitted according to the equation arrived at after a certain theoretical concept.

2.5.3 Consistency Equations

These are based on the fact that quantities such as the fractional composition of a single phase in terms of mass and mole, when added, are equal to one. Some more examples are volume fraction, fraction of mass flow rate and mass fraction of particles used in particle-size distribution. Whatever law is applied, these equations are always satisfied and are called 'consistency equations'. Values not satisfied by these equations indicate either a calculation mistake or unaccounted losses such as leakage. Therefore, the consistency equations are very essential in a model. In many cases, the consistency equations are used to obtain initial guesses for the composition. For example, for a phase consisting of N species, the following will be the consistency equation:

$$\sum_{i=1}^{N} x_i = 1 \tag{2.8}$$

where x_i is the composition of the ith species. The initial guess for x_i may be taken as $x_i = 1/N$ for $i = 1$ to N.

2.5.4 Differential Equations Using Laws of Conservation

The differential equations are obtained when a law applied is valid at all the points in the system. The equation of continuity and the momentum–balance equation are obtained in this manner. The subject is taught at the undergraduate level in the course named Transport Phenomena. These equations are capable of taking into account the generation and depletion of a chemical species or energy generated or consumed in a chemical reaction, viscous dissipation, electrical or electrochemical effects on the flow and other phenomena. The generalised equation of change is known as the Navier–Stokes equation.

2.5.5 Integration over Area, Volume and Time

Many quantities involve integration of a quantity over time, space or both. For example, the flow rate, Q, can be determined by integrating the velocity profile, v, over the cross-sectional area, A.

$$Q = \int\int v\,dA \tag{2.9}$$

Similarly, the shear stress at the wall may be integrated over the entire area to calculate the drag force:

$$F_D = \int\int \tau_w\,dA \tag{2.10}$$

Other equations involving integration are obtained while averaging the parameters. Average velocity is obtained by integrating point velocity over an entire cross-sectional area. For example, time-averaged velocity and number-averaged particle diameter involve integrals.

2.5.6 Population Balance

These equations are obtained while writing population balances. The birth rates may be considered to depend upon the population, food available and other relevant environmental parameters. Similarly, the death rates may be considered to depend on population, with other species using this species as food, availability of food and environmental parameters. Growth in the size of individuals is considered in the model. The growth rate may be considered to depend upon environmental parameters and breakage or agglomeration or coalescence. These models are frequently used in biochemical processes involving microorganisms.

2.6 Simplifying Concepts

The purpose of model development is to simplify the complexity of the process while retaining the relevant characteristics of the process. It is achieved by making appropriate assumptions. Some of the assumptions are so frequently used that they look very obvious. However, choice of a wrong type of assumption will result in prediction of the performance of the process that is quite different from that experimentally observed. Some of the simplifications are made with a point of view of solving the model equations. For example, elimination of dependent or independent variables or reduction to ODE or PDEs, reduction to homogeneous or canonical forms and simplification of geometry and use of dimensionless numbers were discussed by Basmadjian (1999). Assumptions can also be made to ignore one or more of the unimportant mechanisms or phenomena. Some of the frequently used assumptions to simplify a process are discussed here.

2.6.1 Continuum

There are many processes which involve dispersed phases (Figure 2.8). The fluidised beds have solids dispersed in a fluid phase. Bubble columns, mechanically agitated gas–liquid reactors and other gas–liquid contactors consist of gas dispersed in liquid. The spray dryer has liquid dispersed in air. The emulsions contain one liquid dispersed in another liquid. It must be remembered that the laws of conservation are applicable in a continuous phase. Boundary conditions are applied at the interface. Doing it for the entire population of the dispersed phase is quite tedious. It involves writing a separate model for individual populations of the dispersed phase. Let us call this a 'sub-model'. The sub-model may take into consideration the property

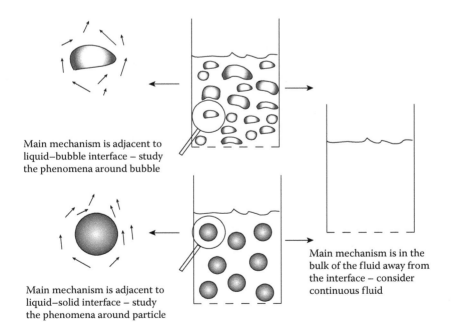

Main mechanism is adjacent to liquid–bubble interface – study the phenomena around bubble

Main mechanism is adjacent to liquid–solid interface – study the phenomena around particle

Main mechanism is in the bulk of the fluid away from the interface – consider continuous fluid

FIGURE 2.8
Dispersed phase and assumption of continuum.

of the 'individual'. The model for the overall process will consider the effect of these sub-models. It makes the model rigorous with enhanced computational effort, which is worth it only if the increased accuracy and widening of the scope of the model are achieved.

Simple models frequently consider the entire dispersion to be a continuous phase. This assumption seems valid for fine sizes of drops, bubbles or particles because the units of the dispersed phase move with the fluid element of the dispersing phase. The laws of conservation for each unit of the dispersed phase can be avoided. Transport properties such as viscosity, thermal conductivity, diffusivity are usually defined for the dispersion. These models, however, are not capable of predicting the details of individuals of the dispersed phase. Therefore, the assumption of a continuum is to be used when the main mechanism is away from the interphase of the dispersed and continuous phases. For example, in the case of bubble columns, if the interest is to find the gas–liquid mass transfer rates, the main mechanism related to it takes place at the gas–liquid interphase. Therefore, the assumption of a continuum is meaningless. In the case of determination of heat transfer coefficient between the wall and the gas–liquid dispersion, the continuum assumption simplifies the model equations. The assumption of continuum should be used cautiously in case the size of an individual of the dispersed phase is large. In such cases, the dispersed phase does not move with the fluid and requires laws of motion for each individual. For example, in cases of fluidised beds with fine particles, the assumption of a continuum is acceptable. It is acceptable in the case of large particles also, since the fluid velocity to fluidise the particles also increases significantly. As a consequence, the concentration of the solids is uniform in the fluidised beds. In the case of slurry bubble columns, it has been observed that the assumption is not valid. The solid concentration decreases from the bottom towards the top of the bed. The assumption of a continuum is incapable of predicting this trend. However, in cases of large particles (>1 mm), the axial solid concentration profile is uniform and the reactor is called 'fluidised beds'.

In short, the assumption of continuum is used to avoid the consideration of an individual entity of the dispersed phase.

2.6.2 Combination of Simple and Rigorous Models

A process model may be broken into various sub-models. The large and complex process models thereafter are simplified by using a combination of simple and rigorous models. It can be done in two ways. Simple models may be used as sub-models, and these sub-models may be combined in a rigorous manner. For example, while modelling mass transfer in a bubble column, a film model may be assumed at the surface of the bubble. While combining the mass transfer of all the bubbles, the bubble-size distribution and axial distribution of the concentration in the bulk may be considered

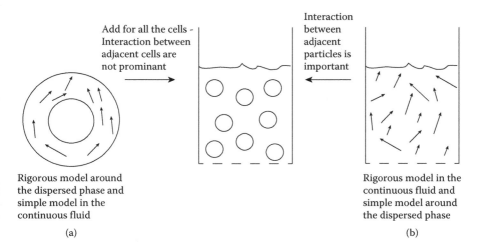

Add for all the cells - Interaction between adjacent cells are not prominant →

Interaction between adjacent particles is important ←

Rigorous model around the dispersed phase and simple model in the continuous fluid

(a)

Rigorous model in the continuous fluid and simple model around the dispersed phase

(b)

FIGURE 2.9
Combination of simple and rigorous models: (a) bubble column and (b) cell model in a fixed bed.

rigorously (Figure 2.9a). There might be another way of combining simple and rigorous models. The sub-models may be rigorous, and these sub-models may be combined using a simple methodology. For example, a cell model for a fixed bed considers many cells, each containing one particle with surrounding fluid. The rigorous model is applied in a cell. The overall performance of the bed is a simple combination of these cells (Figure 2.9b).

2.6.3 Uniform Probability Distribution

In case it is required to consider the individual entity of the dispersed phase, it is assumed that all individual entities of the dispersed phase follow uniform distribution law. It helps in estimation of average properties and evaluation of summation and integrals. For example, the 'penetration' model for transfer processes assumes that all the fluid elements spend the same time at the transfer surface (Figure 2.10). The surface renewal model assumes that the probability of leaving the transfer surface is the same for all the fluid elements present on the transfer surface. This, however, results in exponential probability distribution of the time spent on the surface. In adsorption, the probability of occupying the empty space on the surface is the same. The importance of this assumption will be made clear by examples discussed later in this book (see Chapters 4 and 5).

2.6.4 Parallel Mechanisms

Mass and momentum transfer takes place due to molecular motion and convection. The mechanism of heat transfer is conduction, convection and radiation due to molecular motion, bulk flow and infrared radiation.

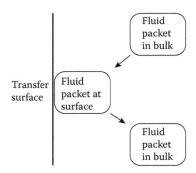

Penetration model: All fluid packets remain at
surface at constant time
Surface renewal: The residence time follows
exponential distribution

FIGURE 2.10
Penetration and surface renewal model.

These mechanisms are considered to work parallel to each other. It looks quite obvious to assume so. However, inherent to this assumption is that the different mechanisms do not interact with each other. The parallel mechanism allows us to treat different mechanisms separately and simplifies the model equations.

2.6.5 Analogy to Electrical Circuits

The flow of various quantities such as fluid flow, rate of heat transfer and rate of mass transfer is considered analogous to the flow of current in electrical circuits (Figure 2.11). The total amount of the quantity transferred can be obtained by integrating with respect to time. The driving force for transfer of these quantities is analogous to electrical current which follows Ohm's law. The rate of diffusive mass and conductive heat transfer follow Fick's and Fourier's laws, respectively. These laws are analogous to Ohm's law. The mass transfer rate by diffusion is proportional to the concentration gradient. The mass transfer rate by convection is also considered proportional to concentration difference. The proportionality constant is called the mass transfer coefficient. Similarly, the heat transfer rate by conduction is proportional to the temperature gradient. The heat transfer rate by convection is also considered proportional to temperature difference. The proportionality constant is called the heat transfer coefficient. The heat transfer due to radiation does not follow Fourier's law. It is described by Stephan–Boltzman's law (i.e. the rate of heat transfer is proportional to the difference of the fourth power of the temperature). Thus, the analogy to electrical circuits is only a simplification of the more complex laws of the transfer processes. However, a resistance to transfer processes is defined in a similar manner, is used to

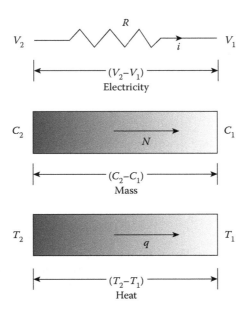

FIGURE 2.11
Analogy of mass and heat transfer with electrical circuits.

TABLE 2.2

Analogy of Heat and Mass Transfer with Electrical Circuit

Quantity	Measured As	Driving Force	Flow Rate	Resistance
Electricity	Charge, Q	Voltage, V	Current, $i = Q/t$	V/i
Heat	Thermal energy, Q	Temperature difference, ΔT	Rate of heat transfer, $q = Q/t$	$\Delta T/q$
Mass	Concentration, mole or mass/volume, c	Concentration difference, Δc	Rate of mass transfer, $J = N/t$	$\Delta c/J$

treat various mechanisms of transfer processes as parallel mechanisms and allows the addition of resistances to transfer processes in parallel or in series in a similar way as was done in the case of electrical circuits. The interpretation of driving force, current and resistance in heat and mass transfer analogous to electricity is presented in Table 2.2.

2.6.6 Film Model

This simple concept has been used extensively. It is based on the assumption that the entire variation of a quantity such as temperature, concentration or velocity varies only in a small region adjacent to the wall (Figure 2.12). This variation is linear. The assumption is helpful in applying the concept of analogy

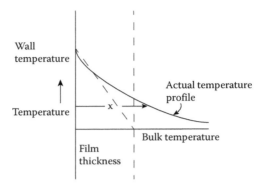

FIGURE 2.12
Film model.

with the electrical circuits. The differential equations are simplified since the derivative is reduced to the ratio, which is a constant. It must always be kept in mind that, in reality, no film exists. The film is a concept only and is often termed a fictitious film. In reality, the concentration or temperature varies continuously from wall to the bulk or centre of the cylindrical vessel at low velocity. At high velocities, it might be better to consider that the variation is only up to a finite distance. The thickness of the film is an important parameter in the model. Since in reality no film exists, the film thickness is not a measureable quantity. It is calculated generally by fitting the experimental data.

2.6.7 Boundary Layer Approximation

It is also based on the assumption that the entire variation of a quantity such as temperature, concentration or velocity varies only in a small region adjacent to the wall (Figure 2.13) and is described by a non-linear function satisfying the boundary conditions. The boundary layer is also fictitious, and in reality it does not exist. The boundary layer extends to a finite dimension known as the 'boundary layer thickness'. The concept used to solve the problems in fluid flow, mass transfer or heat transfer is based on the concepts of momentum, concentration and thermal boundary layers, respectively. All have different thicknesses. Another fact which makes the boundary layer very different from the film is that the boundary layer thickness may vary with the position, as in the case of flat plate it develops from the front of the plate. The boundary layer thickness increases with the distance from the front of the flat plate. In case of an unsteady-state boundary layer, the boundary layer thickness varies with time. In brief, the boundary layer only approximates the real behaviour and does not neglect any of the spatial or temporal variables. The boundary layer may not extend to the bulk and hence requires a buffer layer to describe the variation of the quantity (velocity, concentration or temperature) between the boundary layer and the bulk. In reality, the concentration or temperature

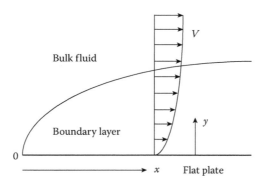

FIGURE 2.13
Boundary layer approximation.

varies continuously from wall to the bulk or centre of the cylindrical vessel at low velocity. Hence, only at high velocities might it be better to consider that the variation is only up to a finite distance. Due to the non-existence of the boundary layer, its thickness is not a measureable quantity and, hence, is calculated by fitting the experimental data.

2.6.8 Order of Magnitude Approximation

The partial differential equation obtained as a result of momentum balance, mass balance or energy balance consists of the three spatial variables, with time as the fourth variable. The equation can be simplified if the main mechanism of mass transfer or heat transfer is by convection in one direction and the main mechanism in other directions is diffusion or conduction, respectively. The ratio of the convective and diffusive terms can be expressed in terms of Peclet number, which is a product of Reynolds number, with Schmidt number or Prandtl number in the case of mass and heat transfer, respectively. If the ratio is high, then the diffusion or conduction in other directions can be neglected. This analysis also points out the importance of dimensionless numbers. The order of magnitude approximation was an important way of simplifying the model equations when the emphasis was to obtain the analytical solutions. Today, CFD software is capable of solving the equations with all the variables. However, the simplification reduces the computational effort and avoids the situation of non-convergence.

2.6.9 Quasi–Steady State

Unsteady-state processes while developing the distributor models involve partial derivatives. The number of equations may be more than one. Solving such problems analytically has not been possible in all cases. Numerical solutions face problems of convergence. The assumption that, during a small

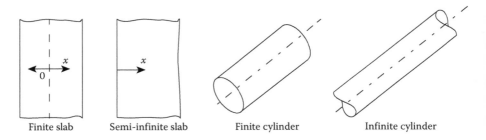

| Finite slab | Semi-infinite slab | Finite cylinder | Infinite cylinder |

FIGURE 2.14
Finite and infinite dimensions.

interval of time, the process is at steady state is known as 'quasi–steady state'. By making such an assumption, the time derivative is separated from the differential equation. The time derivative, however, is present in another equation. As a result, the difficulty level of the solution of model equations is reduced. This equation is equivalent to treat time as a discrete variable.

2.6.10 Finite and Infinite Dimensions

All the physical objects have finite dimensions. However, many mathematical functions approach a fixed value when one of the parameters such as time or distance approaches infinity or zero. When a large value of such a variable is observed, then the experimental values are close to the value at infinity, if not, then at least within experimental error. It results in simplification of the governing model equations or solutions. It is quite common to consider the dynamic behaviour at infinite time the same as that at steady state. A needle may be considered as an infinite cylinder if the tip is not participating in the process. Analytical solutions for semi-infinite slabs, infinite cylinders demonstrate (Figure 2.14) the use of the concept of infinity.

2.7 Summary

The model development involves understanding the complexity of a process, visualising a simplified picture of the process, transforming it into mathematical equations, solving the model equations and then comparing the model predictions with the experimental data. Choice of the software to be used for simulation may depend upon the type of the model. Commonly used assumptions during model development and the types of equations obtained were discussed. A brief overview of the strategy for solving the model equations was presented.

The models developed in future will be more rigorous due to enhanced computational facility, enhanced understanding of the process and expanding scope of application of the chemical processes. For example, the development of microfluidic devices, nanotechnology, microwave processing and molecular dynamic simulations will require models for a wide range of applications.

References

Basmadjian, D. 1999. *The Art of Modeling in Science and Engineering*. Boca Raton, FL: CRC Press.

Chung, C.A. 2004. *Simulation Modelling Handbook: A Practical Approach*. Boca Raton, FL: CRC Press.

Dam, K., Lukszo, Z., Sinivasan, R., and Karimi, I. 2007. Opportunities for agent-based models in computer aided process engineering. In *Proceedings of the European Symposium on Computer Aided Process Engineering ESCAPE 17*. Amsterdam: Elsevier. Retrieved from http://www.nt.ntnu.no/users/skoge/prost/proceedings/escape17/papers/T1-430.pdf

Davis, M.E. 2001. *Numerical Methods & Modeling for Chemical Engineers*. New York: John Wiley.

Dessouky, Y., and Roberts, C.A. 1997. A review and classification of combined simulation. *Comput. Ind. Eng.* 32(2): 251–264.

Dobre, T.G., and Marcano, J.C.S. 2007. *Chemical Engineering*. Weinheim: Wiley-VCH.

Fitt, A.D., Lee, M.E.M., and Please, C.P. 2007. Analysis of heat flow and 'channelling' in a scraped-surface heat exchanger. *J. Eng. Math.* 57: 407–422.

Hussain, M.A. 1999. Review of the applications of neural networks in chemical process control—Simulation and online implementation. *Artif. Intell. Eng.* 13: 55–68.

Korn, G.A. 2007. *Advanced Dynamic-System Simulation-Model replication techniques and Monte Carlo simulation*. Hoboken, NJ: Wiley.

Luyben, W.L. 1996. *Process Modeling, Simulation, and Control for Chemical Engineers*, 2nd ed. New York: McGraw-Hill.

McCabe, W.L., Smith, J.C., and Harriott, P. 1993. *Unit Operations of Chemical Engineering*, 5th ed. New York: McGraw-Hill.

Poling, B.E., Prausnitz, J.M., and O'Connell, J.P. 2001. *The Properties of Gases and Liquids*, 5th ed. New York: McGraw-Hill.

Ramirez, W.F. 1997. *Computational Methods for Process Simulation*, 2nd ed. Oxford: Butterworth-Heinemann.

Ramkrishna, D. 2000. *Population Balances: Theory and Applications to Particulate Systems in Engineering*. San Diego, CA: Academic Press.

Rice, R.G., and Do, D.D. 1995. *Chemical Engineering: Applied Mathematics and Modeling for Chemical Engineers*. New York: John Wiley.

Rodrigues, A.E., and Minceva, M. 2005. Modelling and simulation in chemical engineering: Tools for process innovation. *Comput. Chem. Eng.* 29: 1167–1183.

Rubinstein, R.Y., and Kroese, D.P. 2008. *Simulation and the Monte Carlo Method*. Hoboken, NJ: John Wiley.

Shacham, M. 1982. Equation oriented approach to process flowsheeting. *Comput. Chem. Eng.* 6(2): 79–95.

Shacham, M., and Mah, R.S.H. 1978. A Newton type linearization method for solution of nonlinear equations. *Comput. Chem. Eng.* 2: 64–66.

Soetaert, K., Petzoldt, T., and Setzer, R.W. 2010. Solving differential equations in R. *R J.* 2(2): 5–15.

Stephanopoulos, G., and Han, C. 1996. Intelligent systems in process engineering: A review. *Comput. Chem. Eng.* 20(6–7): 743–791.

Stephanopoulos, G., Johnston, J., Kriticos, T., Lakshmanan, R., Mavrovouniotis, M., and Siletti, C. 1987. DESIGN-KIT: An object-oriented environment for process engineering. *Comput. Chem. Eng.* 11(6): 655–674.

Wainer, G.A. 2009. *Discrete-Event Modeling and Simulation: A Practitioner's Approach.* London: CRC Press.

Zhang, H., and Liu, D. 2006. *Fuzzy Modeling and Fuzzy Control.* Boston: Birkhauser.

3

Models Based on Simple Laws

A large number of models for transfer processes that contain equations are based on laws of conservation of momentum, mass and energy. Some of the models that were developed initially are based on simple laws and were inspired by experimental observations. The laws, though limited in scope, are still used not only as sub-models of various rigorous models but also in various engineering calculations. However, it is essential to understand the simplicity as well as limitations of these models. The simple laws provide us with initial guess values required while using rigorous models. These laws can be used as sub-models or as assumptions if a given process is not very sensitive to a particular parameter. The use of numerical methods is helpful in easily solving some of the problems based on these simple laws. A few commonly used simple laws will be presented in this chapter. They include the ideal gas law, which is used for estimation of thermodynamic properties; the cubic equation of state (EOS), particularly the Peng–Robinson EOS; Newton's law of viscosity and its application in flow problems; Fourier's law for conduction; Fick's first and second laws for diffusion; kinetic rate expressions; the Arrhenius equation and isotherms for adsorption. Definitions of the heat and mass transfer coefficients and resistances to heat and mass transfer are presented. A few applications of the simple laws are presented to demonstrate the importance of these problems.

3.1 Equation of State

The relationship between the pressure, temperature and volume of gases and vapours is called as equation of state. Several thermodynamic properties (e.g. density, enthalpy and vapour–liquid equilibria) can be estimated using methods based on the use of EOSs. These properties determine the performance of various real processes, such as pressure drop in packed beds and interphase transport in multiphase processes involving compressible fluids.

3.1.1 Ideal Gas Law

The properties of the gases and vapours are used in many engineering calculations related to flow of gases, vapour–liquid equilibria, gas–liquid absorption, adsorption of gases etc. The concentration of a chemical species in a mixture of gases is generally expressed in terms of volume fraction. The specific volume of the gas is the reciprocal of density. Good approximations for several properties of real gases are obtained assuming ideal gas laws (Smith et al. 1996). In the ideal gas law, the simplest of the EOS is given as

$$PV = RT \qquad (3.1)$$

where R is the universal gas constant.

The internal energy for a real gas depends upon temperature and pressure. For an ideal gas, the internal energy and enthalpy depend upon temperature only. The latter is frequently used while making energy balances. The enthalpy change for an ideal gas is written as

$$\Delta H = \int_{T_1}^{T_2} C_p \, dT \qquad (3.2)$$

The correlations for specific heat, C_p, as a function of temperature are known for various gases and can be used for engineering calculation.

At ambient conditions, the compressibility factor for gases up to a few bars is close to 1; hence, the ideal gas law is applicable. For real gases, the ideal gas law gives a good initial guess while using numerical techniques to estimate volume from an EOS.

3.1.2 Cubic Equations of State

At elevated pressures and low temperatures, the ideal gas law is not applicable and an EOS for real gas has to be used. Several EOSs are reported in the literature. A class of EOSs for real gases can be expressed in terms of polynomials of the third degree in either volume or compressibility factor, Z. These EOSs have three roots and hence are called cubic EOSs. These EOSs require less calculation effort and are used while modelling the performance of process equipment. All the three roots of the cubic EOS can be obtained analytically. Van der Waals EOS was the first of these. Although many cubic EOSs have been proposed from time to time, only three of the cubic EOSs are given in Table 3.1. The Peng–Robinson EOS can be written in terms of Z, as given here (Peng and Robinson 1976):

$$Z^3 + (1-B)Z^2 + (A-3B^2-2B)Z - (AB-B^2-B^3) = 0 \qquad (3.3)$$

where $A = \dfrac{aP}{R^2T^2}$, $B = \dfrac{bP}{RT}$ and $Z = \dfrac{Pv}{RT}$.

If all the roots are real, then the largest real root corresponds to the gas phase and the smallest corresponds to the liquid phase. The Peng–Robinson EOS provides the roots for both gas and liquid phases more frequently than

TABLE 3.1

Selected Cubic Equation of States for Pure Gases

Investigator	Equation
Van der Waals (1873)	$p = \dfrac{RT}{(V-b)} - \dfrac{a}{V^2}$, $a = \dfrac{27R^2T_c^2}{64P_c}$, $b = \dfrac{RT_c}{8P_c}$
Soave-Redlich-Kwong (1972)	$p = \dfrac{RT}{(V-b)} - \dfrac{a(T)}{V(V+b)T^{0.5}}$,
	$a(T) = 0.42748 \dfrac{R^2T_c^2}{P_c}\left[1 + m\left(1 - T_r^{0.5}\right)\right]^2$,
	$b = 0.08664 \dfrac{RT_c}{p_c}$, $m = 0.48 + 1.574\omega - 0.176\omega^2$
Peng-Robinson (1976)	$p = \dfrac{RT}{(V-b)} - \dfrac{a(T)}{V(V+b)+b(V-b)}$,
	$a(T) = 0.45724 \dfrac{R^2T_c^2}{P_c}\left[1 + \kappa\left(1 - T_r^{0.5}\right)\right]^2$,
	$b = 0.0778 \dfrac{RT_c}{P_c}$,
	$\kappa = 0.37464 + 1.54226\omega - 0.26922\omega^2$
	ω = eccentricity factor

other cubic EOSs. It is an additional advantage. If one real and two imaginary roots are obtained, then the real root corresponds to the gas phase.

In vapour–liquid equilibrium, at least two components are present in the vapour phase. The Peng–Robinson EOS is used to estimate the properties of the mixture using the following mixing rules:

$$a = \sum_i \sum_j x_i x_j a_{ij} \tag{3.4}$$

$$b = \sum_i x_i b_i \tag{3.5}$$

where $a_{ij} = \left(1 - \delta_{ij}\right)a_i^{1/2}a_j^{1/2}$.

The binary interaction coefficient, δ_{ij}, is characteristic of the pair of species, i and j.

For cubic EOSs, some of the properties of the gases can be obtained analytically; for example, for the Peng–Robinson EOS, the fugacity of a pure component is given by

$$\ln\frac{f}{p} = Z - 1 - \ln(Z - B) - \frac{A}{2\sqrt{2}B}\ln\left(\frac{Z + 2.414B}{Z - 0.414B}\right) \tag{3.6}$$

The fugacity for a mixture of gases may be estimated from

$$\ln \frac{f}{x_k p} = \frac{b_k}{b}(Z-1) - \ln(Z-B) - \frac{A}{2\sqrt{2}B}\left(\frac{2\sum_i X_i a_{ik}}{a} - \frac{b_k}{b}\right)\ln\left(\frac{Z+2.414B}{Z-0.414B}\right) \quad (3.7)$$

Similarly, the enthalpy departure of a fluid is given by

$$H - H^* = RT(Z-1) + \frac{T\frac{da}{dT} - a}{2\sqrt{2}b}\ln\left(\frac{Z+2.414B}{Z-0.414B}\right) \quad (3.8)$$

The codes for the estimation of various properties using the Peng–Robinson EOS and other EOSs are reported in the literature using popular software, including MATLAB® (Nasri and Binous 2009). The built-in function 'fsolve' was used to find the roots of the cubic equation. Analytical expressions for other cubic EOSs are available in the literature.

3.2 Henry's Law

Henry's law is used to relate the composition of a chemical species in vapour and liquid at equilibrium. The mole fraction or partial pressure of an ith component in the vapour phase, p_i, is proportional to the mole fraction of that component in liquid, x_i, that is:

$$p_i = H_i x_i \quad (3.9)$$

The proportionality constant, H_i, is known as Henry's constant. For ideal gases:

$$p_i = x_i P \quad (3.10)$$

where P is the total pressure. From Equations 3.9 and 3.10, a linear equilibrium relationship is obtained:

$$\frac{y_i}{x_i} = \frac{H_i}{P} \quad (3.11)$$

In case of adsorption of pure gases at pressure, P, the number of moles adsorbed on the surface, n, is given by the following simple relationship known as Henry's law for adsorption:

$$n = kP \quad (3.12)$$

3.3 Newton's Law of Viscosity

One of the recent trends in solving problems of fluid flow is to use computational fluid dynamics, which is based on the law of conservation of momentum. This approach requires an expression for shear stress, τ_{yx}, in terms of fluid velocity or more precisely shear rate, $\dfrac{dv_x}{dy}$. One of the simplest models for this relationship is Newton's law of viscosity:

$$\tau_{yx} = -\mu \frac{dv_x}{dy} \tag{3.13}$$

where the proportionality constant, μ, is known as the viscosity of the fluid. All the gases and liquids and solutions of low-molecular-weight compounds exhibit Newtonian behaviour (Skelland 1967).

Due to the simplicity of the law, it has been possible to obtain analytical solutions for several flow problems. Dimensionless numbers such as Prandtl number, Pr, and Schmidt number, Sc, as defined in Equations 3.14 and 3.15, have been used in several empirical correlations for heat and mass transfer.

$$Pr = \frac{C_p \mu}{k_f} \tag{3.14}$$

$$Sc = \frac{\mu}{\rho D_{AB}} \tag{3.15}$$

where k_f is the thermal conductivity of the fluid and D_{AB} is the diffusivity of A into B.

All fluids do not follow Newton's law of viscosity. If the shear stress is plotted against shear rate, fluids with different behaviour are observed as shown in Figure 3.1. The curve for Newtonian fluid is a straight line passing through the origin. The fluids for which the line does not pass through the origin but is a straight line are known as Bingham plastic fluids. Fluids for which the curve is not a straight line but is described by a power law are known as power law fluids. If the exponent is less than 1, the fluid is known as a pseudoplastic. If the exponent is larger than 1, then it is called a dilatant. Few fluids exhibit time-dependent behaviour between shear stress and shear rate. These fluids are known as thixotropic or rheopectic fluids depending upon their behaviour with time. Some of the rheological models to describe the non-Newtonian behaviour are given in Table 3.2. Out of various rheological models for non-Newtonian fluids, the power law fluid and the Bingham plastic model can be used with less difficulty than other models. Solutions for obtaining the velocity field in pipes and channels for these fluids following these models are frequently available in the literature (Bird et al. 1960; Skelland 1974). The velocity field is an essential requirement to account for the convective transfer of mass and heat.

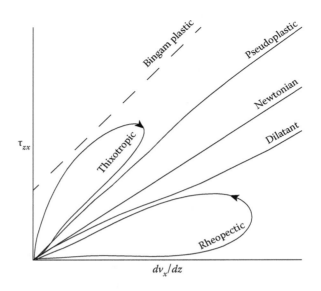

FIGURE 3.1
Rheological models for a few non-Newtonian fluids.

TABLE 3.2

Rheological Models for Non-Newtonian Liquids

Investigator	Equation
Power law or Ostwald–de Waele model	$\tau_{yx} = -m\left\|\dfrac{dv_x}{dy}\right\|^{n-1}\dfrac{dv_x}{dy}$
	If $n < 1$, pseudoplastic; if $n > 1$, dilatant; if $n = 1$, Newtonian.
Eyring model	$\tau_{yx} = -A\ \mathrm{arcsin}\,h\left(-\dfrac{1}{B}\dfrac{dv_x}{dy}\right)$
Ellis model	$-\dfrac{dv_x}{dy} = \left(\phi_0 + \phi_1\left\|\tau_{yx}\right\|^{x-1}\right)\tau_{yx}$
Reiner–Philippoff model	$-\dfrac{dv_x}{dy} = \left(\dfrac{1}{\mu_\infty + \dfrac{(\mu_0 - \mu_\infty)}{1 + (\tau_{yx}/\tau_s)^2}}\right)\tau_{yx}$
Bingham plastic	$\tau_{yx} = -\mu_0\dfrac{dv_x}{dy} \pm \tau_0\ \text{if}\ \left\|\tau_{yx}\right\| > \tau_0$
	$\dfrac{dv_x}{dy} = 0\ \text{if}\ \left\|\tau_{yx}\right\| < \tau_0$

For non-Newtonian fluids, the viscosity is not defined; instead, an apparent viscosity is defined as

$$\mu_a = \frac{\tau_{yx}}{\left|\dfrac{dv_x}{dy}\right|} \tag{3.16}$$

In absence of a well-defined fluid flow field, mostly encountered in multiphase systems, the apparent viscosity cannot be conveniently and confidently determined. These empirical correlations using dimensionless numbers involving viscosity cannot be used easily. A proper rheological model has to be used to find out the flow field first. It is possible only if equations based on laws of conservation of momentum are solved.

3.4 Fourier's Law of Heat Conduction

When heat transfer takes place due to conduction only, then at steady state the heat transfer rate is proportional to the temperature difference (driving force) and area of heat transfer. It is inversely proportional to the distance between the two faces of the transfer surface. Heat transfer rate $\dot{q}$ is written as

$$\dot{q} = kA \frac{(T_2 - T_1)}{\Delta z} = -kA \frac{dT}{dz} \tag{3.17}$$

The differential form is known as Fourier's law of heat conduction. At a steady state, it results in

$$\dot{q} = \frac{k_m A}{\delta_h}(T_2 - T_1) \tag{3.18}$$

The proportionality constant, k_m, is the thermal conductivity of the medium. The distance between the two boundaries maintained at constant temperatures T_1 and T_2 is δ_h.

One may recall Newton's law of cooling which is applicable for convective heat transfer at low temperatures. According to this law, the rate of cooling of the body is proportional to the temperature difference between the body and ambient conditions and the heat transfer area. The heat transfer coefficient is defined in a similar fashion:

$$\dot{q} = hA(T_2 - T_1) \tag{3.19}$$

The effect of flow field on the heat transfer rate is included in the heat transfer coefficient, h.

Heat transfer resistance analogous to electrical resistance for convection as well as conduction can be expressed as follows:

$$\frac{(T_2 - T_1)}{\dot{q}} = \frac{1}{hA} = \frac{\delta_h}{k_m A} \tag{3.20}$$

The resistance can be added depending upon different heat transfer processes taking place in parallel, in series or a combination of both.

Differentiation of Equation 3.17 gives

$$\frac{d}{dz}\left(A\frac{dT}{dz}\right) = 0 \tag{3.21}$$

This equation may be used if the heat transfer area is a variable.

3.5 Fick's First Law

Based on experimental observations, Fick proposed that at a steady state, the mass transfer rate remains constant between two points. The concentration profile is linear. Hence, the mass transfer rate is assumed to be proportional to the concentration gradient. Since it is also proportional to the area of the mass transfer surface, A, the mass transfer rate, N_A, for a steady state may be expressed as

$$N_A = -D_{AB}A\left(\frac{dC_A}{dz}\right) \tag{3.22}$$

Equation 3.20 is known as Fick's first law and should be applied only in cases of steady-state diffusion with no convection. The proportionality constant, D_{AB}, is known as the diffusivity or diffusion coefficient. Since mass transfer results in some amount of convection (diffusional flux), Equation 3.22 is valid only for dilute solutions or in cases of diffusion in solids. In concentrated solutions, the diffusion coefficient depends upon the concentration.

In most of the cases, at least two types of molecules are involved. One is that of the medium, and the second is that of the diffusing species. Therefore, the diffusivity is reported for a pair medium and diffusing species. For example, the diffusion of benzene in air will be different from the diffusion of benzene in nitrogen. If the types of molecules are different, the diffusivity of A into B may be different than that of B into A. But most of the time, they are considered to be the same. In the case of multicomponent systems, Fick's law is applicable for each species since each species has a different diffusivity.

If Fick's law is compared with the definition of mass transfer coefficient,

$$N_A = kA\Delta C_A = J_A A = -D_{AB}A\left(\frac{dC_A}{dz}\right) \tag{3.23}$$

The mass transfer coefficient, k, is equal to the ratio of diffusivity to distance between the points across which the concentration difference exists.

N and J are equal only in the absence of convection. At the wall or interface of two phases, the convection is always zero. If there are at least two species diffusing in opposite directions, it is possible that the sum of all the velocities is zero. In a binary system, such a situation is known as equimolal counter-diffusion. However, it does not mean that N_A and J_A cannot be related otherwise. The molar flux, N_A, and the molar diffusion flux, J_A, are relative to stationary coordinates and coordinates moving with bulk flow with velocity, $\bar{u}$, respectively. Therefore, for a species, i, the two fluxes are given as

$$N_i = C_i u_i \tag{3.24}$$

$$J_i = C_i (u_i - \bar{u}) \tag{3.25}$$

Here, u_i is the velocity of each species, and $\bar{u}$ is the average velocity of all the species.

Fick's first law may also be written in terms of mass flux, n, and mass diffusion flux, j:

$$J_i^* = -cD_{AB} \frac{dx_i}{dz} \tag{3.26}$$

and

$$j = -\rho D_{AB} \frac{dw_i}{dz} \tag{3.27}$$

Mass flux, n_A, relative to fixed spatial coordinates is given by

$$n_i = -\rho D_{AB} \frac{\partial w_i}{\partial z} + w_i \left(\sum_{j=1}^{n} n_j \right) \tag{3.28}$$

and molar flux relative to fixed spatial coordinates is given by

$$N_i = -cD_{AB} \frac{\partial x_i}{\partial z} + x_i \left(\sum_{j=1}^{n} N_j \right) \tag{3.29}$$

Fick's first law does not involve any temporal derivative; hence, it can be applied in steady-state continuous processes. In a batch process, the steady state is never attained. Therefore, understanding the unsteady state is essential for understanding progress in a batch process. It is also helpful for understanding a continuous process during initial periods.

3.6 Fick's Second Law

Let us consider diffusion through a semi-infinite slab. It is assumed that during a short time interval, the system behaves as a steady-state process. Such an assumption is called 'quasi–steady state'. Fick's first law may be differentiated during this period. The rate of accumulation of a chemical species in a volume is equal to the difference between the rate of species entering at z due to diffusion and the rate of species leaving.

$$\text{(Change in conc.)(volume)} = \begin{pmatrix} \text{rate of A entering} \\ - \text{rate of A leaving} \end{pmatrix} \text{(area)(time)} \qquad (3.30)$$

Substitution of terms in Equation 3.30 gives an equation for unsteady-state diffusion:

$$\frac{\Delta C}{\Delta t} = \lim_{\Delta z \to \infty} \frac{\left[\left(D_{AB} \frac{\partial C}{\partial Z} \right)\Big|_{z+\Delta z} - \left(D_{AB} \frac{\partial C}{\partial Z} \right)\Big|_{z} \right]}{\Delta z}$$

or

$$\frac{\partial C}{\partial t} = D_{AB} \frac{\partial^2 C}{\partial z^2} \qquad (3.31)$$

Equation 3.31 is known as Fick's second law and is applicable only in the absence of convection.

Several processes involve cylindrical and spherical objects (e.g. equipment and dispersed phases). In these situations, an appropriate coordinate system should be chosen. Fick's second law in cylindrical coordinates and spherical coordinates, respectively, is given in Equations 3.32 and 3.33:

$$\frac{\partial C_A}{\partial t} = D_{AB} \left(\frac{1}{r} \frac{\partial}{\partial r} \left(r \frac{\partial C_A}{\partial r} \right) \right) \qquad (3.32)$$

$$\frac{\partial C_A}{\partial t} = D_{AB} \left(\frac{1}{r^2} \frac{\partial}{\partial r} \left(r^2 \frac{\partial C_A}{\partial r} \right) \right) \qquad (3.33)$$

Integration of Fick's first law with the boundary condition as a function of time and the solution of Fick's second law are different.

3.7 Film Model

Mass and heat transfer take place between a fluid and a transfer surface. The transfer surface may be a rigid wall or interface of two different phases. The region of interest is near the transfer surface. Mass transfer takes place by diffusion and convection mechanisms. Heat transfer takes place by conduction, convection and radiation, the latter being significant at high temperature. While diffusion and conduction may be estimated using Fick's first law and

Fourier's law, respectively, the convective transport of mass and heat transfer requires knowledge of the convection (i.e. the flow field). The fluid velocity depends upon space variables as well as time. Problems related to diffusion or conduction can be solved much easier than those involving convective transport, as the fluid velocity is absent. At low fluid velocity, the flow is laminar and randomness of the turbulence is absent. However, due to the strong dependence of the flow upon viscosity, the effect of flow field upon transfer rates cannot be ignored and is to be used to estimate mass or heat transfer rates. The flow field adjacent to the transfer surface strongly depends upon the flow field in the bulk of the fluid. It happens in the case of laminar flow.

When the fluid velocities are large, the flow becomes turbulent. The convective transport is higher than the transport due to diffusion or conduction. Therefore, the flow field away from the interface does not seem to have a significant effect on the transfer rates.

Many engineering problems are related to transfer processes at the interface (e.g. fluid–fluid and fluid–solid interfaces). In the bulk of the fluid, the distribution of the temperature or concentration is studied. One of the simplest models to describe mass or heat transfer is the film model or film theory. This model is based on the following assumptions:

1. The entire variation of concentration or temperature takes place within a thin fluid film adjacent to the interface.
2. Within the film, the mass or heat transfer takes place by steady-state one-dimensional diffusion or conduction, respectively.

The first assumption is close to a situation where a high degree of mixing in the bulk fluid is present so that the concentration or temperature is almost uniform in the bulk fluid. Therefore, the effect of fluid flow field on mass or heat transfer is not significant. It also indicates that the film model should not be used in cases of laminar flow. The second assumption is very bold. It ensures that flow field is not considered for the estimation of mass or heat transfer rates. It does not say that there is no flow in the film. But the fluid velocity normal to the interface is absent. The fluid velocity in the vicinity of the interface is laminar and parallel to the interface. Therefore, it is neglected, as the convective transport in the direction of diffusional or thermal transport is normal in regard to the direction of fluid flow.

With these assumptions, mass transfer flux for species A (i.e. N_A, or mass transfer rate per unit area) can be written using Fick's first law:

$$N_A = D_{AB} \frac{(C_{A2} - C_{A1})}{\delta_m} \tag{3.34}$$

where C_{A1} and C_{A2} are the concentrations of the species A at the interface and on the other face of the fictitious film, respectively. The latter is equal to the concentration of A in the bulk fluid.

The diffusional film thickness, δ_m, cannot be measured. It may, however, be estimated. The diffusional flux may be neglected in the case of dilute solutions. In the case of equimolal counter-diffusion, the diffusional flux is zero. In the case of concentrated solutions, the diffusivity depends upon the concentration, and it can vary in the vicinity of the interface.

The mass transfer flux can also be written in terms of mass transfer coefficient:

$$N_A = k(C_{A2} - C_{A1}) \tag{3.35}$$

Comparison of Equation 3.35 with Equation 3.34 provides

$$\delta_m = \frac{D_{AB}}{k} \tag{3.36}$$

Diffusivity is the property of the species in a solution. Mass transfer coefficient is a measurable quantity. It depends upon the hydrodynamics, that is, the fluid flow field in the bulk fluid. The film thickness can now be estimated. The effect of the fluid flow field is included in the mass transfer coefficient.

Analogous equations can be written for heat transfer also:

$$q = k_f \frac{(T_2 - T_1)}{\delta_h} = h(T_2 - T_1) \tag{3.37}$$

Film thickness of thermal film, δ_h, is written as follows:

$$\delta_h = \frac{k_f}{h} \tag{3.38}$$

The relationship for mass transfer flux and heat transfer flux may be compared with the definition of current in the case of flow of electricity (see Section 2.6.5). The current is given as

$$i = \frac{V}{R} \tag{3.39}$$

As voltage is the driving force in the case of electric current, the driving forces for mass transfer and heat transfer are concentration difference and temperature difference, respectively. The current is analogous to mass transfer and heat transfer rates. With this analogy, the resistance to mass and heat transfer has been defined as

$$R = \frac{\Delta V}{i} = \frac{\Delta C_A}{N_A A_s} = \frac{1}{k A_s} = \frac{\Delta T}{q A_s} = \frac{1}{h A_s} \tag{3.40}$$

Here, A_s is the area of the interface or transfer surface. It may be seen that for the estimation of resistance to mass and heat transfer, the values of mass and heat transfer coefficients are required. The resistances are additive in a manner similar to that of resistances in the case of electrical circuits.

3.8 Two-Film Theory

The film model can easily be applied if mass transfer takes place at the solid–fluid surface. In the case of mass transfer across fluid–fluid interfaces, such as in gas–liquid and liquid–liquid systems, the interface separates the transfer processes in the two phases. The properties governing the rate of transfer processes in each phase are different. For example, the viscosity, thermal conductivity and specific heat of the two phases will not be the same. The velocity, concentration and temperature profile does not change continuously from one phase to another. The mechanism of mass transfer across the interface is different. It is governed by the thermodynamic properties of the system. To simplify the transfer process, a simple two-film theory was developed (Figure 3.2). It is based on the following two assumptions:

1. The resistance to mass transfer lies in two different fictitious films adjacent to the interface on both of its sides. These films are similar to that in a film model applied in each phase. Thus, the effect of hydrodynamics in each phase on the film thickness can be considered separately.

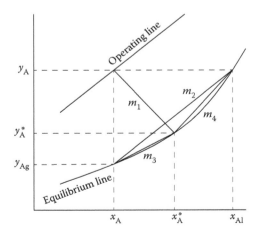

FIGURE 3.2
Two-film model at the interphase.

2. The concentration of the chemical species in both phases at the interface is in equilibrium (i.e. mass transfer resistance of the interface is considered to be insignificant in comparison to the mass transfer resistance in either of the two phases). In a true sense, the equilibrium does not exist since at equilibrium no mass transfer across the interface takes place. However, a significant departure from equilibrium is only at high mass transfer rates (Bird et al. 1960).

The mass transfer coefficient as defined for a single phase cannot be used as such in two-phase systems because the unit of concentration in the two phases is different. The measurable quantities are the concentration in individual fluids. Since the mass transfer coefficient can be determined from hydrodynamics, its value in each phase may be determined. The driving force is the difference between the concentrations at the interface and in the bulk fluid, and it is not known because the concentrations at the interface (from both sides) are unknown. Let us consider a gas–liquid system. The concentrations of species A in the gas and liquid phases are y_A and x_A, and the concentrations at the interface are y_A^* and x_A^* (Figure 3.2). The mass transfer coefficients in the gas and liquid phases are k_g and k_l, respectively. The flux can be written as

$$N_A = k_g \left(y_A^* - y_A \right) \tag{3.41}$$

and

$$N_A = k_l \left(x_A - x_A^* \right) \tag{3.42}$$

The mass transfer coefficients, k_g and k_l, are known as individual mass transfer coefficients in the gas and liquid phases, respectively. To define a mass transfer coefficient in terms of concentrations in the bulk fluids, the units of the concentrations should be the same. If x_{Ag} is in equilibrium with y_A and y_{Al} is in equilibrium with x_A, then the driving force in each phase can be written conveniently. The mass transfer coefficients, K_g and K_l, are defined as

$$N_A = K_g (y_A - y_{Al}) \tag{3.43}$$

$$N_A = K_l (x_{Ag} - x_A) \tag{3.44}$$

The values of K_g and K_l may be called the overall mass transfer coefficients based on units of concentration in the gas and liquid phases, respectively. These can also be expressed in terms of individual mass transfer coefficients as follows:

$$\frac{k_l}{k_g} = -m_1 \tag{3.45}$$

$$\frac{K_1}{K_g} = m_2 \tag{3.46}$$

$$\frac{1}{K_g} = \frac{1}{k_g} + \frac{m_3}{k_1} \tag{3.47}$$

$$\frac{1}{K_1} = \frac{1}{k_1} + \frac{1}{m_4 k_g} \tag{3.48}$$

Here, m_1, m_2, m_3 and m_4 are the slope of the lines as shown in Figure 3.2. Equations 3.47 and 3.48 are the summation of individual resistances to mass transfer to get resistance to overall mass transfer based on respective driving forces as given in Equations 3.43 and 3.44.

It is inconvenient to determine the driving forces in these equations except in a few simple cases. If the mass transfer in the liquid phase is controlling, then $k_g \gg k_1$ or $1/k_1 \gg 1/m_4 k_g$, and hence $K_1 = k_1$. If the mass transfer in the gas phase is controlling, then $k_1 \gg k_g$ or $1/k_g \gg m_3/k_1$, and hence $K_g = k_g$.

If the equilibrium relationship is linear, then $m_2 = m_3 = m_4$. For example, whenever Henry's law is applicable, all these constants are equal to Henry's constant, H. Equations 3.47 and 3.48 can be written as

$$\frac{1}{K_g} = \frac{1}{k_g} + \frac{H}{k_1} \tag{3.49}$$

$$\frac{1}{K_1} = \frac{1}{k_1} + \frac{1}{H k_g} \tag{3.50}$$

Based on Equations 3.49 and 3.50, several models for chemical reaction with mass transfer are reported in the literature.

In cases of the non-linear equilibrium relationship, $y^* = f(x^*)$, the interfacial concentrations can be determined in terms of the individual mass transfer coefficients by the following implicit equation:

$$\frac{y - f(x^*)}{(x - x^*)} = -\frac{k_1}{k_g} \tag{3.51}$$

Once the interfacial concentrations are known, the slopes m_2, m_3 and m_4 are known. Such difficulties do not arise in cases of heat transfer processes since the unit of temperature is the same in both the phases, and at the interface the temperatures on both sides of the interface are assumed to be the same. The interfacial temperature is eliminated while expressing the overall heat transfer coefficient, U, in terms of individual heat transfer coefficients, h_1 and h_2.

$$\dot{q} = h_1(T^* - T_1) = h_2(T_2 - T^*) = U(T_2 - T_1) \tag{3.52}$$

Hence,

$$\frac{1}{U} = \frac{1}{h_1} + \frac{1}{h_2} \tag{3.53}$$

The resistances to heat transfer on both sides are added to get the overall heat transfer resistance.

3.9 Arrhenius' Law

The reaction rates depend upon the concentration of the reactants and the temperature. The rate of reaction can be written as the product of two terms; the first term depends upon the concentration only, and the second term depends upon temperature only (Levenspiel 1999).

$$-r_A = f(C_A, C_B, \ldots).k'\,(T) \tag{3.54}$$

The effect of pressure is considered in the first terms. The reactions may proceed in several steps and may lead to a complex rate expression. There is no general expression which can be used to express all types of concentration-dependent terms in the rate expression. Fortunately, the temperature dependence term can be written as follows (Levenspiel 1999):

$$k'(T) = k'_0 e^{-E/RT} \tag{3.55}$$

The k' is the kinetic rate expression, and k'_0 is the constant. The temperature dependence term is exponential in nature. The constant E is known as the activation energy of the reaction. The exponential behaviour given in Equation 3.53 is known as Arrhenius' law.

The activation energy for a reaction is determined by plotting $\ln(k)$ as a function of $\frac{1}{T}$. The slope of the straight line is $\frac{E}{R}$. If activation energy and the kinetic rate expression k' at a temperature T_1 are known, then the kinetic rate expression at any other temperature T_2 can be determined by the following equation:

$$\ln\left(\frac{r_{A1}}{r_{A2}}\right) = \ln\left(\frac{k'_1}{k'_2}\right) = \frac{E}{R}\left(\frac{1}{T_2} - \frac{1}{T_1}\right) \tag{3.56}$$

3.10 Adsorption Isotherms

Adsorption phenomena have been used as a separation process to remove an unwanted chemical species that is present in a fluid using a solid as the adsorbent. The adsorbate is adsorbed on the surface of the solid. After a long period,

TABLE 3.3

A Few Frequently Used Adsorption Isotherms

Investigator	Isotherm	Remark
Langmuir (1918)	$p = \dfrac{1}{K}\left(\dfrac{\theta}{1-\theta}\right)$	No interaction between the adsorbed molecules
Brunauer–Emmett–Teller (1938)	$\dfrac{q}{q_m} = \dfrac{K_B p_r}{(1-p_r)(1-p_t + K_B p_r)}$	$p_r = \dfrac{p}{p_s}$, q_m = constant
Freundlich (1926)	$q = k_F C^{(1/n_F)}$	Empirical
Radke and Prausnitz (1972)	$q = \dfrac{1}{\left(\dfrac{1}{K_H p} + \dfrac{1}{k_F p^{1/n_F}}\right)}$	Combines Freundlich isotherms and Henry's law

equilibrium is achieved. The concentration of the adsorbate in the fluid, C_A, and the amount adsorbed on the solid, q, is in equilibrium. The equilibrium relationship relating q and C_A is known as an isotherm. There are several simple models which describes isotherms. A few frequently used isotherms are presented in Table 3.3. The isotherms are given in terms of the $K = \dfrac{K_a}{K_d}$ and $\theta = \dfrac{q}{q_0}$, where K_a and K_d are the rates of adsorption and desorption, respectively. The parameter q_0 is the adsorption capacity of the adsorbent. The concentration of a component in the gas phase is expressed in terms of partial pressure of the component in the gas phase p. A detailed description of the isotherms may be found in the literature (Suzuki 1990).

In addition to these, Equation 3.12, known as Henry's law of adsorption, is also used.

3.11 Examples

Several applications of the simple concepts discussed in this chapter are available in textbooks of heat transfer, mass transfer, transport phenomena, and chemical reaction engineering. A few of these will be briefly discussed to illustrate the role of simple models in understanding the behaviour of the phenomena. Ease in applying numerical procedures and, as a consequence, relaxation in consideration of special cases are also hinted at.

3.11.1 Overall Heat Transfer Coefficient in a Composite Cylindrical Wall

Let us consider a tube in which fluid is flowing in a turbulent condition. The tube is made of two concentric tubes of two different materials (Figure 3.3). The heat transfer coefficients on the inside and outside of the tubes are h_1 and h_2, respectively. The thermal conductivity of the inner and outer tubes is

Electrical analogue

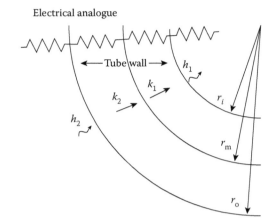

FIGURE 3.3
Heat transfer through a composite cylindrical wall.

k_1 and k_2, respectively. The inner and outer radii of the inner tube are r_i and r_m, respectively, and the radius of the outer tube is r_o.

Using an electrical analogue and the concept of heat transfer resistance, the overall heat transfer coefficient U can be written as

$$\frac{1}{Ur_o} = \frac{1}{h_2 r_o} + \frac{\ln(r_m / r_i)}{k_1} + \frac{\ln(r_o / r_m)}{k_2} + \frac{1}{h_1 r_i} \tag{3.57}$$

3.11.2 Cooling of a Small Sphere in a Stagnant Fluid

A hot solid sphere of diameter D_p at temperature T_s is kept in a stagnant cold fluid maintained at temperature T_f (Figure 3.4). Writing Equation 3.21 in spherical coordinates, we get

$$\frac{1}{r^2}\frac{\partial}{\partial r}\left(r^2 \frac{\partial T}{\partial r}\right) = 0 \tag{3.58}$$

Equation 3.58 is integrated twice to obtain the temperature profile:

$$T = \frac{D_p (T_s - T_f)}{2r} + C_2 \tag{3.59}$$

The following are the boundary conditions:

At $r = \dfrac{D_p}{2}$, $T = T_s$

At $r \rightarrow \infty$, $T = T_f$

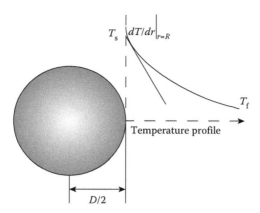

FIGURE 3.4
Heat transfer for a sphere in a stagnant fluid.

Obtaining constants C_1 and C_2 by applying boundary conditions and substituting in Equation 3.59, the following expression is obtained:

$$T = \frac{(T_s - T_f)D_p}{2r} + T_f \tag{3.60}$$

The heat transfer rate can be written as

$$\dot{q} = hA(T_s - T_s) = -k_f A \left.\frac{dT}{dr}\right|_{r=R} = k_f A \frac{(T_s - T_f)D_p}{2(D_p / 2)^2} \tag{3.61}$$

Hence:

$$\mathrm{Nu} = \frac{hD_p}{k_f} = 2 \tag{3.62}$$

3.11.3 Diffusion in a Stagnant Gas Film

Fick's first law can be used to solve several simple problems. As an example, let us consider that a liquid A is evaporating and diffusing into a gas B in a simple device such as a test tube (Figure 3.5). The liquid level of A is maintained at a constant height. The component B is insoluble in A. From Equation 3.29, the diffusion of A in a stagnant gas B is written as follows:

$$N_A = -cD_{AB}\frac{\partial x_A}{\partial z} + x_A(N_A + N_B) = -cD_{AB}\frac{\partial x_A}{\partial z} + x_A N_A \tag{3.63}$$

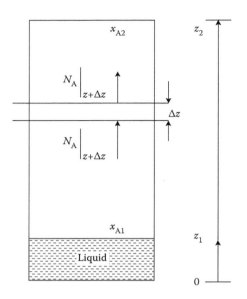

FIGURE 3.5
Diffusion through a stagnant gas film.

The $N_B = 0$ as B is not diffusing. At a steady state, the N_A does not vary with z, that is:

$$\frac{dN_A}{dz} = \frac{d}{dz}\left(\frac{1}{1-x_A}\frac{dx_A}{dz}\right) = 0 \tag{3.64}$$

The boundary conditions are as follows:

At $z = z_1$, $x_A = x_{A1}$

At $z = z_2$, $x_A = x_{A2}$

The solution of Equation 3.64 is given below (Bird et al. 1960):

$$\frac{(1-x_A)}{(1-x_{A1})} = \left(\frac{(1-x_{A2})}{(1-x_{A1})}\right)^{\left(\frac{z-z_1}{z_2-z_1}\right)} \tag{3.65}$$

After getting the concentration profile, the flux can now be written as

$$N_A\big|_{z=z_1} = \frac{pD_{AB}/RT}{(z_2-z_1)(p_B)_{\text{ln}}}(p_{A1} - p_{A2}) \tag{3.66}$$

where $(p_B)_{\text{ln}}$ is the logarithmic mean of p_{B1} and p_{B2}. Experiments to measure the diffusivity are designed based on Equation 3.66.

3.11.4 Diffusion–Reaction Systems

Reaction catalysed by solid takes place at the surface of the catalyst. Hence, porous catalysts are desirable due to their large surface area per unit of the volume. Since the reactants have to diffuse in the pores of the solids, the concentration in the pores is different from that at the outer surface of the catalyst pallet. Let us consider the pores in the catalyst to be cylindrical (Figure 3.6a). The length and radius of the pores are L and r, respectively. The effectiveness factor is defined as

$$\xi_A = \frac{\int_0^L (-r_A)\big|_{C_A(x)} (2\pi r)\,dx}{\int_0^L (-r_A)\big|_{C_A(x=0)} (2\pi r)\,dx} \tag{3.67}$$

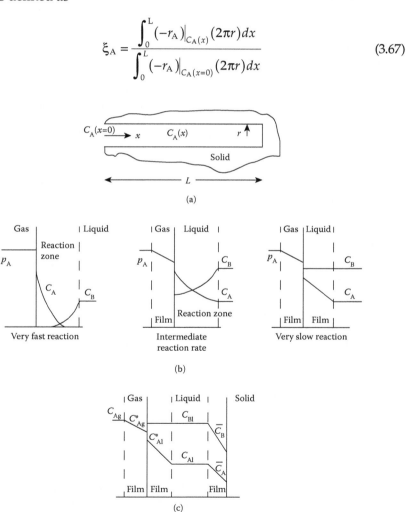

(a)

(b)

(c)

FIGURE 3.6

Diffusion–reaction problems: (a) single catalyst pore, (b) fluid–fluid non-catalysed reactions for very fast, intermediate and slow reactions and (c) fluid–fluid solid-catalysed reaction.

The numerator of equation requires the concentration profile within the pores. It can be obtained by solving the following differential equation:

$$D_A \frac{d^2 C_A}{dx^2} = \frac{2}{r}(-r_A) \qquad (3.68)$$

Equation 3.68 is for cylindrical pores in a slab. It is called a diffusion–reaction equation. The equation can be solved numerically using the ordinary differential equation (ODE) solver of MATLAB for any arbitrary kinetics. However, if more than one species are involved, then the number of differential equations will be more than one. An analytical solution for a first-order reaction is given in almost all of the textbooks. Due to more computational power at the desktop, a wide range of problems can now be solved easily. Gas–liquid non-catalysed reactions also are diffusion–reaction systems. The reactant in the gas phase diffuses into the liquid, where it reacts with another reactant present in the liquid. The observed reaction rate is due to mass transfer on both sides of the interface and the reaction. Using two-film theory, many different cases are possible. Only three of these are presented in Figure 3.6b.

1. When the reaction is very fast (instantaneous reaction), the reaction takes place at a reaction plane or in a reaction zone. As soon as the reactant diffuses into another phase, the reaction starts. Since the reaction is fast, the reactant is consumed before it diffuses into the bulk of the phase. In other words, the reaction is limited to the film (if the film model is considered).

2. For very slow reaction rates, most of the reaction takes place in the bulk of the liquid. The film on the liquid side may or may not be present. The reactants get sufficient time to diffuse into the bulk of the phase. The reactants consumed in the reaction in the film are negligible as compared to the reactant consumed in the bulk. In other words, the reaction takes place mainly in the bulk.

3. For intermediate reaction rates, the reaction takes place in both the film and bulk. Analytical solutions for first-order solutions in all these cases are available in the literature (Levenspiel 1999).

The diffusion–reaction process in the liquid phase can be described by

$$D_A \frac{d^2 C_A}{dx^2} = -r_A \qquad (3.69)$$

This equation can be solved easily by ODE solvers. However, the boundary conditions depend upon the particular case based on the reaction rate. The boundary conditions cannot be determined by a steady-state model. A dynamic model is required to understand the evolution of the steady-state behaviour and the kinetic regime. It will increase the complexity of the model.

3.11.5 Gas–Liquid Solid-Catalysed Reactions

Reactions in gas–liquid systems catalysed by solids are carried out in various types of reactors (e.g. slurry-bubble columns, trickle beds, packed beds and three-phase fluidised beds). One of the reactants is in the gas phase, and the other reactant is in the liquid phase. The reactant in the gas phase diffuses into liquid (Figure 3.6c). Both reactants in the liquid reactant diffuse to the solid surface where the reaction takes place according to the following kinetics:

$$-r_A = k\bar{C}_A^n \bar{C}_B^m \tag{3.70}$$

where $C_{Ag}, C_{Ag}^*, C_{Al}^*$ and C_{Al} are the concentrations of A in the gas phase, on the gas side of the interface, on the liquid phase of the interface and in the bulk liquid, respectively. The average concentrations of A and B in the catalyst surface are $\bar{C}_A$ and $\bar{C}_B$, respectively. k_g, k_l and k_s are the mass transfer coefficients in the gas phase, liquid phase and solid surface, respectively (Figure 3.6c). The interfacial area and the area of catalyst surface per unit volume are a_i and a_c. At a steady state, the mass transfer rate of A, N_A, is written as

$$N_A = k_{Ag}a_i \left(p_{Ag} - p_{Ag}^*\right) = k_{Al}a_i \left(C_{Ag}^* - C_{Al}\right) = k_s a_c \left(C_{Al} - \bar{C}_A\right) = -r_A a_c \tag{3.71}$$

Assuming that the equilibrium at the interface follows Henry's law and the film model holds good, Equation 3.71 can be written as

$$\left(p_{Ag} - \bar{C}_A\right)\left(\frac{1}{k_{Ag}a_i} + \frac{H_A}{k_{Al}a_i} + \frac{H_A}{k_s a_c}\right) = -r_A \tag{3.72}$$

Since the average concentration in the catalyst is not known, the observed rate is defined in terms of the concentration in the easily measurable concentration of A in the gas, p_{Ag}. From Equation 3.66, $\bar{C}_A$ is to be eliminated. It can be done easily if the kinetics is of the first order with respect to A and B and consideration of the effectiveness factor, ξ_A, that is,

$$-r_A = k\bar{C}_A \bar{C}_B \xi_A \tag{3.73}$$

From Equations 3.72 and 3.73:

$$\left(p_{Ag} - \bar{C}_A\right) = N_A = -r_A \left(\frac{1}{k_{Ag}a_i} + \frac{H_A}{k_{Al}a_i} + \frac{H_A}{k_s a_c}\right) = k\bar{C}_A \bar{C}_B \xi_A \left(\frac{1}{k_{Ag}a_i} + \frac{H_A}{k_{Al}a_i} + \frac{H_A}{k_s a_c}\right)$$

$$\tag{3.74}$$

The average concentration of A at the catalyst surface can now be written as

$$\bar{C}_A = \frac{p_{Ag}}{\left(\dfrac{1}{k_{Ag}a_i} + \dfrac{H_A}{k_{Al}a_i} + \dfrac{H_A}{k_s a_c} + \dfrac{1}{k\bar{C}_B \xi_A} \right)} \frac{1}{k\bar{C}_B \xi_A} \tag{3.75}$$

Hence, the observed rate is

$$-r'_A = \frac{p_{Ag}}{\left(\dfrac{1}{k_{Ag}a_i} + \dfrac{H_A}{k_{Al}a_i} + \dfrac{H_A}{k_s a_c} + \dfrac{1}{k\bar{C}_B \xi_A} \right)} \tag{3.76}$$

Equation 3.76 was obtained by eliminating the unknown value of $\bar{C}_A$ (Levenspiel 1999). The value of $\bar{C}_B$ is to be determined by getting the observed reaction rate for B. It is not always possible to eliminate $\bar{C}_A$ for any arbitrary kinetics. However, root-finding methods are implemented in various software which can be used to determine $\bar{C}_A$. For example, Equation 3.74 for the kinetics given by Equation 3.73 can be written as

$$\left(p_{Ag} - \bar{C}_A \right) = k_A \bar{C}_A^n \bar{C}_B^m \xi_A \left(\frac{1}{k_{Ag}a_i} + \frac{H_A}{k_{Al}a_i} + \frac{H_A}{k_s a_c} \right) \tag{3.77}$$

Similarly, for B, the following equation is obtained:

$$\left(C_{Bl} - \bar{C}_B \right) = k_B \bar{C}_A^n \bar{C}_B^m \xi_B \left(\frac{1}{k_{Bs}a_c} \right) \tag{3.78}$$

Equations 3.77 and 3.78 can be solved for $\bar{C}_A$ and $\bar{C}_B$ using iterative procedures that are available with several easily accessible types of software. Theses equations were obtained using the film model. Similar expressions using other theories are available in the literature. The simple model discussed above adds the mass transfer resistances, which is not correct. Whenever mass transfer is coupled with reaction, the resistances cannot be added (Cussler 1998).

3.12 Summary

A few simple models and concepts related to rheological behaviour of the fluid, equations of state, Fourier's law for heat transfer, Fick's laws for mass transfer, Arrhenius' law and adsorption isotherms were presented. The film model for transfer processes and two-film model for the processes taking place at the interface were discussed. Simple laws are frequently used in determining total heat transfer resistances in complex geometry and multi-phase diffusion–reaction problems. The governing equations can be solved numerically even in cases of increased complexity. These concepts were illustrated with the help of a few examples.

References

Bird, R.B., Stewart, W.E., and Lightfoot, E.N. 1960. *Transport Phenomena*. New York: John Wiley.

Cussler, E.L. 1998. *Diffusion-Mass Transfer in Fluid Systems*, 2nd ed. Cambridge: Cambridge University Press.

Levenspiel, O. 1999. *Chemical Reaction Engineering*, 3rd ed. New York: John Wiley.

Nasri, Z., and Binous, H. 2009. Applications of the Peng-Robinson equation of state using MATLAB. *Chem. Eng. Edu.* 43(2): 1–10.

Peng, D.Y., and Robinson, D.B. 1976. A new two-constant equation of state. *Ind. Eng. Chem. Fundam.* 15(1): 59–64.

Skelland, A.H.P. 1974. *Diffusional Mass Transfer*. New York: John Wiley.

Skelland, A.H.P. 1967. *Non-Newtonian Flow and Heat Transfer*. New York: John Wiley.

Smith, J.M., Van Ness, H.C., and Abbott, M.M. 1996. *Introduction to Chemical Engineering Thermodynamics*, 5th ed. New York: McGraw-Hill.

Suzuki, M. 1990. *Adsorption Engineering*. Amsterdam: Elsevier.

4

Models Based on Laws of Conservation

The simple laws discussed in Chapter 3 involve heat transfer and mass transfer, with and without chemical reactions. The mechanisms of heat and mass transfer that were considered were conduction and diffusion, respectively. Another mechanism for mass and heat transfer is convection. To consider the convective transport of heat and mass transfer, knowledge of the flow field in the system is required. Lumped parameter models using average velocity are the exception. Estimations of pressure drop, drag forces, viscous dissipation etc., also require the flow field to be known. The subject of transport phenomena was developed with the aim to find mathematical solutions for the distribution of velocity, temperature and concentration within a system. The development of numerical techniques to solve the governing equation resulted in the development of computational fluid dynamics (CFD). The governing equations describing transfer processes are derived by applying laws of conservation of mass, momentum and energy. The development of these equations and their application in modelling various processes are presented in this chapter.

4.1 Laws of Conservation of Momentum, Mass and Energy

The subject of transport phenomena is based on the fact that during convection, the momentum, mass and thermal energy are conserved in a system. Changes in any of these (i.e. accumulation of these) are equal to changes due to convection and other phenomena. The momentum is a vector; hence, all of its three components are conserved. Accumulation of the momentum is due to the convective transport, the shear stress in a particular direction and the pressure or gravitational force. Similarly, accumulation of mass of each chemical species is due to convective transport, diffusion and chemical reaction. Accumulation of thermal energy is due to convection, conduction, generation of heat due to chemical reaction, viscous dissipation, electrical heating etc. The development of the governing equation is briefly presented here. The detailed derivations of these equations are available in most of the books on transport phenomena (e.g. Bird et al. 1960).

4.1.1 Equation of Continuity

Mass is conserved except in nuclear reactions. It cannot be created or destroyed. It is known as the law of conservation of mass. A substance may consist of various chemical species. If the mass of a particular species is considered, then it may appear or disappear due to a chemical reaction, but the total mass is conserved. Let us apply the law of conservation of total mass over a differential volume element of dimensions Δx, Δy and Δz in Cartesian coordinates (Figure 4.1):

$$\begin{bmatrix} \text{Net rate of} \\ accumulation \\ \text{of mass} \end{bmatrix} = \begin{bmatrix} \text{Rate of mass} \\ \text{entering} \end{bmatrix} - \begin{bmatrix} \text{Rate of mass} \\ \text{leaving} \end{bmatrix} \tag{4.1}$$

For a fluid of density, ρ, Equation 4.1 can be written as

$$\begin{aligned} (\Delta x \Delta y \Delta z)\frac{\partial \rho}{\partial t} &= \left[(\Delta y \Delta z)(\rho v_x)\big|_x + (\Delta x \Delta z)(\rho v_y)\big|_y + (\Delta x \Delta y)(\rho v_z)\big|_z \right] \\ &\quad - \left[(\Delta y \Delta z)(\rho v_x)\big|_{x+\Delta x} + (\Delta x \Delta z)(\rho v_y)\big|_{y+\Delta y} + (\Delta x \Delta y)(\rho v_z)\big|_{z+\Delta z} \right] \end{aligned} \tag{4.2}$$

After rearranging terms, dividing Equation 4.2 by ($\Delta x\,\Delta y\,\Delta z$) and assuming that the limit of Δx, Δy and Δz approached zero, the following equation is obtained:

$$\frac{\partial \rho}{\partial t} + \frac{\partial}{\partial x}(\rho v_x) + \frac{\partial}{\partial y}(\rho v_y) + \frac{\partial}{\partial z}(\rho v_z) = 0 \tag{4.3}$$

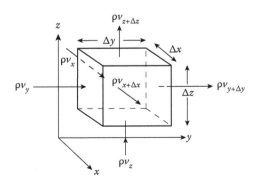

FIGURE 4.1
Fluid flow through a fixed volume element.

Equation 4.3 describes the variation of density of the fluid in terms of spatial velocity distribution. It is applicable to both compressible as well as incompressible fluids. It is helpful in simplifying equations of motion, energy and continuity for each species while deriving these (Bird et al. 1960).

For incompressible fluids, ρ is constant. Equation 4.3 simplifies to the following equation:

$$\frac{\partial v_x}{\partial x} + \frac{\partial v_y}{\partial y} + \frac{\partial v_z}{\partial z} = 0 \tag{4.4}$$

The spatial distribution of velocity should always follow Equation 4.4 for incompressible fluid. For one-dimensional (1D) fluid flow, $v_y = v_z = 0$. Equation 4.4 reduces to

$$\frac{\partial v_x}{\partial x} = 0 \tag{4.5}$$

The velocity in the direction of flow remains constant. For 2D cases, if $v_z = 0$, Equation 4.4 is reduced to the following equation:

$$\frac{\partial v_x}{\partial x} + \frac{\partial v_y}{\partial y} = 0 \tag{4.6}$$

Equation 4.6 is used in two ways. Let the velocities, v_y and v_z, be defined in terms of a new variable, ψ, as follows:

$$v_x = -\frac{\partial \psi}{\partial y}$$

and

$$v_y = \frac{\partial \psi}{\partial x} \tag{4.7}$$

Equation 4.6 is satisfied, but the governing equations to describe the velocity field will have only one variable, ψ, which is called the stream function.

If the velocities, v_y and v_z, are defined in terms of another variable, φ, as follows:

$$v_x = -\frac{\partial \phi}{\partial x}$$

and

$$v_y = -\frac{\partial \phi}{\partial y} \tag{4.8}$$

then Equation 4.6 is again satisfied. The variable φ is known as 'velocity potential'. From Equations 4.7 and 4.8, we get the Cauchy–Riemenn equations:

$$\frac{\partial \psi}{\partial y} = \frac{\partial \phi}{\partial x}$$

and

$$\frac{\partial \psi}{\partial x} = -\frac{\partial \phi}{\partial y} \qquad (4.9)$$

An analytical function $w(z)$ of a complex variable, z, may be chosen such that $w(z) = \phi(z) + i\psi\,(z)$, satisfying Equation 4.8. The velocity components, v_y and v_z, can be obtained as the real and imaginary parts of the following equation:

$$\frac{dw}{dz} = -v_x + iv_y \qquad (4.10)$$

Equation of continuity; stream function, ψ; and velocity potential, φ, in cylindrical and spherical coordinates are given in Tables 4.1, 4.2 and 4.3, respectively. It is obvious that for a steady-state 1D flow problem, the velocity gradient in the direction of flow is always zero irrespective of the coordinate system.

For 2D flows, Equation 4.6 is also used in the following manner:

$$v_y = \int_0^y \frac{\partial v_x}{\partial x}\,dy \qquad (4.11)$$

The limits of integration depend upon the problem.

TABLE 4.1

Equation of Continuity in Cylindrical and Spherical Coordinates

Coordinate System	Equation of Continuity
Cylindrical coordinates	$\dfrac{\partial \rho}{\partial t} + \dfrac{1}{r}\dfrac{\partial}{\partial r}(\rho r v_r) + \dfrac{1}{r}\dfrac{\partial}{\partial \theta}(\rho v_\theta) + \dfrac{\partial}{\partial z}(\rho v_z) = 0$
For incompressible fluids	$\dfrac{1}{r}\dfrac{\partial}{\partial r}(r v_r) + \dfrac{1}{r}\dfrac{\partial v_\theta}{\partial \theta} + \dfrac{\partial v_z}{\partial z} = 0$
Spherical coordinates	$\dfrac{\partial \rho}{\partial t} + \dfrac{1}{r^2}\dfrac{\partial}{\partial r}(\rho r^2 v_r) + \dfrac{1}{r\sin\theta}\dfrac{\partial}{\partial \theta}(\rho v_\theta \sin\theta) + \dfrac{1}{r\sin\theta}\dfrac{\partial}{\partial \phi}(\rho v_\phi) = 0$
For incompressible fluids	$\dfrac{1}{r^2}\dfrac{\partial}{\partial r}(r^2 v_r) + \dfrac{1}{r\sin\theta}\dfrac{\partial}{\partial \theta}(v_\theta \sin\theta) + \dfrac{1}{r\sin\theta}\dfrac{\partial v_\phi}{\partial \phi} = 0$

TABLE 4.2

Stream Function, ψ, and N-S Equation in Cylindrical and Spherical Coordinates

Coordinate System	Stream Function, ψ	N-S Equation
Cylindrical	$v_r = -\dfrac{1}{r}\dfrac{\partial \psi}{\partial \theta}; \; v_\theta = \dfrac{\partial \psi}{\partial r}$	$\dfrac{\partial(\nabla^2\psi)}{\partial t} + \dfrac{1}{r}\dfrac{\partial(\psi,\nabla^2\psi)}{\partial(r,\theta)} = \dfrac{\mu}{\rho}\nabla^4\psi$
Cylindrical ($v_\theta = 0$)	$v_r = \dfrac{1}{r}\dfrac{\partial \psi}{\partial z};$ $v_z = -\dfrac{1}{r}\dfrac{\partial \psi}{\partial r}$	$\dfrac{\partial(E^2\psi)}{\partial t} - \dfrac{1}{r}\dfrac{\partial(\psi,E^2\psi)}{\partial(r,z)} - \dfrac{2}{r^2}\dfrac{\partial\psi}{\partial z}E^2\psi = \dfrac{\mu}{\rho}E^4\psi$
Spherical ($v_\phi = 0$)	$v_r = -\dfrac{1}{r^2 Sin\theta}\dfrac{\partial \psi}{\partial \theta};$ $v_\theta = \dfrac{1}{rSin\theta}\dfrac{\partial \psi}{\partial r}$	$\dfrac{\partial(\tilde{E}^2\psi)}{\partial t} + \dfrac{1}{r^2 Sin\theta}\dfrac{\partial(\psi,\tilde{E}^2\psi)}{\partial(r,\theta)}$ $-\dfrac{2}{r^2 Sin^2\theta}\left(\dfrac{\partial\psi}{\partial r}\cos\theta - \dfrac{1}{r}\dfrac{\partial\psi}{\partial\theta}Sin\theta\right)\tilde{E}^2\psi = \dfrac{\mu}{\rho}\tilde{E}^4\psi$

Note: The operators used are

$$\nabla^2 = \frac{\partial^2}{\partial r^2} + \frac{1}{r}\frac{\partial}{\partial r} + \frac{1}{r^2}\frac{\partial^2}{\partial\theta^2}; \; \nabla^4\psi = \nabla^2(\nabla^2\psi); \quad \frac{\partial(\psi,\nabla^2\psi)}{\partial(r,\theta)} = \begin{vmatrix} \dfrac{\partial\psi}{\partial r} & \dfrac{\partial\psi}{\partial\theta} \\[2ex] \dfrac{\partial\nabla^2\psi}{\partial r} & \dfrac{\partial\nabla^2\psi}{\partial\theta} \end{vmatrix}$$

$$E^2 = \frac{\partial^2}{\partial r^2} - \frac{1}{r}\frac{\partial}{\partial r} + \frac{\partial^2}{\partial z^2}; \; E^4\psi = E^2(E^2\psi); \quad \frac{\partial(\psi,E^2\psi)}{\partial(r,z)} = \begin{vmatrix} \dfrac{\partial\psi}{\partial r} & \dfrac{\partial\psi}{\partial z} \\[2ex] \dfrac{\partial E^2\psi}{\partial r} & \dfrac{\partial E^2\psi}{\partial z} \end{vmatrix}$$

$$\tilde{E}^2 = \frac{\partial^2}{\partial r^2} + \frac{Sin\theta}{r^2}\frac{\partial}{\partial\theta}\left(\frac{1}{Sin\theta}\frac{\partial}{\partial\theta}\right); \; \tilde{E}^4\psi = \tilde{E}^2(\tilde{E}^2\psi); \quad \frac{\partial(\psi,\tilde{E}^2\psi)}{\partial(r,\theta)} = \begin{vmatrix} \dfrac{\partial\psi}{\partial r} & \dfrac{\partial\psi}{\partial\theta} \\[2ex] \dfrac{\partial\tilde{E}^2\psi}{\partial r} & \dfrac{\partial\tilde{E}^2\psi}{\partial\theta} \end{vmatrix}$$

TABLE 4.3

Velocity potential, φ, and Laplace Equation in Cylindrical and Spherical Coordinates

Coordinate System	Velocity Potential, φ	Laplace Equation
Cylindrical, $v_z = 0$	$v_r = -\dfrac{\partial\phi}{\partial r}, \; v_\theta = -\dfrac{1}{r}\dfrac{\partial\phi}{\partial\theta}$	$\dfrac{1}{r}\dfrac{\partial}{\partial r}\left(r\dfrac{\partial\phi}{\partial r}\right) + \dfrac{1}{r^2}\dfrac{\partial^2\phi}{\partial\theta^2} = 0$
Cylindrical, $v_\theta = 0$	$v_r = -\dfrac{\partial\phi}{\partial r}, \; v_\theta = -\dfrac{\partial\phi}{\partial z}$	$\dfrac{1}{r}\dfrac{\partial}{\partial r}\left(r\dfrac{\partial\phi}{\partial r}\right) + \dfrac{\partial^2\phi}{\partial z^2} = 0$
Spherical	$v_r = -\dfrac{\partial\phi}{\partial r}, \; v_\theta = -\dfrac{1}{r}\dfrac{\partial\phi}{\partial\theta}$	$\dfrac{1}{r^2}\dfrac{\partial}{\partial r}\left(r^2\dfrac{\partial\phi}{\partial r}\right) + \dfrac{1}{r^2 \sin\theta}\dfrac{\partial}{\partial\theta}\left(\sin\theta\dfrac{\partial\phi}{\partial\theta}\right) = 0$

4.1.2 Laws of Conservation of Momentum, Mass and Energy

Let us consider a differential volume element of dimensions: Δx, Δy and Δz. Let Q be any of the quantity of momentum, mass or energy. The laws of conservation of Q can be written for this element.

$$
\begin{bmatrix} \text{Accumulation} \\ \text{rate of } Q \\ \text{in the volume} \end{bmatrix} = \begin{bmatrix} \text{Flux of } Q \\ \text{entering due} \\ \text{to convection} \end{bmatrix} - \begin{bmatrix} \text{Flux of } Q \\ \text{leaving due} \\ \text{to convection} \end{bmatrix} + \begin{bmatrix} \text{Molecular} \\ \text{flux of } Q \\ \text{entering} \end{bmatrix} \quad (4.12)
$$
$$
- \begin{bmatrix} \text{Molecular} \\ \text{flux of } Q \\ \text{leaving} \end{bmatrix} + \begin{bmatrix} \text{Generation} \\ \text{rate of } Q \end{bmatrix}
$$

Usually, the molecular transport of momentum, mass and heat is described by Newton's law of viscosity, Fick's first law of diffusion and Fourier's law, respectively. The appropriate substitution of various terms in Equation 4.12 is followed by division by the volume, $\Delta x \Delta y \Delta z$. As $\Delta t \rightarrow 0$, $\Delta x \rightarrow 0$, $\Delta y \rightarrow 0$ and $\Delta z \rightarrow 0$ Equation 4.12 gives the desired differential equations for conservation of Q. Simplifying the equations with the help of the equation of continuity, the equation of change for momentum, mass and thermal energy may be obtained. It is not an easy task to use the shell balance method to model a process in curvilinear coordinates. Bird et al. (1960) suggests starting the problem formulation directly from the equation of motion.

The following equations for momentum transport in Cartesian coordinates are obtained for all types of fluids, whether Newtonian or non-Newtonian, and compressible or incompressible. An absence of derivatives of ρ does not mean that these equations are valid for only incompressible fluids. The derivatives of ρ were eliminated with the help of the equation of continuity.

$$
\rho \left(\frac{\partial v_x}{\partial t} + v_x \frac{\partial v_x}{\partial x} + v_y \frac{\partial v_x}{\partial y} + v_z \frac{\partial v_x}{\partial z} \right) = -\frac{\partial p}{\partial x} + \left[\frac{\partial \tau_{xx}}{\partial x} + \frac{\partial \tau_{yx}}{\partial y} + \frac{\partial \tau_{zx}}{\partial z} \right] + \rho g_x \quad (4.13)
$$

$$
\rho \left(\frac{\partial v_y}{\partial t} + v_x \frac{\partial v_y}{\partial x} + v_y \frac{\partial v_y}{\partial y} + v_z \frac{\partial v_y}{\partial z} \right) = -\frac{\partial p}{\partial y} + \left[\frac{\partial \tau_{xy}}{\partial x} + \frac{\partial \tau_{yy}}{\partial y} + \frac{\partial \tau_{zy}}{\partial z} \right] + \rho g_y \quad (4.14)
$$

$$
\rho \left(\frac{\partial v_z}{\partial t} + v_x \frac{\partial v_z}{\partial x} + v_y \frac{\partial v_z}{\partial y} + v_z \frac{\partial v_z}{\partial z} \right) = -\frac{\partial p}{\partial z} + \left[\frac{\partial \tau_{xz}}{\partial x} + \frac{\partial \tau_{yz}}{\partial y} + \frac{\partial \tau_{zz}}{\partial z} \right] + \rho g_z \quad (4.15)
$$

Equations 4.13 through 4.15 are the x, y and z components of the equation of change, respectively. The stress is a tensor having nine components.

From these equations, the following equations, called 'Navier–Stokes (N-S) equations', are obtained for Newtonian incompressible fluids:

$$\rho\left(\frac{\partial v_x}{\partial t}+v_x\frac{\partial v_x}{\partial x}+v_y\frac{\partial v_x}{\partial y}+v_z\frac{\partial v_x}{\partial z}\right)=-\frac{\partial p}{\partial x}+\mu\left[\frac{\partial^2 v_x}{\partial x^2}+\frac{\partial^2 v_x}{\partial y^2}+\frac{\partial^2 v_x}{\partial z^2}\right]+\rho g_x \quad (4.16)$$

$$\rho\left(\frac{\partial v_y}{\partial t}+v_x\frac{\partial v_y}{\partial x}+v_y\frac{\partial v_y}{\partial y}+v_z\frac{\partial v_y}{\partial z}\right)=-\frac{\partial p}{\partial y}+\mu\left[\frac{\partial^2 v_y}{\partial x^2}+\frac{\partial^2 v_y}{\partial y^2}+\frac{\partial^2 v_y}{\partial z^2}\right]+\rho g_y \quad (4.17)$$

$$\rho\left(\frac{\partial v_z}{\partial t}+v_x\frac{\partial v_z}{\partial x}+v_y\frac{\partial v_z}{\partial y}+v_z\frac{\partial v_z}{\partial z}\right)=-\frac{\partial p}{\partial z}+\mu\left[\frac{\partial^2 v_z}{\partial x^2}+\frac{\partial^2 v_z}{\partial y^2}+\frac{\partial^2 v_z}{\partial z^2}\right]+\rho g_z \quad (4.18)$$

Analytical solutions of these equations may be obtained for equations with one variable and, in a few cases, with two variables. Numerical solutions to equations with two variables can be obtained using various types of software, including MATLAB®. To solve a problem with three or four variables, CFD software is required.

In cases of mass transfer with a chemical reaction in incompressible fluid, the following equation is applicable for each of the chemical species. For example, if there are three chemical species, then three equations are obtained. The diffusivity is assumed to be constant which is true only in the case of dilute solutions:

$$\rho\left(\frac{\partial C_A}{\partial t}+v_x\frac{\partial C_A}{\partial x}+v_y\frac{\partial C_A}{\partial y}+v_z\frac{\partial C_A}{\partial z}\right)=D_{AB}\left[\frac{\partial^2 C_A}{\partial x^2}+\frac{\partial^2 C_A}{\partial y^2}+\frac{\partial^2 C_A}{\partial z^2}\right]-r_A \quad (4.19)$$

In cases of heat transfer, only one equation is obtained:

$$\rho C_P\left(\frac{\partial T}{\partial t}+v_x\frac{\partial T}{\partial x}+v_y\frac{\partial T}{\partial y}+v_z\frac{\partial T}{\partial z}\right)=k\left[\frac{\partial^2 T}{\partial x^2}+\frac{\partial^2 T}{\partial y^2}+\frac{\partial^2 T}{\partial z^2}\right]-\Delta H_{rxn} \quad (4.20)$$

In Equations 4.19 and 4.20, the generation terms (i.e. the last term on the right hand side) are due to chemical reactions only. In the absence of any reaction, $-r_A = 0$ and $\Delta H_{rxn} = 0$.

Several simple models are based on special cases of these equations. The left side has a temporal derivative which is zero in steady-state models. The temporal derivative is considered in dynamic models. The other three terms on the left-hand side of these equations are convective terms. In the absence of convective transport, these terms will be zero. One-dimensional momentum balance for an unsteady-state model gives

$$\rho\left(\frac{\partial v_x}{\partial t}+v_x\frac{\partial v_x}{\partial x}\right)=-\frac{\partial p}{\partial x}+\mu\left[\frac{\partial^2 v_x}{\partial x^2}+\frac{\partial^2 v_x}{\partial y^2}\right]+\rho g_x \quad (4.21)$$

The equation of continuity for 1D flow is given by Equation 4.5. Using it in Equation 4.21 gives

$$\rho \frac{\partial v_x}{\partial t} = -\frac{\partial p}{\partial x} + \mu \frac{\partial^2 v_x}{\partial y^2} + \rho g_x \qquad (4.22)$$

A few simple 1D model equations in Cartesian, cylindrical and spherical coordinate systems are presented in Table 4.4. The solution of these equations depends upon the boundary conditions. The equations in Table 4.4 have been used to model various processes involving multiphase systems. Solutions in Cartesian coordinates are useful in describing the processes occurring at the flat walls and large surfaces. Solutions in cylindrical coordinates are used to describe the processes in tubes, capillaries and cylindrical vessels. Solutions in spherical coordinates are used to describe the processes around the dispersed systems such as bubbles, drops and particles. Often, the non-spherical shapes of the dispersed objects have been approximated by spheres with correction in terms of sphericity.

The well-defined velocities in N-S equations exist in cases of laminar flows only. The random nature of the velocity fluctuations observed in turbulent flows is not described in terms of velocities only.

N-S equations in cylindrical and spherical coordinates are given in Appendix A.

4.1.3 Boundary Conditions

The solution of the differential equations depends upon the boundary conditions (i.e. the conditions at the boundaries of the system). The equations of change based on the laws of conservation are second order in the spatial coordinates and first order in time. The solution requires that two boundary conditions, and one initial condition must be specified to describe the system. Three different types of boundary conditions are commonly used while solving heat and mass transfer problems (Figure 4.2).

4.1.3.1 Boundary Conditions of the First Kind

For heat transfer, the temperature at the boundaries is specified. For mass transfer problems, the concentration of chemical species at the boundary or surface is specified. This type of boundary condition is very frequently used; it is known as a Dirichlet condition, or a boundary condition of the first kind. However, the temperature or concentration may be a function of time and location at the boundary. Such situations are very frequently observed while studying the dynamic response of a system to ramp or sinusoidal input. If the surface is heated or cooled by a fluid going through a phase change, the surface temperature is constant. This kind of boundary condition is applied to heat transfer equipment heated by steam.

TABLE 4.4

A Few Simple Models Based on Laws of Conservation

	Mass Transfer	Heat Transfer
Cartesian Coordinates		
Conditions		
Steady state; no reaction; no convection	$D_{AB} \dfrac{\partial^2 C_A}{\partial x^2} = 0$	$k_f \dfrac{\partial^2 T}{\partial x^2} = 0$
Steady state; with reaction; no convection	$D_{AB} \dfrac{\partial^2 C_A}{\partial x^2} - r_A = 0$	$\dfrac{k_f}{\rho C_p} \dfrac{\partial^2 T}{\partial x^2} - \Delta H_A = 0$
Unsteady state; no reaction; no convection	$\rho \dfrac{\partial C_A}{\partial t} = D_{AB} \dfrac{\partial^2 C_A}{\partial x^2}$	$\dfrac{\partial T}{\partial t} = \dfrac{k_f}{\rho C_p} \dfrac{\partial^2 T}{\partial x^2}$
Cylindrical Coordinates		
Geometry		
Unsteady state; no reaction; diffusion or conduction in radial direction	$\dfrac{\partial C_A}{\partial t} = \dfrac{D_{AB}}{\rho} \left[\dfrac{1}{r} \dfrac{\partial}{\partial r} \left(r \dfrac{\partial C_A}{\partial r} \right) \right]$	$\dfrac{\partial T}{\partial t} = \dfrac{k_f}{\rho C_p} \left[\dfrac{1}{r} \dfrac{\partial}{\partial r} \left(r \dfrac{\partial T}{\partial r} \right) \right]$
Unsteady state; no reaction; diffusion or conduction in axial direction	$\dfrac{\partial C_A}{\partial t} = \dfrac{D_{AB}}{\rho} \dfrac{\partial^2 C_A}{\partial z^2}$	$\dfrac{\partial T}{\partial t} = \dfrac{k_f}{\rho C_p} \dfrac{\partial^2 T}{\partial z^2}$
Unsteady state; with reaction; diffusion or conduction in radial direction	$\dfrac{\partial C_A}{\partial t} = \dfrac{D_{AB}}{\rho} \left[\dfrac{1}{r} \dfrac{\partial}{\partial r} \left(r \dfrac{\partial C_A}{\partial r} \right) \right] - (-r_A)$	$\dfrac{\partial T}{\partial t} = \dfrac{k_f}{\rho C_p} \left[\dfrac{1}{r} \dfrac{\partial}{\partial r} \left(r \dfrac{\partial T}{\partial r} \right) \right] - \dfrac{\Delta H_{rxn}}{\rho C_p}$
steady-state; no reaction; diffusion or conduction in radial direction; convection in axial direction	$v_z \dfrac{\partial C_A}{\partial z} = D_{AB} \dfrac{1}{r} \dfrac{\partial}{\partial r} \left(r \dfrac{\partial C_A}{\partial r} \right)$	$v_z \dfrac{\partial T}{\partial z} = \dfrac{k_f}{\rho C_p} \dfrac{1}{r} \dfrac{\partial}{\partial r} \left(r \dfrac{\partial T}{\partial r} \right)$
steady-state; no reaction; diffusion or conduction in axial direction; convection in axial direction	$v_z \dfrac{\partial C_A}{\partial z} = D_{AB} \dfrac{\partial^2 C_A}{\partial z^2}$	$v_z \dfrac{\partial T}{\partial z} = \dfrac{k_f}{\rho C_p} \dfrac{\partial^2 T}{\partial z^2}$
Spherical Coordinates		
Condition		
Unsteady-state; no reaction; diffusion or conduction in radial direction	$\dfrac{\partial C_A}{\partial t} = \dfrac{D_{AB}}{\rho} \dfrac{1}{r^2} \dfrac{\partial}{\partial r} \left(r^2 \dfrac{\partial C_A}{\partial r} \right)$	$\dfrac{\partial T}{\partial t} = \dfrac{k_f}{\rho C_p} \dfrac{1}{r^2} \dfrac{\partial}{\partial r} \left(r^2 \dfrac{\partial T}{\partial r} \right)$

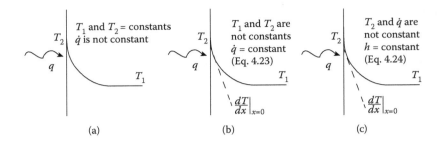

FIGURE 4.2
Types of boundary conditions: (a) Dirichlet, or first kind, (b) Neumann, or second kind and (c) third kind.

4.1.3.2 Boundary Conditions of the Second Kind

A large number of problems require that, at the surface, the heat transfer or mass transfer flux is constant. The flux is specified by Fourier's law or Fick's first law in cases of heat and mass transfer, respectively. This type of boundary condition is specified as given here:

$$q = -k_f \frac{\partial T}{\partial x}\bigg|_{x=0} \tag{4.23}$$

or

$$N_A = -D_{AB} \frac{\partial C_A}{\partial x}\bigg|_{x=0} \tag{4.24}$$

This boundary condition is known as a boundary condition of the second kind or Neumann condition. It is observed in electrically heated heat transfer equipment or solar collectors. Mass transfer across two different phases, with one of them being pure, also has this kind of boundary condition. In cases of adiabatic process, $\dot{q} = 0$.

4.1.3.3 Boundary Conditions of the Third Kind

In the case of convective heat and mass transfer across the surface, the boundary condition of the third kind is observed. It is usually described in terms of heat or mass transfer coefficient.

$$-k \frac{\partial T}{\partial x}\bigg|_{x=0} = h(T_s - T_\infty) \tag{4.25}$$

$$-D_{AB} \frac{\partial C_A}{\partial x}\bigg|_{x=0} = k(C_{As} - C_{A\infty}) \tag{4.26}$$

The temperature at the surface, T_s, or concentration of A, $C_{A\infty}$, need not be constant. It may be a function of time.

4.2 Laminar Flow

In laminar flow, the fluid flows in an orderly manner. No macroscopic inter-mixing of the fluid elements with its neighbour is observed. Laminar flow takes place at low flow rates. From the modelling point of view, a laminar flow can be considered as a well-defined flow which can be expressed in terms of time and space variables.

4.2.1 Velocity Field in Laminar Flow

The laws of conservation provide governing equations which, when solved, give the distribution of velocity, temperature and concentration in a system. These are therefore useful in developing a distributed parameter model. The first terms on the left-hand side of Equations 4.18 and 4.19 are tempo-ral derivative terms. These terms are zero in steady-state models. The other three terms correspond to convection terms.

If the boundary conditions are a function of time or the velocity is a func-tion of time, then the temporal term should be used. However, sometimes the process is so slow that accumulation is negligible, and the temporal term is dropped. This is called a 'quasi-steady-state assumption'.

If the velocity components, v_x, v_y and v_z, are considered to be independent of temperature or concentration, then the equation of motion may be solved independently to get the velocity profile.

The equation of thermal energy and equation of continuity for a chemical species may be solved after substitution of the expression for the velocity profile. This approach is applicable in cases of forced convection. Natural convection is caused by the density difference due to the temperature or concentration gradient. In such cases, the equation of motion and equation of change for thermal energy or the equation of continuity for a chemical species should be solved simultaneously. The models for natural convection are thus somewhat more complex than those for forced convection problems.

4.2.2 Velocity Profile in Simple Geometries

In process industries, fluids flow in circular tubes and cylindrical vessels. The jacketed wall used to exchange heat in a cylindrical vessel is similar to annulus. The fluid flow in many separation devices is similar to flow between two parallel plates and slits. The flow around dispersed systems resemble to flow around spheres. The flow across cylindrical tubes is encountered in heat exchangers. Thus, the most frequently observed geometries are circular tubes, plates, annulus, cylinders and spheres. The flow around these objects has been studied extensively and is available in books of fluid mechanics and transport phenomena. Only a few cases will be discussed to illustrate the use of laws of conservation to obtain the velocity, temperature or concentration profile.

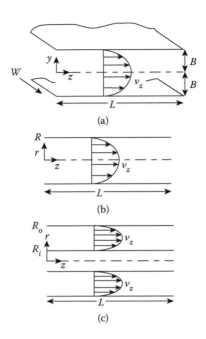

FIGURE 4.3
Flow of Newtonian incompressible fluid: (a) between two parallel plates, (b) in a circular tube and (c) in annulus.

Let us consider the steady-state fully developed flow of Newtonian incompressible fluid between two parallel plates (Figure 4.3a) separated by a distance 2B. The width of the plate W is very large as compared to the gap 2B. Fully developed flow means that there are no end effects. This assumption is true only when the length of plate, L, in the direction of flow is very large. All the convective terms in N-S equations are absent for 1D flow when the equation of continuity is substituted. The gravity has no effect in a horizontal tube or in the case of forced convection. At steady state, the time derivative is zero. After these simplifications and taking $\frac{\partial p}{\partial x} = \frac{\Delta P}{L}$, we get

$$\frac{\Delta P}{L} = \mu \frac{d^2 v_z}{dy^2} \qquad (4.27)$$

The boundary conditions are as follows:

At $y = B$, $v_z = 0$

At $y = 0$, $\frac{dv_z}{dy} = 0$

The velocity profile can be obtained by solving Equation 4.27.

$$v_z = \frac{\Delta P B^2}{2\mu L}\left[1-\left(\frac{y}{B}\right)^2\right] \tag{4.28}$$

For axial flow in a circular tube (Figure 4.3b), we again obtain the similar equation. The variable y is replaced by r. The following boundary conditions are used:

At $r = R$: $v_z = 0$

At $r = 0$: $\dfrac{dv_z}{dr} = 0$

The velocity profile in this case also is parabolic.

$$v_z = \frac{\Delta P R^2}{4\mu L}\left[1-\left(\frac{r}{R}\right)^2\right] \tag{4.29}$$

A fully developed flow of incompressible non-Newtonian fluid in an annulus with R_o and R_i as the radii of the outer and inner tubes, respectively, results in the same equation as in the case of flow in a circular pipe (Figure 4.3c). The boundary conditions are as follows:

At $r = R_o$: $v_z = 0$

At $r = R_i$: $v_z = 0$

The velocity profile is as follows:

$$v_z = \frac{\Delta P R_o^2}{4\mu L}\left[1-\left(\frac{r}{R_o}\right)^2 - \frac{1-(R_i/R_o)^2}{\ln(R_o/R_i)}\ln\left(\frac{R_o}{r}\right)\right] \tag{4.30}$$

Detailed proofs of these problems are not given here as they are available in several books on transport phenomena (Bird et al. 2002). Equation 4.30 reduces to Equation 4.28 for values of $\dfrac{R_i}{R_o}$ close to 1. Under this condition, the flow in an annular flow can be treated as the flow in a slit (Bird et al. 2002). Similar simplifications are used to avoid the equation of change in curvilinear coordinates. The use of N-S equations in rectilinear coordinates is much easier.

In all three problems, the convective terms are zero. The velocity profile still exists as it is related to shear stress.

For flow of non-Newtonian fluids, the starting equations are Equations 4.13 through 4.15. Let us consider flow of a power law fluid in a circular pipe. For axial flow, the z component of the equation of motion is simplified to

$$\frac{\partial p}{\partial z} = -\left(\frac{1}{r}\frac{\partial}{\partial r}(r\tau_{rz})\right) \tag{4.31}$$

Equation 4.31 requires a rheological model to proceed. The correctness of the velocity profile depends on the correct choice of the rheological model (Skelland 1964). For a power law fluid substituting the value of τ_{rz} in Equation 4.31, we get

$$\frac{\Delta p}{L} = \frac{1}{r}\frac{d}{dr}\left\{rm\left(\frac{dv_z}{dr}\right)^n\right\}$$

(4.32)

The boundary conditions are as follows:

At $r = R$: $v_z = 0$

At $r = 0$: $\dfrac{dv_z}{dr} = 0$

Equation 4.32 can now be integrated twice to get the velocity profile.

$$v_z = \frac{\Delta P}{2Lm(n+1)}\left(R^{\frac{n+1}{n}} - r^{\frac{n+1}{n}}\right)$$

(4.33)

4.2.3 Convective Heat and Mass Transfer in Simple Geometries

Let us consider unsteady-state mass transfer in a flat plane in the absence of convection with no reaction. The convection and generation terms are zero, and Equation 4.19 reduces to Equation 3.31. An analogous equation for heat transfer is also obtained. Fick's second law and Fourier's law were not obtained from the law of conservation, but they are used in the equations of continuity, momentum and thermal energy. The problems based on Fick's and Fourier's laws were discussed in Chapter 3 and were in the absence of convection. Heat and mass transfer problems in the presence of convection are discussed in this chapter.

Let us consider that a cold fluid at temperature T_0 is falling down a vertical wall, as shown in Figure 4.4. The hot surface is maintained at a constant temperature, T_s. If the fluid is incompressible non-Newtonian, the velocity profile is given by Equation 4.29. The equation of change of thermal energy for a steady-state heat transfer is obtained by simplifying Equation 4.20:

$$\rho C_p v_z \frac{\partial T}{\partial z} = k \frac{\partial^2 T}{\partial y^2}$$

(4.34)

The boundary conditions are as follows:

$T = T_0$ at $z = 0$ and for $y > 0$
$T = T_0$ at $y = \delta$ and for finite z
$T = T_s$ at $y = 0$ and for $z > 0$

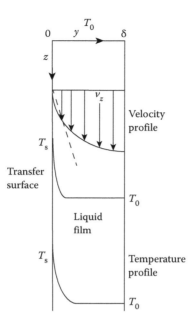

FIGURE 4.4
Heat transfer for fluid falling down a vertical wall.

To obtain an analytical solution, the second boundary condition, $y = \delta$, is replaced by the condition $y \to \infty$. For a short contact time, the parabolic velocity profile is approximated by linear velocity in the vicinity of the surface, that is,

$$v_z = \frac{\rho g \delta}{\mu} y \qquad (4.35)$$

If the heat transfer flux is to be determined, then the temperature gradient at the wall is required. Therefore, the above assumption seems appropriate. The analytical solution of Equation 4.34 is as follows (Bird et al. 1960):

$$\frac{(T - T_0)}{(T_s - T_0)} = \frac{1}{\Gamma(4/3)} \int_{\eta}^{\infty} \exp(-\eta^3) d\eta \qquad (4.36)$$

where

$$\eta = \frac{y}{\left[\left(\dfrac{9 \mu k}{\rho^2 C_p g \delta} \right) x \right]^{\frac{1}{3}}}$$

If the velocity profile is given by Equation 4.29, then Equation 4.34 can be solved numerically. The 'pdepe' solver of MATLAB can be used by replacing the time derivative by spatial derivative. This solver can be used for parabolic partial differential equations for various types of boundary conditions for slabs, cylinders and spheres, and it can consider the generation term also. While using the 'pdepe' solver, use of variables 't' and 'x' only is permitted. The codes for the function, initial condition and boundary conditions are written in separate files and are given in Appendix B. Equation 4.29 was used to describe the velocity profile. The results are presented in Figure 4.5.

For steady-state mass or heat transfer for an incompressible Newtonian fluid, flowing in a circular pipe in the laminar flow regime is a 1D model that can be described by simplifying the equation in cylindrical coordinates, as given in Table 4.4 (e.g. for heat transfer without reaction):

$$\rho C_v v_z \frac{\partial T}{\partial z} = k \left[\frac{1}{r} \frac{\partial}{\partial r} \left(r \frac{\partial T}{\partial r} \right) + \frac{\partial^2 T}{\partial z^2} \right] \tag{4.37}$$

By neglecting the conduction in the direction of flow in comparison to that in the radial direction, we get

$$\rho C_v v_z \frac{\partial T}{\partial z} = k \left[\frac{1}{r} \frac{\partial}{\partial r} \left(r \frac{\partial T}{\partial r} \right) \right] \tag{4.38}$$

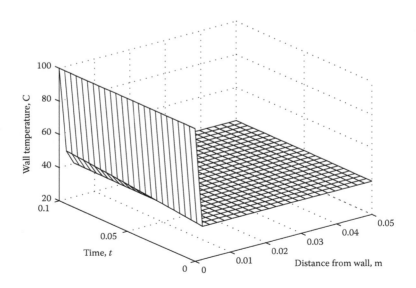

FIGURE 4.5
Numerical solution for steady-state heat transfer with convection in a slab.

Equation 4.38 is known as the Graetz problem. The analytical solution of the Graetz problem is available in the literature (Bird et al. 2002). It can be solved numerically also. The 'pdepe' solver of MATLAB can again be used by taking $m = 1$.

The solution of problems with more variables can be solved using CFD software which is beyond the scope of the present book.

4.3 Boundary Layers: Momentum, Thermal and Diffusional

Various processes in chemical and biochemical systems and environmental sciences are related to mass and heat transfer near a wall, surface or interface. The region of interest is the neighbourhood of the surface.

From the equations of change of momentum, it can be seen that the convective term does not contain viscosity. The viscosity appears only in the terms corresponding to molecular transport. It is true for the equation of continuity of individual species in the case of mass transfer and also for the equation of change of thermal energy in the case of heat transfer. It is easy to realise that, near the wall or surface, the fluid velocity is small in comparison to that in the bulk. Therefore, the molecular phenomena dominate near the surface. This has led to the development of the concept of boundary layers. The entire flow field is divided into two regimes: the first in the vicinity of the surface, where viscosity plays an important role, and another in the bulk, where the flow is not significantly influenced by the existence of the surface. The region adjacent to the wall is called the 'boundary layer'. It is often assumed that the effect of the wall is restricted to the boundary layer only. In the region in which the momentum transfer is considered, the boundary layer is called the 'momentum boundary layer'. Similarly, the region adjacent to the surface in which the entire variation of temperature takes place is called the 'thermal boundary layer'. The 'diffusional boundary layer' is the region adjacent to the surface in which the entire variation of concentration exists. The concept of the boundary layer is an approximation of the physical situation, unlike in film theory in which the film is fictitious.

Let us study the flow of an incompressible Newtonian fluid near a leading edge of a flat plate (Figure 4.6). The flow is 2D. The equation of continuity is Equation 4.6, and the x component of the equation of motion in rectilinear coordinates is

$$\rho\left(v_x \frac{\partial v_x}{\partial x} + v_y \frac{\partial v_x}{\partial y} \right) = \mu\left(\frac{\partial^2 v_x}{\partial x^2} + \frac{\partial^2 v_x}{\partial y^2} \right) \tag{4.39}$$

The contribution of the y-component of the equation of motion is neglected because $v_y \ll v_x$. The limit of integration is from 0 to y, which is due to the

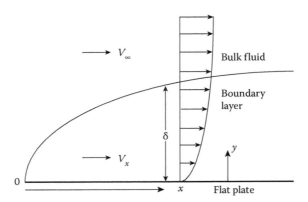

FIGURE 4.6
Boundary layer adjacent to a flat plate.

assumption that all variation of the velocity is limited to a boundary layer of thickness y. The value of v_y from Equation 4.11 is substituted in Equation 4.39. Neglecting the term $\dfrac{\partial^2 v_x}{\partial y^2}$, since it is small as compared to the term $v_x\left(\dfrac{\partial v_x}{\partial x}\right)$ following is obtained.

$$v_x \frac{\partial v_x}{\partial x} + \left(\int_0^y \frac{\partial v_x}{\partial x} dy\right)\frac{\partial v_x}{\partial y} = \frac{\mu}{\rho}\frac{\partial^2 v_x}{\partial y^2} \tag{4.40}$$

The boundary conditions are as follows:

$v_x = 0$ at $y = 0$ (at the surface)
$v_x = v_\infty$ at $y = \delta$ (at the end of the boundary layer)
$v_x = v_\infty$ for all values of y (at the edge)

Equation 4.40 is written in terms of a new variable: $\eta = \dfrac{y}{\delta}$ and $\phi(\eta) = \dfrac{v_x}{v_\infty}$.

$$\left(\int_0^1 \eta\frac{d\phi}{d\eta}d\eta - 2\int_0^1 \phi\eta\frac{d\phi}{d\eta}d\eta\right)\delta\frac{d\delta}{dx} = \frac{\mu}{\rho v_\infty}\int_0^1 \frac{d^2\phi}{d\eta^2}d\eta \tag{4.41}$$

If function $\phi(\eta)$ is chosen such that it satisfies the boundary condition, the expression for δ can be obtained. The following function satisfies the boundary conditions:

$$\phi(\eta) = \frac{3}{2}\eta - \frac{1}{2}\eta^3 \tag{4.42}$$

For function $\varphi(\eta)$, given by Equation 4.42, the boundary layer thickness and velocity distribution profiles are obtained.

$$\delta = \left(\frac{280}{13} \frac{\mu x}{\rho v_\infty} \right)^{1/2} \tag{4.43}$$

and

$$\frac{v_x}{v_\infty} = \frac{3}{2} \frac{y}{\delta} - \frac{1}{2} \left(\frac{y}{\delta} \right)^3 \tag{4.44}$$

A detailed proof is available in textbooks (e.g. Bird 2002). Note that the integration constants in Equation 4.41 depend upon the definition of the function $\varphi(\eta)$. For example, for

$$\phi(\eta) = 2\eta - 2\eta^3 + \eta^4 \tag{4.45}$$

the boundary layer thickness and velocity profiles are given by

$$\delta = \left(\frac{1260}{37} \frac{\mu x}{\rho v_\infty} \right)^{1/2} \tag{4.46}$$

$$\frac{v_x}{v_\infty} = 2\frac{y}{\delta} - 2\left(\frac{y}{\delta} \right)^3 + \left(\frac{y}{\delta} \right)^4 \tag{4.47}$$

To obtain the temperature profile in the boundary layer adjacent to a flat plate, the following equation is also solved together with Equations 4.39 and 4.11:

$$\rho C_p \left(v_x \frac{\partial T}{\partial x} + v_y \frac{\partial T}{\partial y} \right) = k \left[\frac{\partial^2 T}{\partial x^2} + \frac{\partial^2 T}{\partial y^2} \right] \tag{4.48}$$

It is assumed that the velocity and temperature profiles have similar shapes; in other words, a dimensionless temperature profile has the same functional nature as given by Equation 4.45. Also, the thermal boundary layer thickness, δ_T, is proportional to the momentum boundary layer thickness, δ (i.e. $\delta_T = \delta\Delta$).

$$\phi_T(\eta_T) = 2\eta_T - 2\eta_T^3 + \eta_T^4 \tag{4.49}$$

where

$$\eta_T = \frac{y}{\delta_T};$$

and

$$\phi_T(\eta_T) = \frac{T_s - T}{T_s - T_\infty}$$

The only unknown is Δ because δ is given by Equation 4.46, and the temperature profile is given by Equation 4.49. The solution of Equation 4.48 gives a value of Δ which is approximated by $Pr^{-1/3}$ for $\Delta < 1$ (Bird et al. 1960).

Once a functional nature for the variable φ is assumed, the 'pdepe' toolbox of MATLAB can be used to get the boundary layer thickness. However, since a solution to the boundary layer thickness over a flat plate is available in the form of a simple expression, as given by Equation 4.45, the result has been successfully used to model many complex problems.

4.4 Turbulence Models

When the flow is laminar, the laws of conservation provide us the governing equations which can be solved to get the distribution of velocity, concentration or temperature. However, if the flow is not laminar, then the laws of conservation cannot be applied until the turbulent nature is appropriately expressed in terms of the velocity. The modelling of processes involving turbulent flow requires an understanding of turbulent flows.

4.4.1 What Is Turbulence?

It is not easy to define 'turbulence'. Reynolds' experiment of movement of a streak of dye in a circular tube indicated that at high fluid velocity, the motion of the fluid elements in the radial direction is different from that of molecular diffusion. It is responsible for the high degree of mixing which is desirable in many processes. The flow has been called 'turbulent flow'. The heat and mass transfer rates in this regime of flow are high and cannot be explained by molecular transport only. The reason and exact behaviour of the turbulence are not clearly understood; therefore, there is no deterministic method to predict it. Hinze (1975) looked at turbulence as a random process (i.e. the fluid velocity at a point in space changes with time in a random way). This assumption has helped in specifying turbulence in statistical parameters. However, turbulence is not completely random. A chaotic fluctuation of fluid velocity is visualised as turbulence by many other investigators. The turbulence is the instability of the laminar flow; as a result, large eddies are generated. The kinetic energy imparted to the flow is dissipated into smaller eddies, and finally all the energy is dissipated as thermal energy (Figure 4.7a). These eddies are 3D and time dependent. The instabilities are due to interactions between non-linear

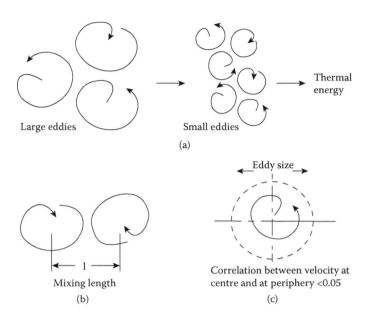

FIGURE 4.7
Turbulence models: (a) dissipation of kinetic energy into thermal energy, (b) mixing length and (c) eddy.

inertial terms and viscous terms in the N-S equation. These interactions are rotational, fully time dependent and mutually connected via vortex stretching that is possible only in 3D space. Therefore, a turbulent flow field can be obtained only by solving N-S equations in three dimensions and with unsteady terms. This is possible only by using CFD codes, and it is beyond the scope of this book.

When N-S equations of motion are solved by taking sufficiently small time and space steps and by using completely defined initial conditions, a solution to turbulent flow can be obtained. This approach is called a direct numerical simulation (DNS). Models for Reynolds stresses are used in the Reynolds-averaged Navier–Stokes (RANS) approach. The k–ε model used in CFD simulations solves two sets of additional differential equations for the conservation of turbulent kinetic energy (k) and the turbulent energy dissipation rate (ε). This increases computational effort and time. Due to an incomplete understanding and excessive computational effort, the turbulence is characterised in many empirical laws which are good for engineering applications. Here, we will look some of the ways in which the processes involving turbulent flow and turbulent transfer process are modelled.

Let the instantaneous fluid velocity be represented by sum of an average velocity, $\bar{v}$, and a randomly fluctuating term, v'. The average velocity is the ensemble average (i.e. it is the average of fluid velocity in several runs measured at same point and time). Often, the time average is used, which will not be the same in the case of an unsteady process. The velocity, v_x, can be written as

$$v_x = \bar{v}_x + v_x' \tag{4.50}$$

The average, $\bar{v}'$, is zero, but the averages, $\overline{v_x'^2}$ and $\overline{v_x' v_y'}$ etc., are non-zero. Substituting Equation 4.50 in Equation 4.3, the equation of continuity becomes

$$\frac{\partial \rho}{\partial t} + \frac{\partial}{\partial x}\left[\rho\left(\overline{v_x} + v_x'\right)\right] + \frac{\partial}{\partial y}\left[\rho\left(\overline{v_y} + v_y'\right)\right] + \frac{\partial}{\partial z}\left[\rho\left(\overline{v_z} + v_z'\right)\right] = 0 \tag{4.51}$$

In this way, the deterministic nature of the equation of continuity and equation of motion are coupled with the statistical nature of the turbulent velocity. Taking the time average of Equation 4.51, we get

$$\frac{\partial \rho}{\partial t} + \frac{\partial}{\partial x}\left(\rho\overline{v_x}\right) + \frac{\partial}{\partial y}\left(\rho\overline{v_y}\right) + \frac{\partial}{\partial z}\left(\rho\overline{v_z}\right) = 0 \tag{4.52}$$

The x component of the equation of motion before using the equation of continuity is as follows:

$$\rho\left(\frac{\partial(\rho v_x)}{\partial t} + \frac{\partial}{\partial x}\left(\rho v_x v_x\right) + \frac{\partial}{\partial y}\left(\rho v_x v_y\right) + \frac{\partial}{\partial z}\left(\rho v_x v_z\right)\right)$$

$$= -\frac{\partial p}{\partial x} + \mu\left[\frac{\partial^2 v_x}{\partial x^2} + \frac{\partial^2 v_x}{\partial y^2} + \frac{\partial^2 v_x}{\partial z^2}\right] + \rho g_x \tag{4.53}$$

Substituting Equation 4.50 in Equation 4.53 provides

$$\frac{\partial}{\partial t}\rho\left(\overline{v_x} + v_x'\right) + \frac{\partial}{\partial x}\rho\left(\overline{v_x} + v_x'\right)\left(\overline{v_x} + v_x'\right) + \frac{\partial}{\partial y}\rho\left(\overline{v_x} + v_x'\right)\left(\overline{v_y} + v_y'\right)$$

$$+ \frac{\partial}{\partial z}\rho\left(\overline{v_x} + v_x'\right)\left(\overline{v_z} + v_z'\right) \tag{4.54}$$

$$= -\frac{\partial(\bar{p} + p')}{\partial x} + \mu\left[\frac{\partial^2\left(\overline{v_x} + v_x'\right)}{\partial x^2} + \frac{\partial^2\left(\overline{v_x} + v_x'\right)}{\partial y^2} + \frac{\partial^2\left(\overline{v_x} + v_x'\right)}{\partial z^2}\right] + \rho g_x$$

The time average of Equation 4.54 gives

$$\frac{\partial}{\partial t}\rho\overline{v_x} + \left(\frac{\partial}{\partial x}\rho\overline{v_x v_x} + \frac{\partial}{\partial y}\rho\overline{v_x v_y} + \frac{\partial}{\partial z}\rho\overline{v_x v_z}\right)$$

$$= -\frac{\partial\bar{p}}{\partial x} + \mu\left[\frac{\partial^2\overline{v_x}}{\partial x^2} + \frac{\partial^2\overline{v_x}}{\partial y^2} + \frac{\partial^2\overline{v_x}}{\partial z^2}\right] + \rho g_x - \left(\frac{\partial}{\partial x}\rho\overline{v_x' v_x'} + \frac{\partial}{\partial y}\rho\overline{v_x' v_y'} + \frac{\partial}{\partial z}\rho\overline{v_x' v_z'}\right)$$

$$\tag{4.55}$$

Comparing Equation 4.55 with Equation 4.53, it is observed that the last term of Equation 4.55 is an additional term that is due to the turbulence. It is the cross-correlation of various velocity components. The other terms use average velocity instead of instantaneous velocity. The last term is known as 'Reynolds stresses'. These are analogous to momentum flux and are also called 'turbulent momentum flux'. In laminar flow, the average and instantaneous velocities are same. Equations similar to Equation 4.55 can be obtained for the y and z components of the equation of motion. If the Reynolds stresses can be expressed in terms of average velocity or its function, then Equation 4.55 can be solved. Various turbulence models are proposed to specify these terms.

Equations of change of thermal energy can be obtained in a similar manner. Instantaneous temperature is assumed to be the sum of an average temperature and a fluctuation term (i.e. $T = \overline{T} + T'$). For turbulent flow problems, Equation 4.20 without the generation term takes the following form:

$$
\rho C_p \frac{\partial \overline{T}}{\partial t} + \left(\frac{\partial}{\partial x} \rho C_p \overline{v_x T} + \frac{\partial}{\partial y} \rho C_p \overline{v_y T} + \frac{\partial}{\partial z} \rho C_p \overline{v_z T} \right)
$$
$$
= k \left[\frac{\partial^2 \overline{T}}{\partial x^2} + \frac{\partial^2 \overline{T}}{\partial y^2} + \frac{\partial^2 \overline{T}}{\partial z^2} \right] - \left(\frac{\partial}{\partial x} \rho C_p \overline{v_x' T'} + \frac{\partial}{\partial y} \rho C_p \overline{v_y' T'} + \frac{\partial}{\partial z} \rho C_p \overline{v_z' T'} \right)
$$
(4.56)

The last term on the right-hand side of Equation 4.46 is due to the fluctuating component of velocity and temperature. It is a cross-correlation of temperature and various velocity components. In the case of mass transfer, the additional term due to turbulence includes the cross-correlation of concentration of the species with velocity components. The equation, in the absence of a chemical reaction, is given here:

$$
\frac{\partial \overline{C_A}}{\partial t} + \left(\frac{\partial}{\partial x} \overline{v_x C_A} + \frac{\partial}{\partial y} \overline{v_y C_A} + \frac{\partial}{\partial z} \overline{v_z C_A} \right)
$$
$$
= D_{AB} \left[\frac{\partial^2 \overline{C_A}}{\partial x^2} + \frac{\partial^2 \overline{C_A}}{\partial y^2} + \frac{\partial^2 \overline{C_A}}{\partial z^2} \right] - \left(\frac{\partial}{\partial x} \overline{v_x' C'_A} + \frac{\partial}{\partial y} \overline{v_y' C'_A} + \frac{\partial}{\partial z} \overline{v_z' C'_A} \right)
$$
(4.57)

The last terms on the right-hand side of Equations 4.55 through 4.57 are due to turbulent flow. The model for these terms requiring no additional equation but only the equation of continuity and the equation of motion to get the velocity field are called 'zero-equation turbulence models'.

In cases of mass transfer with a chemical reaction, more terms involving the fluctuating concentration are generated. In general, the following rate expression:

$$-r_A = f(C_A) = f(C_A + C'_A) \tag{4.58}$$

gives

$$-\overline{r_A} = \overline{f(C_A + C'_A)} \tag{4.59}$$

For example, for a second-order reaction given by

$$-r_A = kC_A^2 \tag{4.60}$$

the fluctuating terms are given by

$$-\overline{r_A} = k\overline{C_A^2} + k\overline{C_A'^2} \tag{4.61}$$

The observed rate depends upon the concentration fluctuation. Only in the case of a first-order reaction does the concentration fluctuation not affect the observed rate of reaction.

4.4.2 Eddy Viscosity, Eddy Diffusivity and Eddy Thermal Conductivity

Reynolds stresses are considered to be stresses due to turbulence (Figure 4.7c), and they are expressed in a manner analogous to Newton's law of viscosity:

$$\rho\overline{v'_x v'_y} = \mu_t \frac{d\overline{v_x}}{dy} \tag{4.62}$$

Since the turbulence diminishes rapidly near the wall, the eddy viscosity is a strong function of the distance from the wall and is not a constant. It is not the property of the fluid but depends upon the turbulence present. In a region adjacent to the wall, the flow is laminar. This region is known as the 'laminar sub-layer'. In this region of the fluid flow, viscous forces predominate. In the region far away from the wall, the viscous forces do not influence the flow field, and the turbulence or eddies play an important role. The flow in this region is known as 'inviscid flow'. The definition of eddy viscosity in Equation 4.55 allows us to add the Reynolds stresses with the viscous forces.

The cross-correlations in Equations 4.56 and 4.57 are also defined in a similar manner:

$$\rho C_v \overline{v'_x T'} = k_t \frac{d\overline{T'}}{dy} \tag{4.63}$$

$$\rho \overline{v_x' C_A'} = D_{AB,t} \frac{d\overline{C_A'}}{dy} \tag{4.64}$$

These terms are known as 'turbulent energy flux' and 'turbulent mass flux', respectively. The constants k_t and $D_{AB,t}$ are 'eddy conductivity' and 'eddy diffusivity', respectively. Since the mass and heat transfer are considered to be analogous, the values of k_t and $D_{AB,t}$ are equal. The turbulence models given by Equations 4.62 through 4.64 have the advantage that these terms can be added directly to the terms involving viscosity, thermal conductivity or diffusivity.

4.4.3 Prandtl Mixing Length

Let us consider that several eddies are moving in the fluid (Figure 4.7b). When two eddies come closer, they are mixed. The mean distance between eddies is called the 'mixing length' which is analogous to the mean-free path in gases. It is the mean distance that an eddy travels before it loses its identity due to interaction with an adjacent fluid. Prandtl obtained the following equation for turbulent momentum flux in terms of the mixing length, l.

$$\rho \overline{v_x' v_y'} = \rho l^2 \left| \frac{d\overline{v_x}}{dy} \right| \frac{d\overline{v_x}}{dy} \tag{4.65}$$

The mixing length is a function of distance from the wall, y. Prandtl used linear dependence (i.e. $l = ky$).

$$\rho C_v \overline{v_x' T'} = \rho l^2 \left| \frac{d\overline{v_x}}{dy} \right| \frac{d\overline{T}}{dy} \tag{4.66}$$

$$\rho \overline{v_x' C_A'} = \rho l^2 \left| \frac{d\overline{v_x}}{dy} \right| \frac{d\overline{C_A}}{dy} \tag{4.67}$$

The mixing length is the same in all three cases. Prandtl used Equation 4.65 to obtain a universal velocity profile in a circular smooth tube. The flow was assumed to take place in three regions consisting of a very thin region adjacent to the wall, called the 'laminar sub-layer'; a turbulent core; and a buffer region in between the two regions. In the laminar sub-layer, only viscous forces were considered. In the turbulent core, viscous forces were neglected. In the buffer layer, both forces were considered. The turbulent momentum flux was described by the 'mixing length'. Few momentum, mass and energy analogies are based on the universal velocity profile.

4.4.4 Turbulence Kinetic Energy and Length and Time Scale

In case of isotropic turbulence it is assumed that the eddy dissipation by the micro-scale eddy moves isotropically and is mainly due to viscous forces. The length scale, η_l; time scale, τ_η; and velocity scale, v_η, of the micro-eddies were expressed in terms of the energy dissipation rate per unit mass, ε, and kinematic viscosity, ν:

$$\eta_l = \left(\frac{\nu^3}{\varepsilon}\right)^{1/4}, \ \tau_\eta = \left(\frac{\nu}{\varepsilon}\right)^{1/2} \text{ and } v_\eta = (\nu E)^{1/4} \tag{4.68}$$

4.5 Surface Renewal Models at High Flux of Momentum, Mass or Heat

The N-S equations can be solved by CFD software. However, CFD models are time consuming, and integration within the process models for the entire plant is still a tough task and requires a large amount of time for obtaining the solutions to flow problems. The use of CFD software is therefore limited to studying a single process. The data obtained are correlated so that they can be used as input to large models. Possibly for this reason, the CFD simulations are sometimes known as 'numerical experiments'. The role of simple models in large models is still important.

An analytical solution to the N-S equation in the case of turbulent flow is not possible unless some simplifying assumptions are made. These result in various types of models. A film model was described in Chapter 3. The film model does not consider any flow in the fictitious film; hence, laws of conservation are not applicable. Other models make use of one or more of these laws.

One of the approaches was to consider that in the turbulent flows, mass or heat between the surface or interface and the bulk of the fluid are exchanged due to the movement of eddies. Earlier models considered that the transfer processes in eddies take place by molecular transport, although, within the eddy, the flow may be laminar. It is believed that most of the transfer is limited to a region adjacent to the transfer surface; the movement of eddies in the bulk need not be considered. The time spent by the eddy at the surface is important. Some of the models based on this philosophy are presented in this section.

4.5.1 Penetration Model (Higbie's Surface Renewal)

It is assumed that from the bulk fluid, packets or eddies arrive at the transfer surface, stay there for some time and then leave the surface to go back to the bulk (Figure 2.10). Eddies have the same temperature or concentration as that of the bulk before their arrival at the transfer surface. During the stay, the mass or heat transfer takes place by 1D unsteady-state diffusion or conduction, respectively. The mass transfer for a chemical species during the stay of eddies on the transfer surface takes place by unsteady-state diffusion from the wall to the eddy (Higbie 1935).

$$\frac{\partial C_A}{\partial t} = D_{AB} \frac{\partial^2 C_A}{\partial x^2} \tag{4.69}$$

with the following boundary conditions:

$C_A = C_A^*; x = 0, t > 0$
$C_A = C_{A\infty}; t = 0, x > 0$
$C_A = C_{A\infty}; x = \infty, t > 0$

This assumption ensures that information about the fluid velocity is not required for the estimation of mass and heat transfer rates. No heat or mass transfer from eddies to the surrounding fluid takes place during the period of travel between the bulk and the transfer surface. It is equivalent to assume that the time taken by eddies to travel from the bulk to the surface and back is zero.

It is interesting to note that the depth of the fluid packet is taken as infinite, indicating that before the concentration or temperature of the boundary of the eddy away from the wall is changed, the eddy leaves the surface. The solution of Equation 4.69 is

$$\frac{C_A^* - C_A}{C_A^* - C_{A\infty}} = erf\left(\frac{z}{2\sqrt{D_{AB}t}}\right) \tag{4.70}$$

Here, t is the time for which the fluid packet stays at the wall. Higbie (1935) assumed that all the fluid packets stay at the surface for equal times, t_e. The average mass transfer to or from the eddy during the exposure time may be given as

$$\frac{1}{t_e}\int_0^{t_e} -D_{AB}\left(\frac{\partial C_A}{\partial z}\right)_{z=0} dt = \frac{1}{t_e}\int_0^{t_e} \left(C_A^* - C_A\right)\sqrt{\frac{D_{AB}}{\pi t}}\, dt = 2\left(C_A^* - C_A\right)\sqrt{\frac{D_{AB}}{\pi t_e}} \tag{4.71}$$

Similarly, the average heat transfer flux can be obtained as

$$q = 2(T_W - T)\sqrt{\frac{k_f \rho_f C_p}{\pi t_e}} \tag{4.72}$$

4.5.2 Surface Renewal Model for Other Residence Time Distributions

All fluid packets need not stay for an equal amount of time on the surface. Any suitable probability distribution function, $\varphi(t)$, may be chosen to describe the exposure time. The average mass transfer rate may be written as

$$\frac{1}{t_e} \int_0^{t_e} \phi(t) \left\{ -D_{AB} \left(\frac{\partial C_A}{\partial z} \right)_{z=0} \right\} dt \tag{4.73}$$

4.5.2.1 Danckwerts' Surface Renewal Model

Danckwerts (1951) assumed that all of the fluid elements present at the surface have the same probability of leaving the surface irrespective of their age. The age distribution, $\varphi(t)$, thus follows the exponential probability distribution:

$$\phi(t) = se^{-st} \tag{4.74}$$

And the average mass transfer rate N_A is obtained as

$$\overline{N_A} = \left(C_A^* - G_{A\infty} \right) \sqrt{s D_{AB}} \tag{4.75}$$

Here, s has been termed the 'fractional rate of surface renewal'. Its value increases with turbulence. However, there is no direct way to estimate it. The average heat flux, q, can also be obtained in a similar manner:

$$q = (T_s - T_b) \sqrt{\frac{k}{\rho_l C_p s}} \tag{4.76}$$

The mass transfer and heat transfer fluxes are proportional to $D_{AB}^{1/2}$ and $k^{1/2}$, respectively, in both the models and do not depend upon the age distribution function.

The drawbacks of the pure surface renewal models are that eddies have finite size and it is not proper to assume unsteady-state transfer of mass and energy when the eddy is present for a very long time, since with progress of time the transfer processes approach towards achieving steady-state mass or heat transfer. Since film theory is applicable at steady state, attempts have been made to combine the concepts of film theory and surface renewals (Figure 4.8).

4.5.3 Surface Renewal Model with a Packet of Finite Length

Toor and Marchello (1958) combined the concepts of surface renewal with film theory. The range of t according to exponential probability distribution

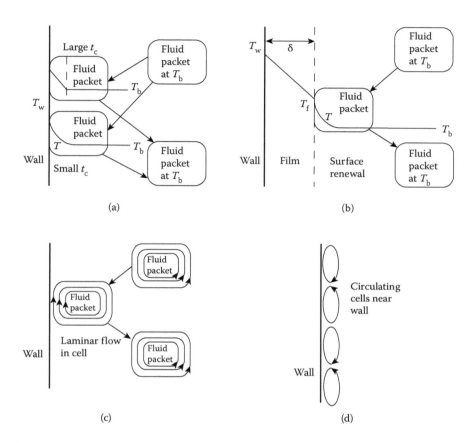

FIGURE 4.8
Various models for high transfer rates. Combination of film theory and surface renewal: (a) applicable at different times of contact, (b) present simultaneously, (c) consideration of laminar flow in the eddy and (d) rolling cell model.

is from 0 to ∞ (i.e. a few elements remain at the surface for a very long time). The boundary condition for Equation 4.69 when time is so large that the conduction and diffusion processes will attain steady state. Therefore, eddies which stay at the surface for a long time attain steady state and are governed by film theory. Eddies staying at the surface for short times follow the surface renewal mechanism. Toor and Marchello (1958) considered the following boundary conditions for the surface renewal mechanism:

$$C_A = C_A^*; x = 0, t > 0$$
$$C_A = C_{A\infty}; t = 0, x > 0$$
$$C_A = C_{A\infty}; x = \delta, t > 0$$

It may be noted that the depth of the eddy is finite, which is reflected in the third boundary condition. At short times, unsteady-state diffusion within the fluid takes place. The instantaneous mass transfer rate is given by the following two equations:

$$n_{A0} = \left(\rho_A^* - \rho_{A\infty}\right)\left(\frac{D_{AB}}{\pi t}\right)^{1/2}\left[1 + 2\sum_{n=1}^{\infty}\exp\left(-\frac{n^2\delta^2}{D_{AB}t}\right)\right] \qquad (4.77)$$

$$n_{A0} = \left(\rho_A^* - \rho_{A\infty}\right)\left(\frac{D_{AB}}{\delta}\right)\left[1 + 2\sum_{n=1}^{\infty}\exp\left(-n^2\pi^2\frac{D_{AB}t}{\delta^2}\right)\right] \qquad (4.78)$$

The first equation converges for $t < \dfrac{\delta^2}{D_{AB}}$, and the second equation converges for $t > \dfrac{\delta^2}{D_{AB}}$. The average mass transfer flux for any arbitrary age distribution function can be estimated as

$$\left(n_{A0}\right)_{av} = \int_0^{\infty} n_A \psi(t)\,dt \qquad (4.79)$$

The average mass transfer flux depends upon the choice of the age distribution function. For equal age of all fluid eddies, as used in Higbie's penetration model:

$$\overline{N_A} = \left(\rho_A^* - \rho_{A\infty}\right)\left(\frac{D_{AB}}{\pi t}\right)^{1/2}\left[1 + 2\sqrt{\pi}\sum_{n=1}^{\infty} ierfc\left(-\frac{n\delta}{\sqrt{D_{AB}t}}\right)\right] \qquad (4.80)$$

$$n_{A0} = \left(\rho_A^* - \rho_{A\infty}\right)\left(\frac{D_{AB}}{\delta}\right)\left[1 + \frac{2}{\pi^2}\frac{\delta^2}{D_{AB}t}\sum_{n=1}^{\infty}\exp\left(-n^2\pi^2\frac{D_{AB}t}{\delta^2}\right)\right] \qquad (4.81)$$

Equations 4.80 and 4.81 correspond to Equations 4.77 and 4.78, respectively. Similarly, expressions for average mass transfer flux corresponding to Danckwerts' surface renewal mechanism are available in the literature (Skelland 1964). The numerical estimation of the series in Equations 4.77 and 4.78 is not a major problem. The estimation of δ, t and s from experimental data is not an easy task. Choice of a suitable age distribution function can also be questioned. For a short time, the film penetration model reduces to penetration or surface renewal models if the age distribution model follows uniform or exponential distribution functions, respectively. For a large contact time, the model reduces to a film model. Therefore, the film penetration model can be considered as applying a penetration model for a short contact time and a film model for a long contact time. Both concepts are used at different times of contact.

4.5.4 Coexistence of Surface Renewal and Film

Wasan and Ahluwalia (1969) also used a combination of film theory and surface renewal theory. In this model, a stagnant film adjacent to the transfer surface is always present. The transfer between the outer edge of the film (away from the wall) and the bulk was considered to take place by the surface renewal mechanism. Thus, the film and surface renewal coexisted. The average heat and mass transfer coefficients were given by

$$h_{av} = \frac{2k_1}{\sqrt{(\pi \alpha t_c)}} + \frac{k_1 \delta}{\alpha t_c}\left[\exp\left(\frac{\alpha t_c}{\delta^2}\right)\left(1 - erf\frac{\sqrt{\alpha t_c}}{\delta}\right) - 1\right] \qquad (4.82)$$

$$k_{av} = \frac{2D}{\sqrt{(\pi D t_c)}} + \frac{\delta}{\alpha t_c}\left[\exp\left(\frac{D t_c}{\delta^2}\right)\left(1 - erf\frac{\sqrt{D t_c}}{\delta}\right) - 1\right] \qquad (4.83)$$

where δ is the thickness of the film and t_c is the contact time. In multiphase systems, the contact time is often related to distance between adjacent dispersed phases. Wasan and Ahluwalia (1969) used their model to correlate the heat transfer coefficient in fluidised beds and used δ as a function of Reynolds number and porosity.

4.5.5 Surface Renewal with Laminar Flow in Eddies

All of the above models considered that the fluid does not move within an eddy (i.e. the convective transport from the wall to the eddy is absent). This assumption seems unrealistic. However, it simplifies the model equations so that an analytical solution can be obtained. The models considering convective transport as well as molecular transport from the surface to the eddy for the period for which it stays at the surface were developed by Ruckenstein (1968). A few more models considering convective transport are also described by Ruckenstein (1987).

Ruckenstein (1968) considered that as soon as an eddy arrives at the surface, the heat or mass transfer takes place by a mechanism which is a combination of steady-state conduction or diffusion, respectively, and convective transport to a fluid element in which the fluid flows parallel to the surface, as given in Figure 4.8. The flow is laminar. The contact time is short; therefore, the change in the temperature or concentration is limited to a small region of the eddy. Thus, the penetration depth is much smaller than the depth of the eddy. In the case of mass transfer, the governing model equations are the equation of continuity, the equation of motion and the equation of change of mass of a chemical species. After simplification, we get Equation 4.38 and the following equation:

$$\rho\left(\frac{\partial C}{\partial t} + v_x \frac{\partial C}{\partial x} + v_y \frac{\partial C}{\partial y}\right) = D_{AB} \frac{\partial^2 C}{\partial y^2} \qquad (4.84)$$

These equations were solved to obtain the following relationship:

$$k \propto \left(\frac{\tau_0}{\rho}\right)^{1/2} \left(\frac{D_{AB}\rho}{\mu}\right)^{2/3} \qquad (4.85)$$

One of the important results was the dependence of the mass transfer coefficient on the two-thirds power of diffusivity. Here, τ_0 is the average shear stress at the wall.

Pinczewski and Sideman (1974) proposed a similar model with the difference that mass transfer to the eddy was by unsteady-state diffusion and convection in the laminar flow parallel to the surface. The model equations consisted of unsteady terms also.

4.6 Analogy between Momentum, Mass and Heat Transfer

To obtain solutions for N-S equations in cases of turbulent flow is very difficult. Although CFD software is now used to study turbulent momentum, mass and energy transfer, the large computation time makes this method not very convenient for engineering purposes or frequent calculations. The zero-equation turbulence models (e.g. eddy viscosity and Prandtl mixing length) are based on the fact that the turbulent flux of momentum, mass and thermal energy is due to a single phenomenon (i.e. the movement of fluid eddies). Earlier attempts to understand the mechanism of transfer processes under turbulent flow conditions were based on the assumption that momentum, mass and energy all are quantities dependent on the hydrodynamics and that the fluid eddy is the carrier of these quantities. Therefore, the momentum transfer, mass transfer and heat transfer were treated in an analogous manner. As a result, several analogies between momentum, mass and energy have been developed. These analogies are simple in a sense that values of velocity, concentration and temperature at the edges of the laminar sub-layer and buffer zone only are required. No detailed concentration or temperature profiles are used. A universal velocity profile in a circular tube or a velocity profile over a flat plate is used. Mechanistic models in each of regions are proposed. Some of the analogies between momentum and mass transfer are described by Skelland (1964). The analogy between momentum and heat transfer can also be obtained in a similar manner.

4.6.1 Universal Velocity Profile

Prandtl used the concept of mixing length and obtained a velocity profile for the flow of a Newtonian fluid in a circular pipe. Let y be the distance from the wall. The shear stress, τ_{rz}, can be written as

$$\tau_{rz} = \mu \frac{dv_z}{dy} + \rho \varepsilon_M \frac{dv_z}{dy} = (\mu + \rho \varepsilon_M) \frac{dv_z}{dy} \tag{4.86}$$

using:

$$\varepsilon_M = l^2 \left(\frac{du}{dy} \right)$$

$$\tau_{rz} = \mu \frac{dv_z}{dy} + \rho l^2 \left(\frac{dv_z}{dy} \right)^2 \tag{4.87}$$

It was assumed that there are three regions in the tube. There is a very thin laminar sub-layer adjacent to the wall. The second term of Equation 4.87 is zero. The velocity profile was assumed to be linear. Hence,

$$\tau_{rz} = \tau_W = \mu \frac{v_z}{y} \tag{4.88}$$

Let us define $v_z^* = \sqrt{\dfrac{\tau_W}{\rho}}$, $y^+ = \dfrac{y \rho v_z^*}{\mu}$ and $v_z^+ = \dfrac{v_z}{v_z^*}$. Substitution of these values in Equation 4.88 gives the velocity profile in the laminar sub-layer extending up to y^+.

$$v_z^+ = y^+ \tag{4.89}$$

In the turbulent core far away from the wall, the first term in Equation 4.87 is neglected because the viscous forces do not play a significant role as compared to the inertial forces. The mixing length was assumed to vary linearly with the distance from the wall (i.e. $l = ay$). The laminar sub-layer and buffer zones are very thin; hence, the shear stress at the interface of the laminar sub-layer and buffer zone and of the buffer zone and turbulent core was assumed to be the shear stress at the wall, τ_w. Equation 4.87 takes the following form:

$$ay \frac{dv_z}{dy} = \sqrt{\frac{\tau_W}{\rho}} = v_z^* \tag{4.90}$$

After integrating and using the dimensionless quantities, we get

$$v_z^+ = 2.5 \ln y^+ + 5.5 \tag{4.91}$$

The constants were determined experimentally. Equation 4.91 is applicable for $y^+ > 30$. The buffer zone lies in between the laminar sub-layer and the turbulent core (i.e. for $5 \leq y^+ \leq 30$). The experimental data were fitted by a line similar to that given in Equation 4.86.

$$v_z^+ = 5.0 \ln y^+ + 3.05 \qquad (4.92)$$

Equations 4.89, 4.91 and 4.92 are called the 'universal velocity profile'. These are used while developing several analogies, as discussed below.

4.6.2 The Reynolds Analogy

Reynolds was the first to propose an analogy between momentum, mass and heat transfer. It was assumed that there is only a turbulent core. The velocity, concentration and temperature are distributed uniformly in the fluid. At the wall, these are zero. The fluid eddies are responsible for the exchange of momentum, mass and thermal energy between the wall and the bulk. A fluid eddy from the bulk, having properties equal to those of the bulk, arrives at the surface and attains the properties of the fluid at the wall. Thereafter, it goes back to the bulk again and attains the properties of the bulk fluid (Figure 4.9). Thus, the resistance to transfer processes in the vicinity of the surface is neglected. Irrespective of the velocity of the fluid, there is a net change of momentum, mass and energy. The velocity at the wall is zero. The analogy between mass, heat and momentum transfer results in the following equation. However, the analogy is valid only when the Prandtl number, Pr, and Schmidt number, Sc, are close to 1. Since for gases Pr and Sc are close to 1, the Reynolds analogy is applicable for transfer processes in the gas phase.

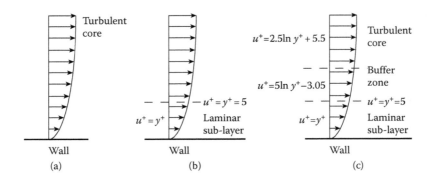

FIGURE 4.9
Analogies between momentum, mass and heat transfer: (a) a turbulent core, (b) a laminar layer and turbulent core and (c) a laminar sub-layer, buffer zone and turbulent core.

$$\frac{Sh}{Sc \cdot Re} = \frac{Nu}{Pr \cdot Re} = \frac{f}{2} \qquad (4.93)$$

where f is the fanning friction factor. The dimensionless numbers are calculated using the fluid velocity in the core, V_∞. The limited scope of this analogy was the result of neglecting regions close to the surface, where the viscous forces dominate and conduction and diffusion play important roles. The resistances due to these are in series with the resistance in the turbulent core and hence should be considered.

4.6.3 Prandtl's Analogy

The range of applicability of the analogy increased when a laminar sub-layer adjacent to the surface was considered by Prandtl. However, it required knowledge of the fluid velocity at the interface of the laminar sub-layer and the turbulent core, v_{yL}. The analogy also exhibits the dependence on the fluid velocity. For mass transfer:

$$\frac{Sh}{Sc \cdot Re} = \frac{f/2}{1 + (v_{yL}/V_\infty)(Sc - 1)} \qquad (4.94)$$

And for heat transfer:

$$\frac{Nu}{Pr \cdot Re} = \frac{f/2}{1 + (v_{yL}/V_\infty)(Pr - 1)} \qquad (4.95)$$

Dividing Equation 4.94 by Equation 4.95, an analogy between heat and mass transfer is obtained:

$$\frac{Sh}{Nu} = \left[\frac{1 + (v_{yL}/V_\infty)(Pr - 1)}{1 + (v_{yL}/V_\infty)(Sc - 1)} \right] \left(\frac{Sc}{Pr} \right) \qquad (4.96)$$

Whereas the Reynolds analogy between heat and mass transfer does not require the value of the friction factor, Prandtl's analogy requires the value of the friction factor and the fluid velocity at the interface of the laminar sub-layer and the turbulent core, v_{yL}. The value of v_{yL} depends upon the geometry of the system in which flow takes place. However, since the laminar sub-layer is normally thin, the curved surface of the pipe may be ignored. Since the velocity profiles for the flow over a flat plate and for the flow in a circular pipe are already known, either of these two may be used to estimate the friction factor at V_∞ and v_{yL}. Prandtl's analogy is better than the Reynolds analogy, but still it deviates at high Sc and Pr values.

4.6.4 Karman's Analogy

To improve the applicability of the analogy between momentum, mass and heat transfer, von Karman assumed the presence of a buffer zone between the laminar sub-layer and turbulent core. Momentum transfer is given by Equation 4.86. The mass transfer flux is given by

$$N_A = D_{AB}\frac{dC_A}{dy} + \rho\varepsilon_D\frac{dC_A}{dy} = (D_{AB} + \rho\varepsilon_D)\frac{dC_A}{dy} \tag{4.97}$$

Neglecting the terms containing ε_M and ε_D in Equations 4.86 and 4.97, respectively, and using Equation 4.88 in Equation 4.86, we get mass transfer flux in the laminar sub-layer.

$$N_{AW} = -D_{AB}\frac{dC_A}{dy} = -\frac{Dv_z^*\rho}{\mu}\frac{dC_A}{dy^+} \tag{4.98}$$

Let the concentration at the wall, the interface of the laminar sub-layer and buffer zone, the interface of the buffer zone and turbulent core and the turbulent core be C_{AW}, C_{AL}, C_{AB} and $C_{A\infty}$, respectively. Integration of Equation 4.98 gives

$$C_{AL} - C_{AW} = -\frac{5Sc}{v_z^*}N_{AW} \tag{4.99}$$

In the buffer zone, Equations 4.86 and 4.97 are applicable. The velocity gradient can be obtained by differentiating Equation 4.92. It is assumed that $\varepsilon_M = \varepsilon_D$. Integrating Equation 4.98 and expressing it in terms of dimensionless quantities give

$$C_{AB} - C_{AL} = -\frac{5[\ln(5Sc + 1)]}{v_z^*}N_{AW} \tag{4.100}$$

In the turbulent core, the viscous forces in Equation 4.86 and diffusion term in Equation 4.97 are neglected. After solving the equations, we get

$$C_{Ab} - C_{A\infty} = -\frac{N_{AW}}{v_z^*}\left\{\frac{V}{v_z^*} - 5(1 + \ln 6)\right\} \tag{4.101}$$

The flux N_{AW} can be expressed in terms of the Sherwood number. The values of C_{AL} and C_{AB} can be eliminated by adding Equations 4.99 through 4.101. Upon arrangement the following equation is obtained:

$$\frac{Sh}{Sc \cdot Re} = \frac{(f/2)}{1 + 5\sqrt{(f/2)}\left[Sc - 1 + \ln\{1 + (5/6)(Sc - 1)\}\right]} \tag{4.102}$$

Similarly, the analogy between heat and momentum transfer is obtained:

$$\frac{\text{Nu}}{\text{Pr} \cdot \text{Re}} = \frac{(f/2)}{1 + 5\sqrt{(f/2)}\left[\text{Pr} - 1 + \ln\{1 + (5/6)(\text{Pr} - 1)\}\right]} \tag{4.103}$$

The analogy between mass and heat transfer can be obtained by dividing Equation 4.102 by Equation 4.103:

$$\frac{\text{Sh}}{\text{Nu}} = \frac{1 + 5\sqrt{(f/2)}\left[\text{Pr} - 1 + \ln\{1 + (5/6)(\text{Pr} - 1)\}\right]}{1 + 5\sqrt{(f/2)}\left[\text{Sc} - 1 + \ln\{1 + (5/6)(\text{Sc} - 1)\}\right]}\left(\frac{\text{Sc}}{\text{Pr}}\right) \tag{4.104}$$

Equation 4.104 also requires the values of the friction factor.

4.6.5 Lin–Moultan–Putnam's Analogy Based on an Eddying Sub-Layer

The two analogies discussed above consider a laminar sub-layer (i.e. in the region adjacent to the wall, the flow is laminar and parallel to the surface). The surface renewal models, except for the one proposed by Wasan and Ahluwalia (1969), consider the presence of eddies. The rolling cell model considers circulatory flow in near proximity to the wall. In reality, for turbulent flow, consideration of the laminar flow near the wall is only an oversimplification. This was termed an eddying sub-layer. However, in this layer, the universal velocity profile cannot be used; another velocity profile is required. Lin et al. (1953) used a velocity profile different from the universal velocity profile and obtained the following analogy for momentum and mass transfer. A detailed proof is given elsewhere (see Lin et al. 1953).

$$\frac{\text{Sh}}{\text{Sc} \cdot \text{Re}} = \left(\frac{f}{2}\right)\frac{1}{\phi_D} \tag{4.105}$$

$$\phi_D = 1 + \sqrt{(f/2)}\left[\frac{14.5}{3}\text{Sc}^{2/3}F(\text{Sc}) + 5\ln\frac{1 + 5.64\text{Sc}}{6.64(1 + 0.041\text{Sc})} - 4.77\right] \tag{4.106}$$

where

$$F(\text{Sc}) = \frac{1}{2}\ln\frac{\left(1 + \frac{5}{14.5}\text{Sc}^{1/3}\right)^2}{1 - \frac{5}{14.5}\text{Sc}^{1/3} + \left(\frac{5}{14.5}\right)^2\text{Sc}^{2/3}} + \sqrt{3}\tan^{-1}\frac{\frac{10}{14.5}\text{Sc}^{1/3} - 1}{\sqrt{3}} + \frac{\pi\sqrt{3}}{6}$$

$$\tag{4.107}$$

The analogy between momentum and heat transfer can be obtained by replacing Sh and Sc by Nu and Pr, respectively, in Equations 4.105 through 4.107. The analogy between the heat and mass transfer can be written as

$$\frac{Sh}{Nu} = \left(\frac{Sc}{Pr}\right)\frac{\phi_H}{\phi_D} \tag{4.108}$$

4.6.6 Chilton-Colburn Analogy: *j* Factors

Karman's analogy and Lin et al.'s analogy seem to be comprehensive. But they are inconvenient to use in multiphase systems. For example, for packed beds, the flow passes through interstitial spaces between the particles. Determination of the friction factor is not an easy task due to its unclear definition in such cases. In the case of momentum transfer, the presence of form drag has no analogy with heat or mass transfer. The analogy of heat and mass transfer with momentum transfer is only due to skin drag component. However, a heat and mass transfer analogy can exist. These are given by Equations 4.104 and 4.108. Another difficulty that one faces is that the empirical correlations for heat and mass transfer do not contain terms such as $\frac{Sh}{Sc \cdot Re}$. Therefore, if heat transfer correlation is available, the mass transfer correlation cannot be obtained easily by using the above-mentioned improved analogies. Chilton–Colburn's analogy is very convenient as the functional nature of the correlations remains same. It is based on the concept of *j* factors. The analogy is based on the similarity between the empirical correlations for heat and mass transfer in similar situations. According to this analogy:

$$J_H = \frac{Nu}{Re \cdot Pr^{1/3}} = J_D = \frac{Sh}{Re \cdot Sc^{1/3}} = a\,Re^{-b} \tag{4.109}$$

The constants *a* and *b* depend upon the type of flow (i.e. the shapes of the objects around which the flow takes place, and the range of Re, Sc and Pr). These constants are determined experimentally. The exponents of the correlation are sometimes rounded so that they take the form given in Equation 4.109.

4.6.7 Heat and Mass Transfer Analogy in Bubble Columns

A Chilton–Colburn analogy has been used frequently to propose correlations. For example, if a correlation for heat transfer is available, an analogous correlation for mass transfer is chosen. The experimental data are used to determine the proportionality constant. Verma (2002) has shown that even the proportionality constant can be determined in the case of heat and mass transfer in bubble columns. The heat transfer coefficient in bubble columns was correlated by (Deckwer 1980)

$$St_H = 0.1 \left(Re\, Fr\, Pr^2 \right)^{-0.25} \qquad (4.110)$$

None of the analogies have a Stanton number, St_H. Therefore, Equation 4.110 has to be written in a form so that an analogy can be applied. We can write Equation 4.110 as

$$St_H = \frac{h}{U\rho_l C_p} = \left(\frac{h d_b}{k_l} \right) \left(\frac{\mu_l}{d_b \rho_l U} \right) \left(\frac{k_l}{C_p \mu} \right) = \frac{Nu}{Pr \cdot Re} \qquad (4.111)$$

where U is the superficial gas velocity, h is the heat transfer coefficient, d_b is the bubble diameter and k_l, μ_l and ρ_l are the thermal conductivity, viscosity and density of the liquid, respectively. Substituting Equation 4.111 in Equation 4.110 and eliminating the Froude number, Fr, we get

$$\frac{Nu}{Re \cdot Pr} = C_H \frac{1}{Pr^{0.5}} \left(\frac{\mu U}{d_b \rho_l g D_c} \right)^{0.25} = \frac{1}{\phi_H} \left(\frac{f}{2} \right) \qquad (4.112)$$

the constant 0.1 in Equation 4.110 is replaced by C_H. Similarly, for mass transfer:

$$\frac{Sh}{Re \cdot Sc} = C_D \frac{1}{Sc^{0.5}} \left(\frac{\mu U}{d_b \rho_l g D_c} \right)^{0.25} = \frac{1}{\phi_D} \left(\frac{f}{2} \right) \qquad (4.113)$$

where C_H is the constant in correlation for the mass transfer coefficient. Dividing Equation 4.112 by Equation 4.113 provides the ratio of the constant in heat and mass transfer in bubble columns.

$$\frac{C_H}{C_D} = \left(\frac{\phi_D}{\phi_H} \right) \left(\frac{Pr}{Sc} \right)^{0.5} \qquad (4.114)$$

To evaluate the functions ϕ_D and ϕ_H, the values of the friction factor and Pr and Sc numbers are required. The average value of Pr and Sc can be used. The friction factor for pipe flow was used by taking the circulation velocity in the bubble column as given by Zehner (1986). Equation 4.114 reveals two important findings. The ratio of the constants in heat and mass transfer depends upon the friction factor, and it also depends upon the range of Pr and Sc, even though the average properties are used. The ratio does not depend upon the definition of the Reynolds number.

4.6.8 Limitations of Analogies

Each of the momentum, mass and heat transfer processes has specific features which make them different from each other and are responsible for limited use of the analogy. These features are as follows:

1. For flow past objects such as in packed beds and fluidised beds, the drag force is composed of two terms: 'skin drag' and 'form drag'. There is no parallel to form drag in cases of mass and heat transfer. Therefore, if the form drag is significant, the momentum transfer is no more analogous to heat and mass transfer. However, the analogy between heat and mass transfer can still exist.

2. The diffusion flux (i.e. the velocity due to diffusion) is observed in mass transfer. It is absent in momentum and heat transfer. The diffusion flux is significant in concentrated solutions. Under this condition, the analogy of mass transfer with momentum and heat transfer breaks down.

3. At high temperature, a significant contribution to heat transfer by radiation makes it different from mass and momentum transfer processes. The heat transfer does not remain analogous to mass and momentum transfer at high temperatures.

When applying empirical correlations based on momentum, mass and heat transfer analogies, the above points should be considered.

4.7 Simple Models for Reactors and Bioreactors

Chemical and biochemical processes are carried out in various reactors and bioreactors. The development of a few simple process models will now be presented. The common reactors are completely mixed-tank reactors (CSTRs), plug flow reactors (PFRs) and batch reactors. The other non-ideal reactors and bioreactors commonly used are modelled. The effect of the flow within the non-ideal reactors can, however, be lumped into residence time distribution (RTD). The conversion can be expressed as a combination effect of RTD and kinetic expressions. Model equations will be described for some of the ideal and simple non-ideal reactor models. Since kinetic rate expression is an important term in the model equations and is generally non-linear, the model equations should be written and solved for every reaction.

4.7.1 Stirred-Tank Reactors

Let us consider a stirred-tank reactor in which an exothermic first-order reaction, $A \rightarrow B$, is being carried out. The feed with concentration, C_{A0},

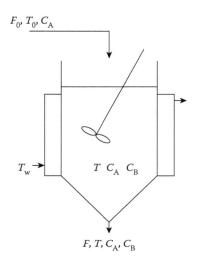

$$F_0, T_0, C_A$$

$$T_w \rightarrow \quad T \ C_A \ C_B$$

$$F, T, C_A, C_B$$

FIGURE 4.10
Stirred-tank reactor.

and temperature, T_0, enters at the mass feed rate, F_0. The product with concentration, C_A and C_B, and temperature, T, leaves at the mass feed rate, F. The volume of the reactor, V, may change with time. To remove the heat from the reactor due to the heat of reaction, ΔH_{rxn}, cooling water at temperature T_w is passed through the jacketed wall (Figure 4.10).

It is desired to develop a dynamic model to predict the performance of the reactor. Let us assume that the stirred-tank reactor behaves as an ideal CSTR. This assumption implies that the concentration and temperature of the contents in the CSTR are equal to those of the leaving stream. Also, there is no spatial variation of these parameters, and hence a lumped parameter model will be developed.

The model equations can now be written using laws of conservation of mass and energy, which are material and energy balances in cases of lumped parameter models. Material balances for A and B give the following (and see Equation 4.1):

$$\frac{d}{dt}(\rho V C_A) = F_i C_{A0} - F_o C_A - (-r_A) \tag{4.115}$$

$$\frac{d}{dt}(\rho V C_B) = F_i C_{B0} - F_o C_B + (-r_A) \tag{4.116}$$

The overall energy balance gives

$$\frac{d}{dt}(\rho V H T) = F_i H_0 - F_o H + \Delta H_{rxn}(-r_A)V - q \tag{4.117}$$

When $F_i = F_o$, a steady-state model is obtained. The model equations are not yet complete since the generation terms $-r_A$ and ΔH_{rxn} are not yet defined. The rate of reaction can be written using the kinetic expression. For example, for a first-order irreversible reaction, the following rate expression will be used:

$$-r_A = f(C_A) = kC_A = k_0\, e^{(-E/RT)}\, C_A \tag{4.118}$$

Note that the temperature in the reactor is affecting the rate of reaction. The heat of reaction is known or can be estimated from the heat of formation or the heat of combustion. It is a constant. The heat removed by the cooling water can be estimated in terms of the overall heat transfer coefficient, U.

$$q = UA_s\,(T-T_w) \tag{4.119}$$

If the temperature of the cooling or heating medium changes, the wall temperature, T_w, also changes. In a steady-state model, the value of T_w is constant.

4.7.2 Batch and Semi-Batch Reactors

The model equations for batch and semi-batch (also called 'fed-batch') reactors are similar to the model for CSTR. In the batch reactor, there is no inlet and outlet; hence, Equations 4.115 through 4.117 take the following form after substituting $F_i = F_o = 0$:

$$\frac{d}{dt}(\rho VC_A) = -(-r_A) \tag{4.120}$$

$$\frac{d}{dt}(\rho VC_B) = (-r_A) \tag{4.121}$$

The overall energy balance gives

$$\frac{d}{dt}(\rho VHT) = \Delta H_{rxn}(-r_A)V - q \tag{4.122}$$

The volume of the reactor may change due to a chemical reaction. It can be expressed as

$$V = V_0\left[1+\left(\frac{V_f}{V_0}-1\right)\frac{C_A}{C_{A0}}\right] \tag{4.123}$$

Here, V_f is the final volume if the reaction is complete, and it can be determined from the stoichiometry. In reality, the reaction may never be complete in cases of reversible reaction. Equations 4.120 through 4.122 with Equations 4.118 and 4.119 form the set of model equations for a batch reactor.

In a semi-batch reactor, there is no outlet. If the reactants are added continuously, then model equations are obtained by substituting $F_o = 0$ in Equations 4.115 through 4.117 and including Equations 4.118 and 4.119 to the set of equations.

Equation 117 may also be used in a model for CSTR if there is volume change due to a chemical reaction.

In a PFR, all of the fluid elements move with same velocity and leave the reactor at the same time. No intermixing takes place. Such flows are closed to the flow observed in tubular reactors under turbulent flow conditions. The concentration in all fluid element changes with time due to chemical reactions as the fluid elements move in the reactor. The conversion in the PFR is given by Equation 2.2.

The PFR, CSTR and batch reactors are ideal reactors. The real reactors exhibit behaviour different from that exhibited by ideal reactors. The distributions of velocity, concentration and temperature in CSTRs and batch reactors are uniform. In the PFR, the residence times for all fluid elements are equal. The real reactors neither show uniform distribution of concentration and temperature nor exhibit the same residence time for all fluid elements. The conversion in non-ideal reactors can be described by the following equation:

$$\frac{C_A}{C_{A0}} = \int_0^\infty E(-r_A)\,dt \qquad (4.124)$$

Here, $E\Delta t$ is the probability of a fluid element having residence time between t and $t+\Delta t$ measured at the exit. Equation 4.124 combines the knowledge of hydrodynamics and kinetics obtained from separate studies. All of the hydrodynamics effects are included in residence time distribution, E. The reactors can be modelled by combining models for RTD, and kinetic model are available. The hydrodynamics and kinetics can be modelled simultaneously also, as discussed next. A few approaches to model RTDs are presented below.

4.7.3 Axial Dispersion Model

In a plug flow system that includes packed beds, fluidised beds and bubble columns, the concentration of a chemical species changes in the axial direction. This variation is due to either depletion or generation of the chemical species due to chemical reaction or the unsteady nature of the concentration at the inlet. Due to the concentration gradient developed, diffusion in the axial direction will also take place (Figure 4.11b). At the same time, a turbulent mass flux will also be present. Assuming that there is no radial variation of concentration, the equation of continuity for the chemical species A will be

$$\frac{\partial C_A}{\partial t} + v_z \frac{\partial C_A}{\partial z} = (D_{AB} + \varepsilon_D)\frac{\partial^2 C_A}{\partial z^2} \qquad (4.125)$$

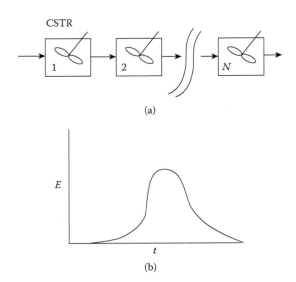

(a)

(b)

FIGURE 4.11
RTD from (a) a tank in series model and (b) an axial dispersion model.

If the length of the reactor is L, then the defining dimensionless variables are $\zeta = (z+tv_z)/L$, $\zeta = z/L$ and $\theta = t / \bar{t}$, and we get

$$\frac{\partial C_A}{\partial \theta} + \frac{\partial C_A}{\partial \zeta} = \frac{\hat{D}}{v_z L}\frac{\partial^2 C_A}{\partial \zeta^2} \tag{4.126}$$

The parameter $\left(\hat{D}/v_z L\right)$ is called the vessel dispersion coefficient. It uses the concept of eddy diffusivity. Equation 4.126 can be solved numerically using MATLAB's 'pdepe' function (see the Appendix B). An analytical solution for residence time distribution is available in the literature only for small values of the dispersion coefficient (<0.01) (Levenspiel 1999).

$$E = \frac{1}{2\sqrt{\pi\left(\hat{D}/v_z L\right)}}\exp\left\{-\frac{(1-\theta)^2}{4\left(\hat{D}/v_z L\right)}\right\} \tag{4.127}$$

Equation 4.127 can be used in Equation 4.124 to get the conversion in a PFR with axial dispersion. To consider hydrodynamics and kinetics together, add the generation term (= $-r_A$) to the left side of Equation 4.125 to get

$$\frac{\partial C_A}{\partial t} + v_z \frac{\partial C_A}{\partial z} = \hat{D}\frac{\partial^2 C_A}{\partial z^2} - (-r_A) \tag{4.128}$$

It has been shown that both approaches give different results (Levenspiel 1999). This difference is due to the fact that, while obtaining RTD, the reaction at the local level was neglected. Therefore, the latter approach, which is distributed parameter models, is more reliable than the RTD approach. Equation 4.128 may easily be solved numerically. An analytical solution for Equation 4.128 for the first-order reaction is as follows (Levenspiel 1999):

$$\frac{C_A}{C_{A0}} = \frac{4a \exp\left(\frac{1}{2}\frac{v_z L}{\hat{D}}\right)}{(1+a)^2 \exp\left(\frac{a}{2}\frac{v_z L}{\hat{D}}\right) - (1-a)^2 \exp\left(-\frac{a}{2}\frac{v_z L}{\hat{D}}\right)} \tag{4.129}$$

where

$$a = \sqrt{1 + 4k\tau\frac{\hat{D}}{v_z L}}$$

4.7.4 Tank in Series Model

In real reactors, backmixing is usually present and is a major cause for deviation from an ideal reactor performance. One method to model the effect of backmixing is to use the axial dispersion model. Another approach is to consider that there are several CSTRs in series. This model is called the 'tank in series model'. The entire reactor of volume, V, is equivalent to N equal-sized tanks, each having a volume of $V_i = V/N$ (Figure 4.11a). Writing a material balance for the pulse tracer in the first tank:

$$\frac{dC_A}{dt} = -F_o C_A \tag{4.130}$$

Equation 4.130 is obtained by dropping the generation term $(-r_A)$ in Equation 4.115. After putting integration of Equation 4.130 gives

$$\frac{C_{A1}}{C_{A0}} = \overline{t_1} E_1 = \exp\left(-t / \overline{t_1}\right) \tag{4.131}$$

Here, $\overline{t_1}$ is the average residence time in the first tank. For the second tank, the material balance gives

$$\frac{dC_A}{dt} = F_o C_{A1} - F_o C_A \tag{4.132}$$

The value of C_{A1} is substituted from Equation 4.131. The solution of Equation 4.132 gives the tracer concentration after the second tank. Similarly, for N tanks in series, the RTD is given by

$$E = \frac{\left(t_i / \bar{t}\right)^{N-1}}{(N-1)!} \exp\left(t_i / \bar{t}\right) \tag{4.133}$$

4.7.5 Modelling of Residence Time Distribution

Real reactors and processes have complicated flow behaviour. The non-ideal behaviour is the result of malfunctions such as channelling, backmixing and dead zones, in addition to the molecular transport and turbulence. The RTDs for plug flow, axial dispersion, mixed tanks etc. are known and are discussed in this chapter. If a vessel is thought to contain one or more of such units connected in series or parallel in a reasonable way, then the RTD of the entire vessel may be modelled. This approach has been called the 'compartmental model'. It is assumed that the entire system is connected compartments, each having a well-defined RTD. The RTD of the entire system depends upon the latter's volume, type and interconnections. Levenspiel (1999) considered three types of compartments: the mixed-flow region, plug flow region and dead region. In the literature, a plug flow with axial distribution region has also been considered as a compartment.

1. *Mixed-flow region*: A fraction of the total volume may behave as a region which may be modelled as a CSTR. For example, in a multistaged tower, each stage may be a well-mixed region (Figure 4.12a). Joshi and Sharma (1979) proposed a circulation cell model, assuming that there are several circulatory cells in bubble columns. In mechanically agitated contactors used to mix viscous liquids, more than one agitator is used (Figure 4.12b), each mixing a portion of the fluid. It may be considered to consist of many mixed-flow regimes. The volume of each mixed-flow regime should be specified.

2. *Plug flow region*: The channelling in the system is normally modelled as regions behaving as a plug flow system. The length and volume of this kind of region are the parameters to be considered (Figure 4.12).

3. *Plug flow with axial distribution region*: This may be used in place of the plug flow region to model channelling (Claudel et al. 2003).

4. *Dead region*: In a system, there might be regions which are stagnant, such as behind baffles, around nozzles extending inside or in wakes behind objects. These are considered regions which do not take part in the reaction or which do not contribute towards RTD. The volume of the dead region can be subtracted from the total volume. Few models consider the exchange of contents between the active region and dead region. The active region is the region which is not the dead region.

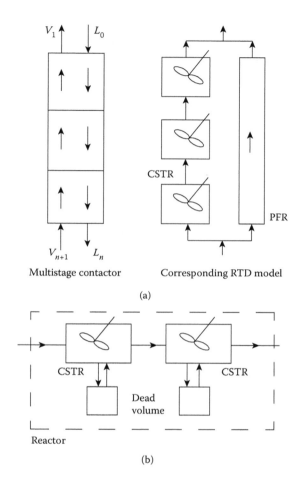

FIGURE 4.12
Compartmental model: (a) CSTR in series and a parallel PFR and (b) CSTR in series parallel arrangement with a dead region.

The interconnection of the units is also important. The units may be connected in series or in parallel. The interconnections may be unidirectional or bidirectional. Unidirectional interconnection tells us that there is no backmixing between two compartments. The recycle of a steam is also taken into consideration. In some sense, the tank in series model is a kind of compartmental model.

Since there is no outlet from a dead region, it is attached only with CSTRs. It is not possible to attach to PFRs. However, it is possible to exchange material between CSTRs and dead regions. PFRs and CSTRs may be connected in series or in parallel. A couple of compartmental models are shown in Figure 4.12a and 4.12b.

The compartmental models are developed by fitting the proposed model with the experimental value. The RTD is characterised by its moments, mean value or expectation, standard deviation, or the second moment and sometimes the third moment. Sometimes, evolutionary methods such as genetic algorithms are used to minimise error (Claudel et al. 2003). Theoretical estimation of the moments may be estimated from moments for the individual compartments using the concepts of probability. The moments can also be determined by probabilistic simulation, as discussed in Chapter 7.

A compartmental model has to be proposed, and then a theoretical RTD should be matched with experimental RTD behaviour (i.e. the moments). A few general concepts are helpful. The observation of peaks is an indication that channelling is taking place; hence, a PFR compartment may be used. Periodicity of the peaks indicates recycle streams.

4.8 Summary

The laws of conservation of momentum, mass and energy provide a set of model equations in many problems related to fluid flow, mass and heat transfer. Equations of continuity, change of momentum and thermal energy and continuity for an individual chemical species are presented. A few simplifications of N-S equations in rectilinear, cylindrical and spherical coordinates are presented.

The N-S equations are extended for turbulent flow. The Reynolds stresses thus introduced may be modelled as eddy viscosity, eddy diffusivity and thermal conductivity, Prandtl's mixing length etc.

At high flux, surface renewal types of models or models combining features of film models and surface renewal models may be applied.

Another approach to explain mass and heat transfer phenomena is to consider them analogous to the viscous flow. The universal velocity profile used in most analogies is presented. The analogies are based on an assumption of a laminar or eddying layer near the wall, a buffer zone and a turbulent core.

The surface renewal models and analogy equations are based on the laws of conservation.

The development of models for stirred-tank, batch and fed-batch reactors is illustrated. A model for plug flow with axial dispersion was also presented.

One of the approaches to study reactor performance is to study residence time distribution. Tank in series and compartmental models were briefly discussed.

References

Bird, R.B., Stewart, W.E., and Lightfoot, E.N. 1960. *Transport Phenomena*. New York: John Wiley.

Bird, R.B., Stewart, W.E., and Lightfoot, E.N. 2002. *Transport Phenomena*. 2nd ed. New York: John Wiley.

Claudel, S., Fonteix, C., Leclerc, J.P., and Lintz, H.G. 2003. Application of the possibility theory to the compartment modelling of flow pattern in industrial processes. *Chem. Eng. Sci.* 58: 4005–4016.

Danckwerts, P.V. 1951. Significance of liquid film coefficients in gas absorption. *Ind. Eng. Chem.* 43: 1460–1467.

Deckwer, W.D. 1980. On the mechanism of heat transfer in bubble column reactors. *Chem. Eng. Sci.* 35: 1341–1346.

Higbie, R. 1935. Rate of absorption of a gas into a still liquid during short periods of exposure. *Trans. AIChE J.* 31: 365–389.

Hinze, J.O. 1975. *Turbulence*, 2nd ed. New York: McGraw-Hill.

Joshi, J.B., and Sharma, M.M. 1979. A circulation cell model for bubble columns. *Chem. Eng. Res. Des.* 57: 244–251.

Levenspiel, O. 1999. *Chemical Reaction Engineering*. New York: John Wiley.

Lin, C.S., Moulton, R.W., and Putnam, G.L. 1953. Mass transfer between solid wall and fluid stream. *Ind. Eng. Chem.* 45: 636–640.

Pinczewski, W.V., and Sideman, S. 1974. A model for mass(heat) transfer in turbulent tube flow. Moderate and high Schmidt (prandtl) numbers. *Cehm. Eng. Sci.* 29(9): 1969–1976.

Ruckenstein, E. 1968. A generalised penetration theory for unsteady convective mass transfer. *Chem. Eng. Sci.* 23: 363–371.

Ruckenstein, E. 1987. Analysis of transport phenomena using scaling and physical models. In *Advances in Chemical Engineering*, vols. 13, ed. J. Wei, 11–112. Orlando, FL: Academic Press.

Skelland, A.H.P. 1964. *Diffusional Mass Transfer*. New York: John Wiley.

Skelland, A.H.P. 1967. *Non-Newtonian Flow and Heat Transfer*. New York: John Wiley.

Toor, H.L., and Marchello, J.M. 1958. Film-penetration model for mass and heat transfer. *AIChE J.* 4(1): 97–101.

Verma, A.K. 2002. Heat- and mass-transfer analogy in a bubble column. *Ind. Eng. Chem.* 41(4): 882–884.

Wasan, T., and Ahluwalia, M.S. 1969. Consecutive film and surface renewal for heat or mass transfer from a wall. *Chem. Eng. Sci.* 24: 1535–1542.

Zehner, P. 1986. Momentum, mass and heat transfer in bubble columns. Part 2. Axial blending and heat transfer. *Int. Chem. Eng.* 26: 29–35.

5

Multiphase Systems without Reaction

Many chemical and biochemical reactions are heterogeneous in nature, involving all three phases of solid, liquid and gas. Filtration, sedimentation and dispersion of solids in a liquid involve two phases. One phase is fluid, and the other phase is solid particles. The solid particles may be stationary as in fixed beds, or it may be moving as in fluidised beds or mechanically stirred vessels. Gas–liquid contactors involve the gas phase in the form of bubbles dispersed in liquid or liquid dispersed as droplets in gases. Many processes are carried out in which all three phases (solid, liquid and gas) are present (e.g. slurry bubble column, three-phase fluidised beds and slurry reactors). In bioprocesses, microorganisms or enzymes supported on solids may be present as the solid phase in bioreactors. Some multiphase contactors are presented in Figure 5.1.

The laws of conservations presented in Chapter 4 are applicable in a single phase only. This does not account for a mass transfer across a boundary between two phases, since the mechanisms of mass transfer across boundaries are different from those of convection or diffusion. Mass transfer across the interface is governed by an equilibrium relationship. The equation of continuity for individual chemical species is applicable in each individual phase. For this purpose, each dispersed unit should be considered an individual phase since each of these has an interface with a continuous phase.

Phenomena such as heat transfer between heat transfer surface and the dispersed medium involve a wall of the vessel or an outer surface of the tubes that is used to supply or remove heat. The interface does not offer any resistance to heat transfer. Mass transfer in electrochemical systems also involves a surface, either of the vessel or of the immersed body.

The movement of individual phases (e.g. settling of particles and bubble velocity in bubble columns) is governed by the forces of buoyancy, gravity and drag force acting on the individual dispersed phase.

Various applications of laws of conservation (discussed in Chapter 4) and simple laws (discussed in Chapter 3) to model processes in multiphase systems in the absence of chemical reactions are presented in this chapter.

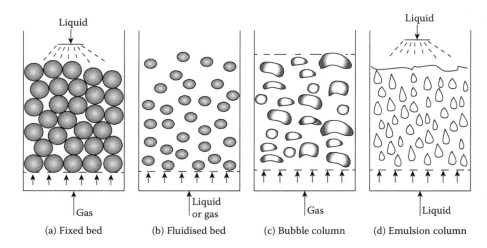

(a) Fixed bed   (b) Fluidised bed   (c) Bubble column   (d) Emulsion column

FIGURE 5.1
Various multi-phase systems: (a) fixed bed, (b) fluidised bed, (c) slurry bubble column and (d) emulsion column.

5.1 Consideration of a Continuous-Phase Axial Solid Profile in a Slurry Bubble Column

Let us take a case in which the region of interest is away from the transfer surface, which may be the interface of the phases or an external surface such as a wall or immersed objects. The multiphase system may be considered in such cases as a continuous phase.

A number of reactions require the solids to be in suspension form. If the bubbling of gas is used to keep the solids in suspension, the reactor is known as a 'slurry bubble column' (SBC). These columns find uses in F-T synthesis, coal liquefaction, hydrogenation of oil, wastewater treatment and many other chemical and biochemical reactions.

The SBC consists of a vertical column filled with slurry, in which the gas is bubbled through a suitably designed distributor into the column. It may be operated in batch or in continuous modes. In the fluidised beds, the solids are uniformly distributed in the column. But in SBC, the solids are not uniformly distributed. An axial distribution of solids is observed in SBC. It has been modelled by a one-dimensional sedimentation–dispersion model.

Gas is introduced through a gas distributor placed at the bottom of the SBC. As the gas bubbles move up, the slurry also moves up with it. The solids in the slurry remain suspended due to the local liquid velocities. The frequently changing velocity of the fluid resulting in back-mixing may be modelled, considering the solid dispersion coefficient analogous to the diffusion and eddy diffusion. Since the particles tend to settle, another term takes this into account. The model based on this concept is known as the sedimentation–dispersion model. It is based on the assumption of 'continuum' (i.e. the slurry is treated as a fluid whose properties may be estimated from the properties of solid and fluid). This assumption is useful in cases in which the region near the interface is not important.

Let us assume that complete radial mixing exists in the column (i.e. the concentration of solids is the same at all radial positions at a given height). This assumption results in a one-dimensional (1D) model. Writing material balance around a differential strip at a position z from the bottom gives (Figure 5.2).

$$
\begin{bmatrix} \text{Net rate of} \\ \textit{accumulation} \\ \text{of solids} \end{bmatrix} = \begin{bmatrix} \textit{Accumulation rate} \\ \text{of solids due to} \\ \text{solid dispersion} \end{bmatrix} + \begin{bmatrix} \textit{Accumulation rate} \\ \text{of solids due to} \\ \text{settling} \end{bmatrix}
$$
$$
+ \begin{bmatrix} \textit{Accumulation rate of} \\ \text{solids due to slurry} \\ \text{flow} \end{bmatrix}
\tag{5.1}
$$

With appropriate signs, substituting the various terms as shown in Figure 5.2, the following equation was obtained (Suganama and Yamanishi 1966; Farkas and Leblond 1969; Yamanaka et al. 1970; Smith and Ruether 1985):

$$
\frac{\partial}{\partial t}(1-\varepsilon_g)C_s = \frac{\partial}{\partial z}\left[(1-\varepsilon_g)E_s\frac{\partial C_s}{\partial z}\right] + \frac{\partial}{\partial z}\left[(1-\varepsilon_g)U_cC_s\right] - U_{sl}\frac{\partial C_s}{\partial z}
\tag{5.2}
$$

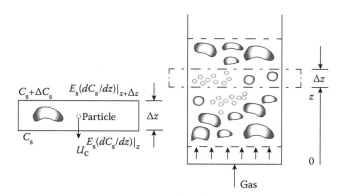

FIGURE 5.2
A sedimentation–dispersion model for an axial solids concentration profile in a slurry bubble column.

Assuming gas holdup, ε_g; solid dispersion coefficient, E_s; and hindered settling velocity, U_c, to be independent of z, Cova (1966) obtained the following equation to describe the solid distribution in an SBC:

$$\frac{\partial C_s}{\partial t} + (U_c - U_{sl})\frac{\partial C_s}{\partial z} = E_s \frac{\partial^2 C_s}{\partial z^2} \tag{5.3}$$

Equation 5.3 is applicable for a co-current operation. For counter-current operations, the superficial velocity of the slurry, U_{sl}, may be used. Smith and Ruether (1985) represented settling velocity in terms of slip velocity as $U_s\,\varepsilon_l$. In batch mode operation ($U_{sl} = 0$), the solids concentration does not change with time. Therefore, Equation 5.3 reduces to

$$E_s \frac{\partial^2 C_s}{\partial z^2} + U_c \frac{\partial C_s}{\partial z} = 0 \tag{5.4}$$

The general solution of Equation 5.4 is

$$C_s = A + Be^{(-U_c z/E_s)} \tag{5.5}$$

The following are the boundary conditions:

At $z = 0$: $C_s = C_s^0$

At $z = 0$: $E_s \left.\frac{\partial C_s}{\partial z}\right|_{z=0} + U_c C_s^0 = 0$

Integration constants, A and B, were obtained and substituted in Equation 5.5.

$$\frac{C_s}{C_s^0} = \exp\left(-\frac{U_c}{E_s}z\right) \tag{5.6}$$

The following boundary equation can also be used (Farkas and Leblond 1969):

At $z = 0$: $C_s = C_s^0$

At $z \rightarrow \infty$: $C_s \rightarrow \infty$

The solution is the same as given in Equation 5.6, although the above boundary condition is not valid as the column extends up to $z = L$ only and there is no slurry beyond that.

Equation 5.6 is exponential in nature and represents the axial distribution of solids in an SBC when operated in batch mode. The concentration is maximal at the bottom and decreases as the axial distance from the bottom increases. It has the least value at the top of the column.

TABLE 5.1

A Few Correlations for Hindered Settling Velocity, U_c, and Dispersion Coefficient for the Solids, E_s

Investigators	U_c	E_s
Imafuku et al. (1968)	$U_c = 1.45 U_{t,\infty}^{0.65} \varepsilon_g^{4.65}$	$E_s = E_l$
Kato et al. (1985)	$U_c = 1.33 U_{t,\infty} \left(U_g / U_{t,\infty} \right)^{0.25} \varepsilon_g^{2.5}$	$Pe_{pl} = 13 Fr \left(1 + 0.009 Re Fr^{-0.8} \right) / \left(1 + 8 Fr^{0.85} \right)$
Smith and Ruether (1985)	$U_c = 1.44 U_{t,\infty}^{0.78} U_g^{0.23} \varepsilon_g^{3.5}$	$Pe_{pl} = 9.6 \left(Fr^6 / Re_g \right)^{0.1114} + 0.019 Re_p^{1.1}$

Equation 5.6 can be expressed as the concentration profile in terms of the ratio of the solids concentration to the average solids concentration, $\overline{C}_s$, as

$$\frac{C_s}{\overline{C}_s} = \frac{U_c L}{E_s} \frac{\exp\left(-\dfrac{U_c z}{E_s} \right)}{\left[1 - \exp\left(-\dfrac{U_c L}{E_s} \right) \right]} \tag{5.7}$$

Although two constants, U_c and E_s, appear in the model, in fact there is only one constant, the ratio $U_c:E_s$ (Imafuku et al. 1968; Murray and Fan 1989; Zhang 2002). This ratio can be determined by plotting $ln(C_s)$ versus z. The slope of the line is U_c-E_s, and the intercept on the y axis is C_s^0.

Several investigators used the free settling velocity of solids as U_c (Cova 1966; Imafuku et al. 1968; Farkas and Leblond 1969; Kojima and Asano 1981; Kojima et al. 1986). Correlations for U_c are also available. Cova (1966) has taken the solid dispersion coefficient equal to the liquid dispersion coefficient (i.e. the values for E_s for particles were taken as equal to those of liquid, E_l). A few correlations for U_c and E_s are presented in Table 5.1. The Peclet number, Pe_{pl}, was based on U_g^c, the critical velocity (i.e. the minimum gas velocity to suspend the solids). Fr is the Froude number.

5.2 Single Interface: The Wetted Wall Column

A wetted wall column is used to study the mass transfer coefficient at the gas–liquid interface due to a well-defined interfacial area and velocity profile in the liquid. The liquid falls down over the column wall under the influence of gravity. The gas flows through the core of the tube. There is only one interface between the gas and liquid (Figure 5.3). Other equipment

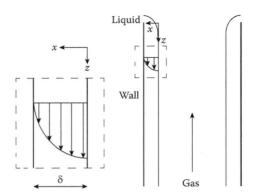

FIGURE 5.3
Mass transfer in a wetted wet column.

using the concept of falling film (e.g. as a falling film evaporator or falling film absorbers) behaves similar to a wetted wall column. The liquid may flow inside or outside of the tube. In all of this equipment, the interfacial area, a, is equal to πD. The transfer rates can be estimated from experimental data or by using an appropriate model due to the well-defined interfacial area.

For incompressible Newtonian fluid, the equation of change for momentum and equation of continuity give

$$\mu \frac{d^2 v_z}{dx^2} = \rho g \tag{5.8}$$

The boundary conditions are as follows:

At $x = \delta$: $v_z = 0$

At $x = 0$: $\dfrac{dv_z}{dx} = 0$

The solution of Equation 5.8 gives the velocity profile

$$v_z = \frac{\rho g \delta^2}{2\mu}\left[1 - \left(\frac{x}{\delta}\right)^2\right] \tag{5.9}$$

Assuming that the liquid film is very thin in comparison to the column diameter, D, the average fluid velocity, $\bar{v}_z$, can be obtained as

$$\bar{v}_z = \frac{\displaystyle\int_0^{\pi D}\int_0^{\delta} v_z\, dx\, dy}{\displaystyle\int_0^{\pi D}\int_0^{\delta} dx\, dy} = \frac{\rho g \delta^2}{3\mu} \tag{5.10}$$

And liquid film thickness, δ, is given by

$$\delta = \frac{Q}{\pi D \bar{v}_z} = \left(\frac{3\mu Q}{\rho g \pi d}\right)^{1/3}$$

(5.11)

where Q is the volumetric flow rate. The average time of contact, $\bar{t}_c$, can be written as

$$\bar{t}_c = \frac{L}{\int_0^\delta v_z \, dx} = \frac{2L}{3}\left(\frac{\pi D}{Q}\right)^{2/3}\left(\frac{3\mu}{\rho g}\right)^{1/3}$$

(5.12)

Equations 5.8 through 5.12 are valid in all vertical column walls, irrespective of the geometry, in light of the assumed thin film thickness. For mass transfer, the equation of continuity for chemical species, after substituting the velocity profile from Equation 5.9 into it, gives us

$$\frac{\rho g \delta^2}{2\mu}\left[1 - \left(\frac{x}{\delta}\right)^2\right]\frac{\partial C_A}{\partial z} = D_{AB}\frac{\partial^2 C_A}{\partial x^2}$$

(5.13)

The boundary conditions are as follows:
At $z = 0$: $C_A = 0$
At $x = 0$: $C_A = C_{A0}$

At $x = \delta$: $\dfrac{\partial C_A}{\partial x} = 0$

The solution to Equation 5.13 can be obtained numerically. Many models are based on the assumption of short tubes, which means that the length of the tube is so short that the concentration of A does not penetrate the liquid film even at the lower end. It is mathematically equivalent to impose the condition that $\dfrac{x}{\delta} \ll 1$. For short tubes, Equation 5.13 is reduced to

$$\frac{\rho g \delta^2}{2\mu}\frac{\partial C_A}{\partial z} = D_{AB}\frac{\partial^2 C_A}{\partial x^2}$$

(5.14)

The solution of Equation 5.14 is (Bird et al. 1960)

$$\frac{C_A}{C_{A0}} = 1 - erf\left(\sqrt{\frac{\rho g \delta^2}{8 D_{AB}\mu}}\frac{x}{\sqrt{z}}\right)$$

(5.15)

The total mass transfer rate of A at the interface is

$$W_A = \int_0^W \int_0^L N_{Ax}\big|_{x=0}\, dz\, dy = W C_{A0}\delta\sqrt{\frac{2 D_{AB} L \rho g}{\mu}}$$

(5.16)

In cases of short contact time, the region in which the solution is required is near the surface, where the velocity is almost constant. The time of contact can be written as $t_c = \dfrac{z}{v_z}$, and hence $\partial t = \dfrac{\partial z}{v_z}$. Equation 5.14 is converted to Equation 4.69. The solution for constant concentration at the wall (boundary condition of the first kind) is given by Equation 4.70.

If the surface concentration is not constant and the gas phase also offers resistance to mass transfer, then the mass transfer coefficient in the gas should also be taken into account. Zhi and Kai (2009) studied the absorption of CO_2 by monoethanolamine (MEA) in a wetted wall column. Unsteady-state diffusion of CO_2 into the liquid film, accompanied by a pseudo-first-order reversible chemical reaction, was assumed. Equation 4.69 is modified as

$$\frac{\partial C_A}{\partial t} = D_{AB}\frac{\partial C_A^2}{\partial z^2} - k\left(C_A - C_{eq}\right) \tag{5.17}$$

The boundary conditions are as follows:

$C_A = C_{eq}$ at $t = 0$ for all z

$C_A = C_{A,0}$ at $z = 0$ for all t

$C_A = C_{A0}$ at $z = 0$ for $t \rightarrow \infty$

The solution of Equation 5.17 is

$$\begin{aligned}
\left(C_A - C_{eq}\right) = \frac{1}{2}\left(C_{A0} - C_{eq}\right)&\left[\exp\left(-x\sqrt{\frac{k}{D_{AB}}}\right).erfc\left(\frac{x}{2\sqrt{D_{AB}t}} - \sqrt{kt}\right)\right.\\
&\left.+\exp\left(x\sqrt{\frac{k}{D_{AB}}}\right).erfc\left(\frac{x}{2\sqrt{D_{AB}t}} + \sqrt{kt}\right)\right]
\end{aligned} \tag{5.18}$$

The mass transfer rate is given by

$$N_A = -D_{AB}\frac{dC_A}{dz}\bigg|_{z=0} = \left(C_{A0} - C_{eq}\right)\left[\sqrt{\frac{D_{AB}}{\pi t}}e^{-kt} + \sqrt{kD_{AB}}\,erfc\left(-\sqrt{kt}\right)\right] \tag{5.19}$$

All of these results look elegant, but in practice it is very difficult to maintain a well-defined laminar flow at the wall. Often, a wavy surface is encountered which not only increases the interfacial area but also enhances the mass transfer rate due to induced flow near the interface.

5.3 Stationary Dispersed-Phase Systems (Gas–Solid Systems)

Packed beds are used as fluid solid contactors and are used as adsorbers, absorbers, distillation columns etc. The solid particles, which may be inert, are packed in a column. The fluid, which may be gas or liquid, enters from one end of the column and leaves from the other end. The packed beds are also used as reactors, where the solids act either as catalysts or as a catalyst support. In all of these applications, the particles remain stationary, and it is the fluid phase which is moving. It is not only a formidable task to model the performance of a packed bed, considering the detailed flow around each particle, but also a not very fruitful exercise because the particles are randomly packed in the bed.

The processes in packed beds, and in general for all multiphase systems, can be divided into two major classes based on the region of interest. In many cases, the processes around the particles are not very important and may be neglected in simple models. But if catalysts are actively involved, then the region of interest is in the neighbourhood of the particles. This classification is valid only for simple models. In rigorous models, processes in both regions are considered. Heat transfer between the wall and packed bed is now being considered to illustrate some of the modelling concepts.

5.3.1 Heat Transfer in Packed Beds

Let us understand the modelling problems in packed beds. One can recall the complexities presented in Chapter 1. Wall-to-bed heat transfer takes place through two mechanisms. The heat transfer is mainly between the wall and the fluid flowing through the interstitial space near the wall. The mechanism of heat transfer is convection. This term is called 'fluid convection'. The fluid conduction may be very small compared to the convective transport, except at a low Reynolds number. The particles, which are adjacent to the wall and are in its contact, are heated (or cooled) by the fluid present around it. Due to the thermal conductivity of the particle being higher than that of the gas, the heat transfer between the particle and gas takes place at a higher rate than in the fluid convection. Let us call this 'particle convection' (Figure 5.4a).

Heat transfer models may be classified as 'one-temperature models' and 'two-temperature models' (Figure 5.4b). The one-temperature models are frequently called 'pseudo-homogeneous models'. In these models, the solids and fluids are assumed to be a continuum phase. The laws of conservation are written as if only one phase is present. The steady-state model for heat transfer considered steady-state conduction to a homogeneous phase having properties estimated from the properties of solids, fluids and fluid velocity.

The simplest model for heat transfer in packed beds assumes that the heat transfer by conduction is negligible as compared to the convective heat

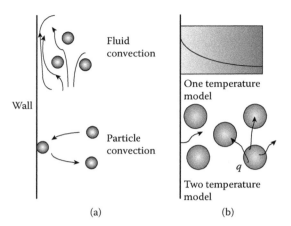

(a)   (b)

FIGURE 5.4
Mechanisms of heat transfer in fluidised beds: (a) fluid convection and particle convection and (b) one-temperature and two-temperature models.

transfer in the axial direction. This assumption is valid only at high fluid velocity. The bed temperature at any cross-section is constant except in very thin regions adjacent to the wall. This condition also is possible at high fluid velocity. The laws of conservation of thermal energy give

$$U_g \rho_g C_{pg} \frac{dT}{dz} = h(T_W - T) \qquad (5.20)$$

Equation 5.20 is an ordinary differential equation which can be solved analytically as well as numerically by using the fourth-order Runga–Kutta method. For constant wall temperature and a constant heat transfer coefficient, the solution of Equation 5.20 is

$$\frac{(T_W - T)}{(T_W - T_0)} = \exp\left(\frac{-h}{U_g \rho_g C_{pg}} \frac{z}{L} \right) \qquad (5.21)$$

The solution of Equation 5.20 for constant heat flux, q, is

$$(T_W - T) = \frac{h(T_W - T)}{U_g \rho_g C_{pg}} \frac{z}{L} \qquad (5.22)$$

The wall temperature, T_W, is a function of the axial position, z.

The pseudo-homogeneous models considering radial thermal diffusivity for gas flowing in packed beds may be written as

$$U_g \rho_g C_{pg} \frac{\partial T}{\partial z} = k_r \frac{1}{r} \frac{\partial}{\partial r}\left(r \frac{\partial T}{\partial r} \right) \qquad (5.23)$$

If axial thermal diffusivity is also considered, then the model equations will be

$$U_g\rho_g C_p \frac{\partial T}{\partial z} = k_r \frac{1}{r}\frac{\partial}{\partial r}\left(r\frac{\partial T}{\partial r}\right) + k_a \frac{\partial^2 T}{\partial z^2} \tag{5.24}$$

The solution of Equation 5.24 is

$$\frac{T-T_0}{T_W-T_0} = 1 - 2\sum_{n=1}^{\infty} \frac{J_0(\lambda_n y)}{J_0(\lambda_n)} \exp\left[\frac{(U_g\rho_g C_{pg})(B_n-1)z}{2k_a}\right] \tag{5.25}$$

If a gas–liquid mixture is flowing through the packed bed, then the term on the left-hand side is modified accordingly:

$$(U_g\rho_g C_{pg} + U_l\rho_l C_{pl})\frac{\partial T}{\partial z} = k_r \frac{1}{r}\frac{\partial}{\partial r}\left(r\frac{\partial T}{\partial r}\right) + k_a \frac{\partial^2 T}{\partial z^2} \tag{5.26}$$

The boundary conditions are the following:

At the centre of the column (i.e. at $r = 0$): $\dfrac{\partial T}{\partial r} = 0$

Far away from the entrance (i.e. as $z\to\infty$): $T \to T_W$

One more boundary condition is required. For constant wall temperature, the boundary conditions of the first kind will be used:

At $r = 0$, $T = T_W$

The solution of Equation 5.26 is given as (Moreira et al. 2006)

$$\frac{T-T_0}{T_W-T_0} = 2\sum_{n=1}^{\infty} \frac{J_0\left(a_n \dfrac{r}{R}\right)\exp\left[\dfrac{(U_g\rho_g C_{pg} + U_l\rho_l C_{pl})(1-B_n)z}{2k_a}\right]}{J_0(a_n)} \tag{5.27}$$

where

$$B_n = \left[1 + \frac{4a_n^2 k_r k_a}{R^2(U_g\rho_g C_{pg} + U_l\rho_l C_{pl})}\right]; \text{ and}$$

$J_0(a_n) = 0$.

If the boundary condition of the third type is used:

At $r = 0$: $-k_r \dfrac{\partial T}{\partial r} = h(T_{r=R} - T_W)$

$$\frac{T - T_0}{T_W - T_0} = 2\sum_{n=1}^{\infty} \frac{\text{Bi}(1 + B_n)J_0\left(a_n\frac{r}{R}\right)\exp\left[\frac{(U_g\rho_g C_{pg} + U_l\rho_l C_{pl})(1 - B_n)z}{2k_a}\right]}{B_n\left(\text{Bi}^2 + a_n^2\right)J_0(a_n)} \tag{5.28}$$

and

$$\text{Bi}J_0(a_n) + a_n J_1(a_n) = 0$$

where Bi is the Biot number based on k_r.

These models are one-temperature models. Since the fluid and solids are considered one phase, no heat transfer between these two phases is written.

In the two-temperature models, fluid and solids are considered to be at different temperatures. The heat transfer between the fluid and solids can also be taken into account. This model is applicable at a low fluid velocity at which the rate of heat transfer between the fluid and solid is small, thus causing a temperature difference between the two phases. The fluids and solids are considered separate phases.

A two-temperature model for 2D packed beds consisting of spheres was proposed by Laguerre et al. (2006). The position of each particle was specified by subscripts i and j in the horizontal and vertical directions, respectively. The temperatures of air and particle are T_a and $T_{p(i,j)}$, respectively. Equating the accumulation of heat in the particle as equal to the heat transferred to the air and the surrounding sphere, the temperature of the particle is described by the following equation:

$$\rho_s C_{ps}\frac{\pi d_p^3}{6}\frac{dT_{p(i,j)}}{dt} = h_j\pi d_p^2\left(T_{a(i,j)} - T_{p(i,j)}\right)$$
$$+ C\left[T_{p(i-1,j)} + T_{p(i+1,j)} + T_{p(i,j-1)} + T_{p(i,j+1)} - 4T_{p(i,j)}\right] \tag{5.29}$$

Only four spheres were considered for the 2D cubic arrangement. For the variables, the heat transfer coefficient between the air and jth particle, h_j, and the conduction between two spheres, C, the following correlations were used:

$$\frac{h_j d_p}{k_a} = \left[2 + 1.09\left(\frac{d_p U \rho_a}{\mu_a}\right)^{0.53}\left(\frac{C_{pa}\mu_a}{k_a}\right)^{1/3}\right]\left[1 + 0.2\exp\left(-\frac{j-1}{1.22}\right)\right] \tag{5.30}$$

The temperature of the air is given as

$$\varepsilon\rho_a C_{pa}\left(\frac{\partial T_a}{\partial t} + v_z\frac{\partial T_a}{\partial z}\right) = h_p S_p\left(T_{p(i,j)} - T_{a(i,j)}\right)$$
$$+ \rho_a C_{pa}\left(D_r\frac{\partial^2 T_{a(i,j)}}{\partial r^2} + D_z\frac{\partial^2 T_{a(i,j)}}{\partial z^2}\right) \tag{5.31}$$

The unsteady-state term is the 'thermal inertia of air'. It was neglected in comparison to the thermal inertia of the solids. For spherical particles, the surface per unit volume, $S_p = \dfrac{\pi}{d_p}$. These partial differential equations were solved by discretising with $\Delta z = \Delta r = d_p$. After discretising, the following equation is obtained:

$$\left(T_{p(i,j+1)} - T_{a(i,j)}\right) = \frac{h_p \pi}{\rho_a C_{pa} v_z}\left(T_{p(i,j)} - T_{a(i,j)}\right)$$

$$+ \frac{D_z}{v_z d_p}\frac{\left(T_{a(i,j+2)} - T_{a(i,j+1)} - T_{a(i,j)} + T_{a(i,j-1)}\right)}{2}$$

$$+ \frac{D_r}{v_z d_p}\frac{\left(T_{a(i-1,j)} - 2T_{a(i,j)} + T_{a(i+1,j)}\right)\left(T_{a(i-1,j+1)} - 2T_{a(i,j+1)} + T_{a(i+1,j+1)}\right)}{2}$$

$$(5.32)$$

Equations 5.29 through 5.32 were solved numerically.

The assumption of continuum results in a simple model providing only average axial variations of temperature. A 2D model describes the radial temperature profile. A two-temperature model considering the temperature difference between the fluid and particles is much more complex, and the use of empirical equations cannot be avoided. The model equations are to be solved numerically after discretising the equations.

5.3.2 Mass Transfer in Packed Beds

The equation of continuity presented in Chapter 4 is based on the assumption that the diffusivity or thermal conductivity is the same in all directions. However, many industrial operations are carried under turbulent flow conditions in which the diffusivity or thermal conductivity in axial and radial directions may be different. In cases of mass transfer in a packed bed under turbulent flow conditions, the effective diffusivity is taken to be $D_{eff} = D_{AB} + \varepsilon_d$. Let us assume that the packed bed behaves as a continuous single fluid (i.e. the assumption of continuum is considered). In reality, the bed is not homogeneous and is known as pseudo-homogeneous. The concentration profile may be obtained by simplifying the equation of continuity for the species in cylindrical coordinates, if the diffusivity is isotropic, as follows:

$$v_z \frac{\partial C_A}{\partial z} = D_{eff}\left[\frac{1}{r}\frac{\partial}{\partial r}(rC_A) + \frac{\partial^2 C_A}{\partial z}\right] \qquad (5.33)$$

Due to different values of eddy diffusivity in radial and axial directions, the effective diffusivity will also be different. The equation of continuity for the species will be

$$v_z \frac{\partial C_A}{\partial z} = D_{effr}\left[\frac{1}{r}\frac{\partial}{\partial r}(rC_A)\right] + D_{effz}\left[\frac{\partial^2 C_A}{\partial z}\right] \qquad (5.34)$$

In terms of the dimensionless numbers, Equation 5.34 takes the following form:

$$\frac{N}{2}\frac{\partial C^*}{\partial Z} = \frac{1}{Pe_{mr}}\left[\frac{1}{r^*}\frac{\partial C^*}{\partial r^*} + \frac{\partial^2 C^*}{\partial r^{*2}}\right] + \frac{1}{Pe_{mz}}\left[\frac{\partial^2 C^*}{\partial Z}\right] \qquad (5.35)$$

where

$$Pe_{mr} = \frac{v_z d_p}{D_{effr}}; \ Pe_{mz} = \frac{v_z d_p}{D_{effz}}; \ C^* = \frac{(C_A - C_0)}{(C_W - C_0)}; \ Z = \frac{z}{r}; \text{and } r^* = \frac{r}{R}$$

Note that if N is the ratio of the column diameter to the particle diameter then $D_T = Nd_p$. The following boundary conditions were used by Dixon et al. (2003):

At the centre of the tube (i.e. at $y = 0$): $\dfrac{d\theta}{dy} = 0$

At the wall (i.e. at $y = 1$): $\dfrac{d\theta}{dy} + Bi\theta = Bi$ for $x > 0$

At the entrance (i.e. at $y = 1$): $x = 0$, $\dfrac{d\theta}{dy} + Bi\theta = 0$

As $x \rightarrow \infty$, $\theta \rightarrow 1$.

And, as $x \rightarrow -\infty$, $\theta \rightarrow 1$

The analytical solutions for Equation 5.35:

$$\theta = 1 - \sum_{n=1}^{\infty} \frac{Bi(1+\beta_n)J_0(\lambda_n y)}{(Bi^2 + \lambda_n^2)\beta_n J_0(\lambda_n)} \exp\left[\frac{Pe_{mz}}{2}\frac{N}{2}x(1-\beta_n)\right] \qquad (5.36)$$

where

$$\beta_n = \sqrt{1 + \left(\frac{16\lambda_n^2}{N^2 Pe_{mz} Pe_{mr}}\right)} \qquad (5.37)$$

and λ_n are the roots of the equation.

$$\lambda_n J_1(\lambda_n) - Bi J_0(\lambda_n) = 0 \qquad (5.38)$$

The heat and mass transfer in packed beds are generally based on the assumption of continuum. This assumption simplifies the equations to an extent that analytical solutions are possible for different types of boundary conditions. Since the detained temperature or concentration distribution around the particles is not of interest, therefore, these solutions are adequate.

5.4 Moving Dispersed Systems (Gas–Solid Systems): Wall-to-Bed Heat Transfer in a Fluidised Bed

Fluidised beds are used in various chemical, biochemical and environmental processes involving reactions and transfer processes. The understanding of wall-to-bed heat and mass transfer is necessary for the design of fluidised beds. There is a voluminous literature on the mechanism of wall-to-bed heat and mass transfer. It has been reviewed by many investigators from time to time (Botterill 1975; Saxena 1989).

Fluidised beds show different characteristics under different operating conditions. Various regimes of fluidisation are presented in Figure 5.5. It is almost impossible to propose a single model to describe the heat transfer or hydrodynamic parameters in all regimes of fluidisation. There are mainly two types of fluidisation: particulate and aggregative. In particulate fluidised beds, the particles move as individuals. However, this can also be observed in gas fluidised beds consisting of large particles, beds operating at elevated pressures, beds operating near incipient fluidisation, lean fluidised beds and the freeboard region.

The particles move as aggregates in aggregative fluidisation. These aggregates are often called 'packets' or 'emulsion'. In fixed beds, gas flows through the interstitial spaces between the particles. As the gas velocity is increased, the drag force acting on the particles balances the weight of the particles. This is the condition of incipient fluidisation. The fluid velocity is called 'minimum fluidisation velocity'. As the gas velocity is increased further, the particles starts moving as packets, because a fraction of the gas moves as bubbles that have almost no particles. The region outside the bubble consists of particles, and a fraction of gas still flows through the interstitial space. This indicates the assumption that the properties of the emulsion at any superficial gas velocity are the same as those of emulsion at incipient fluidisation. Aggregative fluidisation is observed in a few liquid fluidised beds (e.g. the fluidisation of lead shots by water).

As the gas velocity is increased further, the bubble size increases. The bubble coalescence and breakup become more frequent at a high velocity. It becomes difficult to identify individual bubbles. The bed is called

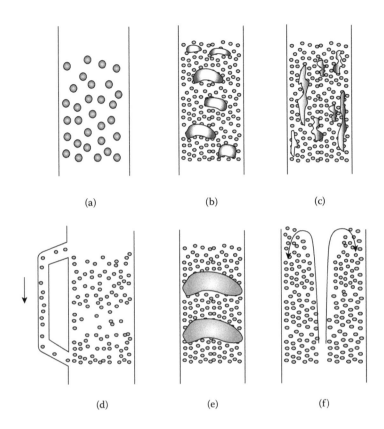

FIGURE 5.5
Regimes of fluidisation: (a) particulate fluidisation, (b) aggregate fluidisation, (c) turbulent fluidisation, (d) circulating fluidised bed, (e) slugging fluidised bed and (f) spouted bed.

a 'turbulent-fluidised bed'. A further increase of gas velocity will result in the entrainment of solids, but still it can be operated by circulating the solids up to a certain superficial gas velocity. This is called 'fast fluidisation'. A further increase of gas velocity will result in pneumatic conveying.

If the column diameter is small, then by increasing gas velocity, the bubble size becomes so large that it is equal to the column diameter. Under this condition, the bubbles move in the form of slugs. These beds are known as 'slugging fluidised beds'.

Because a model is not applicable in all regimes of fluidisation, it becomes necessary to know the conditions under which a fluidised bed exhibits its characteristics. Geldart classified powders into four groups: A, B, C and D. They are based on the particle size and density of the particle. Other classifications are also reported in the literature (Saxena 1989). These classifications help researchers in choosing an appropriate model for the process under consideration.

It has been generally considered that heat transfer between a wall and bed takes place through two parallel mechanisms: the particle-convective transfer and the fluid-convective transfer. In the former model of transfer, the particles arrive at the transfer surface, pick up energy and return to the core of the bed, where it give up its energy, thus contributing to heat transfer. In the latter, the particles scour the thermal layer on the transfer surface and thus enhance the fluid-convective transfer.

Ziegler and Brazelton (1964) observed that the particle convection accounts for 80%–90% of the total heat transfer. Baskakov and Suprun (1972) found that the relative contribution of the fluid-convective transfer increased from 5%–90% as the particle size increased from 0.16 to 4.0 mm. Denloye and Botterill (1978) considered that the gas convective transfer is equal to heat transfer in a quiescent bed. The contribution of the gas convective component increased with particle size and pressure. Saxena et al. (1978) pointed out that the fluid-convective component is significant under the following conditions in gas fluidised beds:

1. Beds consisting of large particles
2. Beds operating at elevated pressures
3. Beds operating near incipient fluidisation
4. Lean fluidised beds
5. Surface geometries that result in regions of either high voidages or stagnant particles

Washmund and Smith (1967) have carried out heat transfer studies in liquid fluidised beds with particles of the same size and density but different thermal diffusivities. Since the difference between the heat transfer coefficients was insignificant at high porosities, it was concluded that the fluid convection is the main mechanism. Patel and Simpson (1977) studied heat transfer in aggregative and particulate fluidised beds, and they found that in both situations, the principal mechanism is the fluid convection.

Thus, the fluid-convective term is equally important in certain cases of gas fluidised beds, and it is the main mechanism in liquid fluidised beds. Because the particles themselves cannot carry mass, the fluid-convective transfer is the sole mechanism of mass transfer. Therefore, consideration of the fluid-convective term is necessary for the development of models for heat and mass transfer in fluidised beds.

Earlier models for heat transfer, when the fluid-convective term was not clearly understood, were based on the nature of the scouring of the thermal boundary layer on the transfer surface by the particle. Notable among these models are those of Dow and Jacob (1951), Leva et al. (1949) and Levenspiel and Walton (1954).

In fluidised bed combustors, the particle sizes larger than 1 mm employed and hence the relative contribution of the fluid-convective term are important. As these combustors came into prominence recently, there is renewed interest to understand the fluid-convective transfer.

Botterill and Denloye (1978) developed a model for an emulsion-phase gas convective term, considering it to be the same as that of a quiescent bed. However, it neglects any possible dependence of a gas convective component on superficial fluid velocity. Zebrodsky et al. (1981) stressed the need to consider the dependence of a convective term on fluid velocity and assumed it to be proportional to $U^{0.2}$ in the expression presented.

Ganzha et al. (1982) have evaluated the convective term, assuming that the turbulent boundary layer is disrupted at the front of the particle and reformed within the wake. The conductive term was evaluated by solving the unsteady-state conduction equation for an infinite composite layer of gas and solids. The model reproduced the dependence of the heat transfer coefficient on the Reynolds number.

A number of models have been proposed to describe the heat transfer coefficient. This was due to the fact that earlier modelling efforts were made to obtain an analytical solution of the model equations. However, today, numerical tools and software to solve model equations are available. An emphasis on obtaining an analytical solution can be considered as a constraint only.

Four types of mechanism for wall-to-bed heat transfer in fluidised beds have been reported in the literature:

1. The film model considers that the entire resistance to heat transfer lies in the gas film. The role of the particles is to scour the film (i.e. the particles reduce the film thickness and enhance the heat transfer rates). In other words, the particles act as turbulence promoters only. The fluid convection is the main mechanism.

2. A thermal boundary layer is formed. It is discontinued at places where the particle touches the wall.

3. The particles take active part in transferring the heat. They are heated at the wall, and they carry heat to the bulk. This has been called 'particle convection'. The heat transfer is between the particle and the surrounding fluid. The surface renewal by particles as individuals or in groups (packet) in particulate and aggregative fluidisation, respectively, was considered.

4. Both particle convection and fluid convection contribute towards heat transfer.

The models based on various simple theories and those based on laws of conservation are described in this section.

5.4.1 Models Based on Film Theory

Early attempts to model were based on film theory. Levenspiel and Walton (1954) proposed a model assuming that the main mechanism of heat transfer is fluid convection. A vertical boundary layer at the wall is formed (Figure 5.6a). The boundary layer is disrupted due to the presence of the particles. The boundary layer is formed in between two successive horizontal layers of equally spaced particles. The distance between two successive layers of particles is

$$D_r = \frac{\pi d_p}{6(1-\varepsilon)} \tag{5.39}$$

The average film thickness is given as

$$\delta_{av} = \frac{10}{3}\left(\frac{D_r \mu}{U \rho_f}\right)^{1/2}\left[\left(1+\beta^2\right)^{3/2} - \beta^3\right] \tag{5.40}$$

where

$$\beta = 0.041(1-\varepsilon)\sqrt{\frac{D_r U \rho_f}{\mu}} \tag{5.41}$$

The heat transfer coefficient can be estimated from

$$h_W = \frac{k_f}{\delta_{av}} \tag{5.42}$$

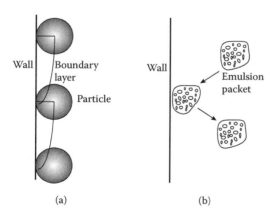

(a)   (b)

FIGURE 5.6
Models for wall-to-bed heat transfer based on (a) film theory (Levenspiel and Walton 1954) and (b) surface renewal (Mickley and Fairbanks 1955).

The wall-to-bed heat transfer increases as the superficial gas velocity increases; it bypasses through a maxima and then decreases. The model based on film theory does not predict this trend.

5.4.2 Models Based on the Surface Renewal Concept

Several models for heat transfer in fluidised beds are based on the concept of surface renewal. The surface is renewed by the individual particles or emulsion packets. Models based on both approaches are discussed now to clarify the different assumptions and results obtained.

Mickley and Fairbanks (1955) developed a model based on surface renewal. It was assumed that a packet of particles arrives at the surface, stays there for sometime during which unsteady-state heat transfer takes place and then leaves the surface (Figure 5.6b). The instantaneous heat transferred at time t is given by

$$q(t) = (T_W - T_b)\sqrt{\frac{k_{eff}\rho_{eff}C_{peff}}{\pi t_e}} \tag{5.43}$$

Equation 5.43 is valid for constant wall temperature. A local instantaneous heat transfer coefficient, $h(t, z)$, is given by

$$h(t,z) = \sqrt{\frac{k_{eff}\rho_{eff}C_{peff}}{\pi t_e}} \tag{5.44}$$

The properties of the packets are those of gas–solid dispersion at incipient fluidisation. This is therefore independent of fluidisation velocity. The average heat transfer can be written as

$$h = \sqrt{\frac{k_{eff}\rho_{eff}C_{peff}}{\pi}}\int_0^L\int_0^\infty t_e^{-1/2}\psi(t_e)dt_e dz \tag{5.45}$$

To evaluate the time, t_e, two cases were considered. If the packets are moving in the vertical direction, then $t_e = \dfrac{L}{U}$. The average heat transfer coefficient is obtained by integrating Equation 5.45:

$$h = 2\sqrt{\frac{k_{eff}\rho_{eff}C_{peff}}{\pi}\left(\frac{U}{L}\right)} \tag{5.46}$$

Ziegler et al. (1964) proposed a model based on the arrival and departure of particles. The following assumptions were made.

All the particles are spherical and move as individuals. This assumption may be valid in particulate fluidised beds.

The particles from the bulk arrive at the surface, where the particle stays for some time. During this period, the heat transfer between the particle and the surrounding fluid takes place. The heat transfer by conduction between the particle and the surface is negligible. At temperatures lower than 600 °C, the radiant heat transfer is insignificant.

The equation of change of thermal energy in terms of dimensionless parameters gives

$$\frac{\partial T^*}{\partial \tau} = \frac{1}{r^{*2}} \frac{\partial}{\partial r^*}\left(r^{*2} \frac{\partial T^*}{\partial r^*}\right) \tag{5.47}$$

The boundary conditions are as follows:

Initial condition (i.e. at $\tau = 0$ and all r^*): $T^* = 1$

At the centre of the particle (i.e. $r^* = 1$ and all τ): $\dfrac{\partial T^*}{\partial r^*} = 0$

At the surface of the particle (i.e. $r^* = 1$ and all τ): $\dfrac{\partial T^*}{\partial r^*} = -Nu T^*$

where $T^* = \dfrac{(T_W - T)}{(T_W - T_b)}$; $r^* = \dfrac{r}{(d_p/2)}$; $\tau = \dfrac{k_s \theta}{\rho_s C_{ps}(d_p/2)^2}$; and $Nu = \dfrac{h_p(d_p/2)}{k_s}$

The solution of Equation 5.47 is

$$T^* = \frac{2Nu}{r^*}\sum_{n=1}^{\infty}\exp\left(\lambda_n^2\tau\right)\left[\frac{\left\{\lambda_n^2 + (Nu-1)^2\right\}\sin\lambda_n}{\lambda_n^2\left\{\lambda_n^2 + Nu(Nu-1)\right\}}\right]\sin\left(\lambda_n^2 r^*\right) \tag{5.48}$$

where λ_n^2 is the root of

$$\lambda_n \cot \lambda_n = 1 - Nu \tag{5.49}$$

It was argued that the particles are small and the fluid velocity around the particle is quite lower. It is a situation similar to the cooling of a small sphere in a stagnant fluid. The value of Nu is 2, as shown in Equation 3.55. Only the first eigenvalues were taken as significant. Using a Taylor series expansion of sin in Equation 5.48 is reduced to

$$T^* = e^{-3Nu\tau} \tag{5.50}$$

The instantaneous heat transfer rate is given by

$$q = h\left(\pi d_p^2\right)(t_W - T) = h\left(\pi d_p^2\right)k_g(t_W - t_b)e^{-3Nu\tau} \tag{5.51}$$

The residence time distribution was described by the gamma distribution function:

$$f(\tau) = \frac{1}{\alpha!\beta^{\alpha+1}} \tau^\alpha e^{-\tau/\beta} \tag{5.52}$$

It was chosen based on the argument that the residence time will be distributed on both sides of a preferred value. This function gives value in the range from zero to infinity.

The average heat transfer rate can be written as

$$q_{av} = \int_0^\infty qf(\tau)\,d\tau = \frac{2\pi d_p k_g (t_W - t_b)}{(1+3Nu\beta)^{\alpha+1}} \tag{5.53}$$

The parameters α and β are related with the average time, $\bar{\tau}$, and the time at which maximum value occurs, τ_{max}, as

$$\bar{\tau} = \beta(\alpha+1) \text{ and } \tau_{max} = \beta\alpha.$$

The concentration of particles on the surface was estimated by assuming that the surface is fully packed with particles (i.e. the spherical particles are arranged on the hexagonal arrangement). Therefore, the average heat transfer rate multiplied by the concentration of the particles on the surface is equal to $h(t_W - t_b)$. The heat transfer coefficient, h, can hence be written as

$$h = \frac{4\pi/\sqrt{3}}{\left(1 + \frac{12k_g\bar{\theta}}{\rho_s C_s d_p^2 (\alpha+1)}\right)^{\alpha+1}} \tag{5.54}$$

Here, the average resistance time of the particle at the surface, $\bar{\theta}$, was back calculated using the experimental data. A simple equation based on the terminal velocity of the particles along the height of the heat transfer surface was given, but it gave a higher estimate of $\bar{\theta}$.

The surface renewal model with the coexistence of film proposed by Wasan and Ahluwalia (1969) was developed for heat transfer in fluidised beds. The heat transfer coefficient is given by Equation 4.83. However, to use this equation for the prediction of heat transfer coefficients, the film thickness, δ, and contact time, t_c, should be determined. The inter-particle distance, d_r, in the fluidised bed can be expressed in terms of the porosity, ε, if a particular arrangement for the particles can be assumed. The average position of the particles was assumed to be at the edges of a cube. The inter-particle distance, d_r, was given by

$$d_r = \left\{\frac{\pi}{6(1-\varepsilon)}\right\}^{1/3} d_p \tag{5.55}$$

The average contact time for the fluid, t_c, is the time in which the fluid travels from one particle to another. It was given by

$$t_c = \left\{ \frac{\pi}{6(1-\varepsilon)} \right\}^{1/3} \frac{d_p}{U} \tag{5.56}$$

The film thickness in a fluidised bed was given by Levenspiel and Walton (1954). For laminar flow of the fluid:

$$\delta = 0.0716 d_p \frac{A_l}{B_l} \tag{5.57}$$

where

$$B_l = 0.02144 \left(\frac{d_p}{U t_c} \right)^{1/2} \left(\frac{d_p \rho_f U}{\mu_f} \right)^{1/2} \quad \text{and} \quad A_l = \left(1 + B_l^2 \right)^{3/2} - B_l^3 \tag{5.58}$$

After substitution of t_c and δ, an expression for a heat transfer coefficient was obtained:

$$\frac{h d_p}{k_l} = 1.257 \, \text{Re}^{1/2} \, \text{Pr}^{1/2} (1-\varepsilon)^{1/6} + 3.715 \, \text{Re}^{1/2} \, \text{Pr} (1-\varepsilon)^{1/6}$$
$$A_l \left\{ e^{L_l} \left(1 - \text{erf} \sqrt{L_l} \right) - 1 \right\} \tag{5.59}$$

where

$$L_l = 0.0898 \, \text{Pr}^{-1} \, A_l^2 \tag{5.60}$$

For the turbulent flow, the film thickness is given by

$$\delta = 0.0597 d_p \frac{A_t}{B_t} \tag{5.61}$$

where

$$B_t = 0.2902 \left(\frac{d_p}{U t_c} \right)^{4/5} \left(\frac{d_p \rho_f U}{\mu_f} \right)^{1/5}$$
$$A_t = \left(1 + B_t^{5/4} \right)^{9/5} - B_t^{9/4} \tag{5.62}$$

After substitution of t_c and δ, the heat transfer coefficient can be estimated using Equation 4.77:

$$\frac{h d_p}{k_l} = 1.257 \, \text{Re}^{1/2} \, \text{Pr}^{1/2} (1-\varepsilon)^{1/6} + 0.2148 \, \text{Re}^{4/5} \, \text{Pr} (1-\varepsilon)^{1/15}$$
$$A_t \left\{ e^{L_t} \left(1 - \text{erf} \sqrt{L_t} \right) - 1 \right\} \tag{5.63}$$

where

$$L_1 = \frac{26.873(1-\varepsilon)^{-1/5}}{\Pr \mathrm{Re}^{3/5} A_t^2} \tag{5.64}$$

Wasan and Ahluwalia (1969) observed that data of Levenspiel and Walton (1954) and Dow and Jacob (1951) were comparable with Equation 4.83 when the film thickness in laminar flow conditions was used. The data of Mickley and Trilling (1949) and Washmund and Smith (1967) were comparable by using an expression for the film thickness in turbulent flow conditions.

There are several modifications to these models by either considering the presence of a gas film adjacent to the heat transfer surface or using a variety of surface renewal rates. In summary, there is no analytical model which is applicable for all types of fluidisation regimes and types of particles.

A model based on surface renewal by the packet for a constant heat flux was proposed by Huang and Levy (2004). The boundary conditions are as follows:

$T = T_b; z > 0, t = 0$

$T = T_b; z \to \infty, t > 0$

$-k_d \dfrac{\partial T}{\partial z} = q_w; z = 0$

The temperature profile in the bed is given by

$$T - T_b = \frac{q_w}{k_s}\left[2\left(\frac{t}{\pi k_d \rho_d C_{pd}}\right)^{1/2} \exp\left(-\frac{\rho_d C_{pd} x}{4k_d t}\right) - x\left\{1 - erf\left(\frac{x}{2}\sqrt{\frac{\rho_d C_{pd}}{k_d t}}\right)\right\}\right] \tag{5.65}$$

From the discussion above, it is clear that in the absence of radiation, the mechanism of heat transfer is fluid convection as well as particle convection. The overall heat transfer coefficient, h, can be written as

$$h = h_f + h_p \tag{5.66}$$

where h_f and h_p are the fluid-convective and particle-convective heat transfer coefficients, respectively. In the presence of radiation, heat transfer due to radiation can be added to these. The particle-convective terms have been determined either by surface renewal by individual particles in particulate fluidisation, circulating fluidised beds (Al-Bosoul 2002), lean fluidised beds etc. or by conductive heating of particles by the surrounding fluid (Botterill 1975; Adams and Welty 1979). In bubbling aggregative fluidisation, it can be determined by any of the models based on surface renewal of emulsion or packets of particles.

The radiative heat transfer coefficient is given by

$$h_r = \frac{\sigma\left(T_p^4 - T_b^4\right)}{\left[\dfrac{1}{\varepsilon_p} + \dfrac{1}{\varepsilon_b} - 1\right]\left(T_p - T_b\right)}$$

(5.67)

5.5 Moving Dispersed Systems (Gas–Liquid Systems): Transfer Processes in Bubble Columns

The bubble columns are gas liquid contactors in which gas is introduced at the bottom of a column filled with liquid or slurry. Bubbles move vertically upward. The movement of the bubbles induces fluid to flow, and a liquid circulation flow is developed. Due to vigorous mixing by the liquid circulation flow, high heat transfer rates are observed. The heat transfer from wall to bulk in bubble columns has been modelled using various concepts discussed in Chapters 3 and 4 (e.g. surface renewal, boundary-layer theory and a combination of film and surface renewal). A few of these models are discussed below. The region of interest is around the transfer surface, not the gas–liquid interface.

5.5.1 Models Based on the Boundary Layer Concept

Zehner (1986) proposed a model based on the concept of heat transfer through a boundary layer at the wall. The heat transfer surface was considered to be analogous to flow over a flat plate. The flow over the surface was assumed to have an eddy velocity, U_f, given as

$$U_f = \left[\frac{1}{25}\left(\frac{\rho_l - \rho_g}{\rho_l}\right)gDU\right]^{1/3}$$

(5.68)

where U is the superficial gas velocity, D is the column diameter and ρ_g and ρ_l are the density of gas and liquid, respectively. The flat plate is not infinite. It is assumed that the boundary layer thins at places where the bubble penetrates the thermal boundary layer (Figure 5.7a). It can be determined from the known mean bubble diameter and gas holdup. The length of the flat plate, l, is therefore equal to the inter-bubble distance and can be estimated as

$$L = d_{av}\sqrt{\frac{\pi}{6\varepsilon_g}}$$

(5.69)

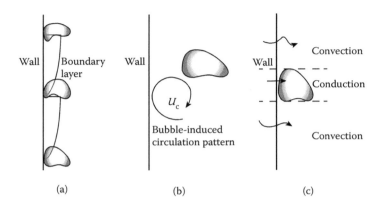

FIGURE 5.7
Models for heat transfer in bubble columns: boundary layer. (a) Based on Zehner 1986, (b) based on Joshi et al. (1980) and (c) based on Verma (1989).

where d_{av} is the average bubble diameter. Heat transfer for flow over a flat plate is assumed to be analogous to this case. Zehner (1986) used the following equation:

$$\frac{hL}{k_1} = 0.18 \left(\frac{U_f L}{h} \right)^{2/3} \left(\frac{C_p \mu_1}{k_1} \right)^{-1/3} \qquad (5.70)$$

At the places where the bubble penetrates boundary layer, heat transfer is inhibited owing to the thermal conductivity of the gas being much lower than that of the liquid. This fact was taken into account by assuming that the heat transfer coefficient is proportional to the liquid holdup $(1-\varepsilon)$, resulting in

$$h = 0.18 (1-\varepsilon) \left(\frac{k_1 \rho_1^2 C_p U_f^2}{L \mu_1} \right)^{1/3} \qquad (5.71)$$

In the absence of the data on length of the boundary layer, L, a constant value of 0.007 was assumed. The value of the h estimated using Equation 5.71 becomes constant above a certain value of U, which is in agreement with the experimental results.

The boundary layer concept of Zehner (1986) gives an estimate of the heat transfer coefficient, which is in much better agreement with experimental data compared to that given by the surface renewal concept of Deckwer (1980).

5.5.2 Models Based on the Surface Renewal Concept

Deckwer (1980) assumed that in the vicinity of the wall surface, there is a steady but irregular flow of the fluid eddies from the bulk fluid to the wall and vice versa. The fluid elements stay for a certain contact time at the surface and then leave it and enter the bulk fluid again. During its stay, heat transfer takes place by unsteady-state heat conduction. The temperature distribution in the fluid element is described by the unsteady-state conduction in the fluid.

$$\frac{\partial T}{\partial t} = \left(\frac{k_1}{\rho_1 C_p}\right)\frac{\partial^2 T}{\partial z^2} \tag{5.72}$$

The boundary conditions are given as:

$T = T_s; z = 0, t \leq 0$

$T = T_b; z > 0, t = 0$

$T = T_s; z \to \infty, t > 0$

Solving this equation and using temperature for the computation of heat flux, i:

$$q = 2\sqrt{\frac{k_1 \rho_1 C_p}{\pi \theta}}\,(T_s - T_b) \tag{5.73}$$

Comparing this equation with the definition of the heat transfer coefficient:

$$h \propto \sqrt{\frac{k_1 \rho_1 C_p}{\theta}} \tag{5.74}$$

The contact time is not known. In turbulent flow it is often considered that the eddy dissipation by the macroscale eddy moves isotropically and comprises mainly viscous forces. The length scale, η_l, and velocity scale, v, of the micro-eddies can be expressed in terms of energy dissipation rate per unit mass, E, and kinematic viscosity, v:

$$\eta_l = \left(\frac{v^3}{E}\right)^{1/4}$$

and

$$v = (vE)^{1/4} \tag{5.75}$$

It was argued that because turbulence intensity remains constant in the radial direction, the role of micro-eddies was responsible for the radial flow as proposed by Kast (1962). Hence, it was assumed that the average contact time, θ, is related to η_l and v in the following way:

$\theta = \eta_l / v$.

Introducing the value of θ in the equation:

$$h \propto \sqrt{\frac{k \rho_1 C_p E^{1/2}}{v^{1/2}}} \tag{5.76}$$

The entire energy dissipation rate is given by the product of the gas, volume flow rate and pressure drop:

$$P = U\rho_l g H_s S \qquad (5.77)$$

Therefore, the energy dissipation rate per unit mass of the liquid is given by

$$E = Ug \qquad (5.78)$$

Using the value of E in Equation 5.76 and rearranging in terms of dimensionless numbers, the following equation was obtained:

$$St = 0.1\left(Re\,Fr\,Pr^2\right)^{-0.25} \qquad (5.79)$$

5.5.3 A Unified Approach Based on Liquid Circulation Velocity

Joshi et al. (1980) considered that there is liquid circulation induced by the motion of bubbles (Figure 5.7b). The liquid circulation patterns are responsible for high heat transfer rates, as in the case of mechanically stirred vessels. The liquid circulation velocity, U_c, in bubble columns is given by

$$U_c = 1.31\left[gD(U - \varepsilon U_{b\infty})\right]^{1/3} \qquad (5.80)$$

The empirical relationships available for heat transfer in mechanically agitated contactors and also for flow-through pipes were used to predict the heat transfer in bubble columns. In the correlation for the Nusselt number for mechanically agitated contactors, the fluid velocity was replaced by the liquid circulation velocity, U_c, to give expression for the Nusselt number in the case of bubble columns.

$$\frac{hD}{k_l} = 0.031\left[\frac{D^{1.33}g^{0.33}(U - \varepsilon U_{b\infty})^{1/3}\rho_l}{\mu_l}\right]^{0.66}\left(\frac{C_p\mu_l}{k_l}\right)^{1/3}\left(\frac{\mu_l}{\mu_w}\right)^{0.14} \qquad (5.81)$$

However, the estimation of bubble velocity, $U_{b\infty}$, is not an easy task. An equation similar to this equation was used to predict the heat transfer coefficient in bubble columns. This equation has a limited scope as it did not predict a constant heat transfer coefficient for $U > 01$ m·s^{-1} (Saxena and Verma 1989). The main difficulty seems to be in estimating the exact amount of energy to be transferred to microscale eddies as the gas velocity increases. Also, at low gas velocities, the term $(U - \varepsilon U_{b\infty})$ became negative.

5.5.4 Modified Boundary-Layer Model

Verma (1989) considered surface renewal mechanisms to be more logical due to vigorous mixing in the bubble column. It was pointed out that whenever a gas bubble comes in the proximity of the wall, a thin liquid layer is formed. In this layer, heat transfer takes place due to conduction only (Figure 5.7c). Therefore, the fraction of the heat transfer rate at these places on the transfer surface, where the air bubble is present, is negligible compared to that at the rest of the transfer surface. A proposed mechanism must take this into consideration. The fraction of the transfer surface not covered by the gas phase may be approximated by $(1-\varepsilon)$. Zehner's model (1986) has this term. Consideration of this term in Deckwer's model also predicts the trend of the heat transfer coefficient in a wide range of air velocities. Introduction of the term $(1-\varepsilon)$ changed the constant from 0.1 to 0.12. The Stanton number, St, is given by

$$St = 0.12(1-\varepsilon)\left(Re\,Fr\,Pr^2\right)^{-0.25} \tag{5.82}$$

5.5.5 Surface Renewal with an Adjacent Liquid Layer

Lewis et al. (1982) proposed a mechanism based on a combination of steady-state conduction through a liquid layer adjacent to a transfer surface and by unsteady-state conduction from the liquid layer to the layer containing eddies. The concept is similar to that of Wasan and Ahluwalia (1969). On eliminating temperature at the outer face of the film, the heat transfer coefficient for a flat plate was found to be

$$h = \frac{h_p}{a^2}\left[erfc(a).\exp\left(a^2\right)-1\right]+h_p \tag{5.83}$$

where

$$a = \frac{2h_f}{\sqrt{\pi}h_p}, h_f = \frac{k_1}{\delta}$$

and

$$h_p = 2\left[\frac{k_1\rho_1C_1}{\pi\theta}\right]$$

Equation 5.83 is the same as Equation 5.82. The contribution of Lewis et al. (1982) to estimate the contact time, θ, as

$$\theta = \frac{L}{U_c'} \tag{5.84}$$

The liquid circulation velocity, U_c', was calculated by the following equation:

$$U_c' = 1.36\left[H_s g\left(U - \varepsilon U_s\right)\right]^{1/3} \tag{5.85}$$

Considering the packet heat transfer resistance in series with film heat transfer resistance:

$$h = \left[\frac{\delta}{k_1} + \left(\frac{\pi L}{4k_1 \rho_1 c_p U_c'}\right)^{1/2}\right]^{-1} \tag{5.86}$$

Considering the laminar sub-layer in turbulent flow, the value of δ can be evaluated as

$$\delta = \frac{5\sqrt{2\mu_1}}{\rho_1 U_c' \sqrt{f}} \tag{5.87}$$

where f is the fanning friction factor. Lewis et al. (1982) have shown that heater dimensions have an effect on the heat transfer coefficient if vertical dimensions are less than 0.08 m. Zehner (1986) used the inter-bubble distance as vertical dimensions. The value used by him was 0.07 m, which is close to the value of the heater dimension used by Lewis et al. (1982). Other models can be used confidently if the length of the transfer surface is larger than 0.07 m, because the flow may not be fully developed for small values of length of the transfer surface. However, it is not a severe limitation because the heat transfer surface in real situations is usually than this value.

5.6 Regions of Interest Adjacent to the Interface

Many phenomena take place around the interface (e.g. the mass transfer between the dispersed phase and the continuous medium, or the frictional force on the dispersed drop, bubble or particle). These processes are governed by the molecular or convective phenomena in the near vicinity of the interface. Sometimes, the processes away from the interface are neglected while developing a model to describe these processes. Here, a couple of examples are chosen to demonstrate the use of physical laws to model a few processes which otherwise are complex.

5.6.1 Terminal Velocity of Bubbles

The gas holdup and gas–liquid mass transfer rates depend on the rise velocity of the bubbles. The bubble velocity in the presence of other bubbles is called the 'swarm velocity'. The terminal velocity of an isolated bubble depends upon the shape of the bubble also because the drag force on the bubble depends upon the bubble shape. There are three different regimes of terminal velocity of an isolated bubble.

Small bubbles are spherical or nearly spherical. The size and hence the terminal velocity are dominated by the viscous and buoyancy forces. The terminal velocity increases with the bubble diameter. In the intermediate region, the bubbles are not spherical. The terminal velocity is dominated by surface tension and inertia forces. The terminal velocity may decrease with the bubble diameter. It may be due to surface contamination, wake structure or oscillating bubble shape. The terminal velocity of very large bubbles is governed by inertial and buoyancy forces. The bubbles are of a spherical-cap shape. The terminal velocity increases with increasing bubble diameter. Estimation of bubble velocity, while taking into account changes in shape and size, is not an easy task. At one stage, it is felt that an empirical correlation should be used for closure of the problem.

5.6.1.1 Terminal Velocity in a Viscosity-Dominated Regime

Bubbles are spherical. The forces acting on the bubble are buoyancy, gravitation and drag. Taking the force balance:

$$(\rho_l - \rho_g)gV_b = (\pi a^2)\frac{1}{2}\rho_l U_{b\infty}^2 f = (\pi R_0^2)\frac{1}{2}\rho_l U_{b\infty}^2 C_D \qquad (5.88)$$

where a is half of the major axis of the bubble, f is the friction factor and C_D is the drag coefficient. The terminal velocity of an isolated bubble is

$$U_{b\infty} = \sqrt{\frac{4}{3}\frac{gd_b(\rho_l - \rho_g)}{C_D \rho_l}} \qquad (5.89)$$

The drag coefficient for the bubble is not equal to that for a rigid sphere. Therefore, an appropriate expression for the drag coefficient is required. The difference is due to the internal circulation in the bubble and the change in the shape of the bubble. The shape of the bubble is also not known. Wallis (1974) gave an expression for C_D to cover a wide range of Reynolds numbers. It takes into account the deformation of the bubble in terms of the Evötös number, Eo.

$$C_D = \max\left[\min\left\{\max\left(\frac{16}{Re},\frac{13.6}{Re^{0.8}}\right),\frac{48}{Re}\right\}, \min\left(\frac{Eo}{3},0.47Eo^{0.25}We^{0.5},\frac{8}{3}\right)\right] \qquad (5.90)$$

The dimensionless number Reynolds, Re; Weber, We; and Evötös, Eo, are defined as follows:

$$Re = \frac{d_e U_{b\infty} \rho_l}{\mu_l} \tag{5.91}$$

$$Eo = \frac{g(\rho_l - \rho_g)d_e^2}{\sigma} \tag{5.92}$$

$$We = \frac{\rho_l U_{b\infty} d_e}{\sigma} \tag{5.93}$$

5.6.2 Gas–Liquid Mass Transfer Coefficient

A gas–liquid interface exists when either gas is dispersed as bubbles in the liquid or liquid is dispersed as droplets in gas. Let us consider the case when gas is dispersed as bubbles. These bubbles move in the liquid, get deformed while moving and may change path in helical or zigzag manners. The mass transfer at the interface is governed by the fluid in the vicinity of the bubble. If the bubbles are small, the flow around the bubble will be laminar and the fluid velocity in close proximity of the interface will be very small. The molecular diffusion will dominate under this condition. For large bubbles moving with large velocity, the flow around the bubble is turbulent, making the convective transport the main mechanism of mass transfer. Different bubble shapes, an absence of universal drag coefficients and the turbulent structure are some of the complexities which make it difficult to obtain a universal expression for a gas–liquid mass transfer coefficient. A few modelling efforts are presented here.

Small bubbles are spherical and behave like rigid spheres. If the fluid around the bubbles is considered to be stagnant, then the mass transfer coefficient, k_L, is given by

$$\frac{k_L d_b}{D_{AB}} = 2 \tag{5.94}$$

The relationship is analogous to that of Equation 3.62.

In a gas–liquid contactor, the bubbles are not stationary and rigid; hence, the mass transfer coefficient is always higher than this. The following semi-empirical equation, Equation 5.95, tries to correct this value (Garner and Suckling 1958):

$$\frac{k_L d_b}{D_{AB}} = 2 + 0.6\,Re^{1/2}\,Sc^{1/3} \tag{5.95}$$

For large bubbles moving at large velocity, it is reasonable to assume that the turbulent flow conditions will exist. For large bubbles, the mass transfer is assumed to take place due to eddies around the interface. The scale of these eddies is very small as compared to the bubble diameter. The turbulent mass transfer is often modelled using a surface renewal mechanism or eddy cell model.

Calderbank and Moo Young (1961) used a dimensionless approach to obtain an expression for the mass transfer coefficient. The Reynolds number was defined using the root mean square value of a fluctuating component. The fluctuating velocity is given by

$$\overline{u_d^2} \propto \left(\frac{P}{V}\right)^{2/3} \left(\frac{d}{\rho_l}\right)^{2/3} \tag{5.96}$$

where $\dfrac{P}{V}$ is the power dissipated per unit volume. In cases of mass transfer at the interface of the bubbles, the following was obtained:

$$\frac{k_L L}{D_{AB}} = 0.42 \left[\frac{\mu_l(\rho_l - \rho_g)g}{\rho_l^2}\right]^{1/3} \left(\frac{\mu_l}{\rho_l D_{AB}}\right)^{1/2} \tag{5.97}$$

Equation 5.97 is valid for the large bubble ($d_b > 2.5$ mm). For small bubbles ($d_b < 1.0$ mm), the following equation was obtained (Linek et al. 2005):

$$\frac{k_L L}{D_{AB}} = 0.31 \left[\frac{\mu_l(\rho_l - \rho_g)g}{\rho_l^2}\right]^{1/3} \left(\frac{\mu_l}{\rho_l D_{AB}}\right)^{2/3} \tag{5.98}$$

Lamont and Scott (1970) proposed a model for a mass transfer coefficient based on the eddy cell model. It was observed that the mass transfer coefficient is due to viscous motion as well as the inertial motion of eddies.

The mass transfer coefficient in terms of rate of energy dissipation, ε, for the fluid surface is given by

$$k_L \propto \left(\frac{\mu_l}{\rho_l D_{AB}}\right)^{-1/2} \left(\frac{\varepsilon \mu_l}{\rho_l}\right)^{1/4} \tag{5.99}$$

The wave number, n, is the number of cycles per unit length. Turbulence energy spectrum, E, as a function of wave number can be written as follows (Hinze 1959):

$$E = 0.45 \varepsilon^{2/3} n^{5/3} \left[1 - \frac{0.6\mu_l}{\rho_l \varepsilon^{1/3}} n^{4/3}\right] \tag{5.100}$$

The mass transfer coefficient is

$$k_L \propto \int_{n_B}^{n_0} D_{AB}^{1/2} \varepsilon^{1/5} n^{1/3} E(n) \, dn \tag{5.101}$$

where n_0 is the wave number at which $E(n) = 0$ and n_B is the wave number corresponding to the bubble dimensions. Substituting Equation 5.100 into Equation 5.101 and integrating, we get

$$k_L = c_1 \left(\frac{\varepsilon \mu_1}{\rho_1} \right)^{1/4} \left(\frac{\mu_1}{\rho_1 D_{AB}} \right)^{-1/2} \tag{5.102}$$

In a bubble column, there is no moving part; hence, the energy dissipated is $\rho g U_b$, and the rate of energy dissipation is $g U_b$. In agitated vessels, the energy is dissipated by the agitator which should be added. It may be determined experimentally.

Higbie's surface renewal model has also been used to describe the mass transfer coefficient:

$$k_L = 1.13 \sqrt{\frac{U_{sl} D_{AB}}{d_b}} \tag{5.103}$$

The proof is similar to the derivation given in Chapter 4. It is known that, after some time, the chemicals accumulate at the interface. As a result, the bubbles start behaving as rigid spheres. The mass transfer coefficient under such conditions is given by

$$k_L = 0.6 \left(\frac{U_{sl}}{d_b} \right) D_{AB}^{2/3} \left(\frac{\mu_1}{\rho_1} \right)^{-1.6} \tag{5.104}$$

5.6.3 Mass Transfer over a Flat Plate: Boundary Layer

The mass transfer coefficient can also be determined based on the concept of a boundary layer. To understand it, let us consider the flow over a flat plate (Figure 5.8). The velocity profile is given by Equation 4.47. Similarly, it can be shown that the concentration profile is given by

$$\frac{C_A}{C_{A\infty}} = \frac{3}{2} \frac{y}{\delta_c} - \frac{1}{2} \left(\frac{y}{\delta_c} \right)^3 \tag{5.105}$$

The equation of continuity of species A is given by

$$v_x \frac{\partial C_A}{\partial x} + v_y \frac{\partial C_A}{\partial y} = D_{AB} \left(\frac{\partial^2 C_A}{\partial x^2} + \frac{\partial^2 C_A}{\partial y^2} \right) \tag{5.106}$$

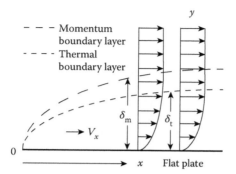

FIGURE 5.8
Momentum and diffusional boundary layer over a flat plate.

Assuming that the convection dominates in the x direction and the diffusion dominates in the y direction, Equation 5.106 can be written as

$$v_x \frac{\partial C_A}{\partial x} = -D_{AB} \frac{\partial^2 C_A}{\partial y^2} \tag{5.107}$$

It is integrated once and after integration to get

$$\frac{\partial}{\partial x} \int_0^{\delta_c} v_x C_A \, dy = -D_{AB} \frac{\partial C_A}{\partial y}\bigg|_{y=0} \tag{5.108}$$

where δ_c is the thickness of the diffusional boundary layer. The equation for δ_c can be obtained as

$$\frac{D_{AB}\rho}{\mu} = \frac{4}{3} \frac{d}{dx}\left(\frac{\delta_c}{\delta}\right)^3 + \left(\frac{\delta_c}{\delta}\right)^3 \tag{5.109}$$

A detailed proof is given elsewhere (Bird 1960). Integrating with boundary conditions

at $x = 0$, $\dfrac{\delta}{\delta_c} = 0$ and substituting Equation 4.46, we get

$$\left(\frac{\delta_c}{\delta}\right)^3 = \left(\frac{D_{AB}\rho_1}{\mu_1}\right) + \frac{C}{x^{-3/4}} \tag{5.110}$$

The term containing x is neglected because even as $x \to \infty$, $\dfrac{\delta}{\delta_c}$ is finite. Therefore:

$$\left(\frac{\delta_c}{\delta}\right)^3 = \frac{D_{AB}\rho_1}{\mu_1} \tag{5.111}$$

Using Equation 4.46 for the value of δ, we get

$$\frac{\delta_c}{x} = 4.64\left(\frac{\mu_1}{x\rho_1 v_\infty}\right)\left(\frac{D_{AB}\rho_1}{\mu_1}\right)^{1/3} \tag{5.112}$$

The mass transfer is determined as

$$k_L = \frac{N_A}{(C_{AW}-0)} = \frac{-D_{AB}\dfrac{\partial C_A}{\partial y}\Big|_{y=0}}{C_{AW}} = \frac{3D_{AB}C_{AW}}{2\delta_c} \tag{5.113}$$

Substitution of the value of δ_c gives the relationship for the mass transfer coefficient for a flat plate:

$$\frac{k_L L}{D_{AB}} = 0.646\left(\frac{L v_\infty \rho_1}{\mu_1}\right)^{1/2}\left(\frac{\mu_1}{\rho_1 D_{AB}}\right)^{1/3} \tag{5.114}$$

5.6.4 Simultaneous Heat and Mass Transfer: Drying of Solids

The models discussed in this chapter consist of model equations which are solved in a sequential manner. The equations of motions are solved first to get the velocity profile. The expression for the velocity is obtained by substituting the velocity profile into the equation of change of thermal energy to get the temperature profile. The concentration profile is obtained by substituting the velocity profile in the equation of continuity for each chemical species. Processes involving phase change (e.g. drying, boiling, melting and crystallisation) cannot be modelled in this way. The temperature and concentration profiles affect each other; hence, the model equations are to be solved simultaneously.

Let us consider drying of an object whose shape can be considered as an infinite flat plate. It is exposed to air (Figure 5.9). Drying is an unsteady-state process because the moisture content changes with time. The heat transfer in the plate takes place due to conduction only. There is no convection term.

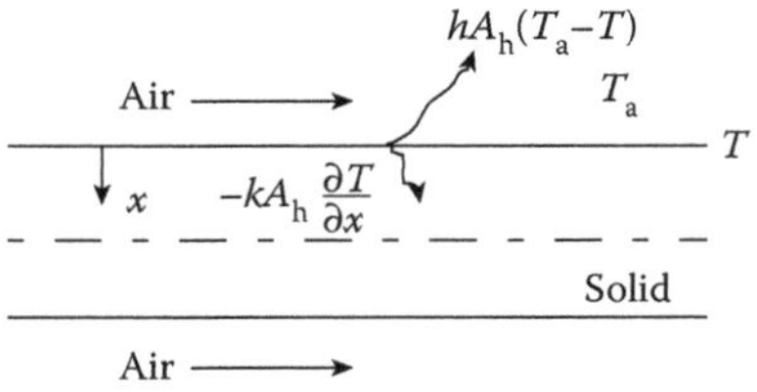

FIGURE 5.9
Drying of an infinite slab containing moisture exposed to air.

The latent heat of vaporisation is to be considered in the generation term. The equation of change of thermal energy under such conditions is given as follows (Kanevce et al. 2002):

$$\rho_s C_{ps} \frac{\partial T}{\partial t} = \frac{\partial}{\partial x}\left(k_s \frac{\partial T}{\partial x}\right) + \alpha\lambda_1 \frac{\partial(\rho_s X_m)}{\partial t} \tag{5.115}$$

The second term on the right-hand side of Equation 5.115 is the generation term. Here, α is the ratio of rate of evaporation of the moisture to the rate of change of moisture content, and X_m and λ_1 are the moisture content on a dry basis and the latent heat of vaporisation, respectively. Because thermal conductivity of the material, k_s, changes with moisture content, a variation of k_s with the distance from the centre of the slab, x, is considered in the spatial derivative.

The rate of change of moisture content is obtained by writing the equation of continuity for moisture. The accumulation is due to diffusion and thermal diffusion.

$$\frac{\partial(\rho_s X_m)}{\partial t} = \frac{\partial}{\partial x}\left(D_{AB}\rho_s \frac{\partial X_m}{\partial x} + D_{AB}\rho_s \zeta \frac{\partial T}{\partial x}\right) \tag{5.116}$$

In the absence of thermal diffusion, neglecting the thermo-gradient coefficient, ζ, we get

$$\frac{\partial(\rho_s X_m)}{\partial t} = \frac{\partial}{\partial x}\left(D_{AB}\rho_s \frac{\partial X_m}{\partial x}\right) \tag{5.117}$$

The shrinkage of the material due to removal of moisture content in the solids results in moving the boundary condition, which was resolved by defining a new variable, $\frac{x}{L}$. However, in the absence of shrinkage, it is not required. Let us assume that the specific volume, v_s, varies linearly with the moisture content:

$$v_s = \frac{1}{\rho_s} = \frac{(1 + \beta X_m)}{\rho_{b0}} \tag{5.118}$$

The density of the fully dried material is ρ_{b0}, and the coefficient of thermal expansion is β. Substituting Equation 5.118 in Equations 5.115 and 5.117, we get

$$\frac{\partial T}{\partial t} = \frac{k_s}{\rho_s C_{ps}} \frac{\partial T^2}{\partial x^2} + \frac{\alpha\lambda_1}{C_{ps}} \frac{\rho_s}{\rho_{b0}} \frac{\partial X_m}{\partial t} \tag{5.119}$$

$$\frac{\partial X_m}{\partial t} = D_{AB} \frac{\rho_{b0}}{\rho_s} \frac{\partial X_m^2}{\partial x^2} + \frac{\rho_{b0}}{\rho_s^2} \frac{\partial(D_{AB}\rho_s)}{\partial x} \frac{\partial X_m}{\partial x} \tag{5.120}$$

The boundary conditions are

$$t = 0, T = T_0, X_m = X_m^0$$

The energy balance at $x = L$ gives

$$\left[-k\frac{\partial T}{\partial x} + h(T_a - T) - (1-\alpha)\lambda k_m (C_a - C)\right]\Bigg|_{x=L} = 0 \qquad (5.121)$$

Equating change in moisture content to mass transfer, we get the following boundary condition:

$$D_{AB}\rho_s \frac{\partial X_m}{\partial x}\Bigg|_{x=L} + k_m (C_a - C) = 0 \qquad (5.122)$$

Equations 5.119 and 5.120 are coupled equations and are to be solved simultaneously. The solutions can be obtained numerically.

5.6.5 Membrane Processes: Model for Pervaporation

'Pervaporation' is a membrane separation technique in which an organic compound having low miscibility with water is separated on the basis of the differences in permeability of the solute and the solvent. The membrane is usually supported on a porous plate. The feed enters on one side of the membrane. The permeate side is kept at low pressure to maintain a high concentration gradient across the membrane. Pervaporation is usually used to separate aroma from food material.

A resistance-in-series model to predict the permeate flux was proposed by Liu et al. (1996). It was assumed that the separation involves the following steps:

1. The compounds are adsorbed on one side of the membrane layer.
2. The adsorbed components diffuse through the membrane and then the membrane support.
3. On the other side of the membrane, the components evaporate due to low pressure and are removed.

The model is based on the mass transfer resistances in series. These resistances are due to convection in the bulk, diffusion in the membrane, diffusion in the membrane support and convection on the other side of the membrane (Figure 5.10).

Let C_{Ab} and C_{Ai} be concentration in the bulk and at the interface on the feed side, respectively. If it is assumed that the convective flux of a solute

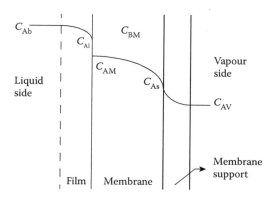

FIGURE 5.10
Mass transfer resistances in a membrane pervaporation.

is negligible and diffusivity in the bulk is constant, Fick's first law can be applied because diffusion is the sole mechanism of mass transfer. The flux on the feed side is

$$J_A = k_L \left(C_{Ab} - C_{Ai} \right) \tag{5.123}$$

Fick's law is applicable within the membrane also; however, due to the high concentration of the species, the diffusion coefficient will depend on the concentration. An empirical expression for diffusivity for a binary system is

$$D_{AM} = D_{AM}^0 \exp\left(\phi_{AA} C_{AM} + \phi_{AB} C_{BM} \right) \tag{5.124}$$

where C_{AM} and C_{BM} are the concentrations of species A and B in the membrane. ϕ_{AA} and ϕ_{BA} are the swelling parameters.

The mass transfer flux within the membrane is given by

$$J_A = -D_{AM} \frac{\partial C_{AM}}{\partial z} = \frac{D_{AM}^0}{\delta_M} \left(C_{AM} - C_{AS} \right) \beta_A \tag{5.125}$$

where:

$$\beta_A = \frac{1}{C_{AM} - C_{AS}} \int_{C_{AM}}^{C_{AS}} \exp\left(\phi_{AA} C_{Az} + \phi_{AB} C_{Bz} \right) dC_{Az} \tag{5.126}$$

Similar expressions for the species B should be written. The concentrations C_{AM} and C_{AS} may be determined from thermodynamic equilibrium. The model of Liu et al. (1996) used the following equations:

$$C_{AM} = \psi_{AM} \frac{p_{AM}}{p_A^0 (T_1)} \tag{5.127}$$

and

$$C_{AS} = \psi_{AS} \frac{p_{AS}}{p_A^0(T_2)} \exp\left[-\frac{V_A}{RT_2}\left\{P_A - p_A^0(T_2)\right\}\right] = \psi_{AS} \frac{p_{AS}}{p_A^0(T_2)} \qquad (5.128)$$

where ψ_{AM} and ψ_{AS} are partition coefficients. The exponential term is close to 1 and can be neglected. The mass transfer flux of A and B is

$$J_A = \frac{D_{AM}^0}{\delta_M}\left(\psi_{AM}\frac{p_{AM}}{p_A^0(T_1)} - \psi_{AS}\frac{p_{AS}}{p_A^0(T_2)}\right)\beta_A \qquad (5.129)$$

$$J_B = \frac{D_{BM}^0}{\delta_M}\left(\psi_{BM}\frac{p_{BM}}{p_B^0(T_1)} - \psi_{BS}\frac{p_{BS}}{p_B^0(T_2)}\right)\beta_B \qquad (5.130)$$

The mass transfer flux for A and B through the membrane support layer is given by

$$J_A = \frac{D_{AS}}{\delta_S RT}(p_{AS} - p_{AV}) = \frac{D_{AS}}{\delta_S RT}\left(p_{AS} - P_V\frac{J_A}{J_A + J_B}\right) \qquad (5.131)$$

$$J_B = \frac{D_{BS}}{\delta_S RT}\left(p_{BS} - P_V\frac{J_B}{J_A + J_B}\right) \qquad (5.132)$$

If the porosity and tortuosity of the membrane support are ε and τ, respectively, the diffusivity of A in the membrane support, D_{AS}, can be expressed in terms of the molecular diffusivity, D_{AB}, and the Knudsen diffusion coefficients, D_{AK}:

$$\frac{1}{D_{AV}} = \frac{\tau}{D_{AB}\varepsilon} + \frac{1}{D_{AK}} \qquad (5.133)$$

There are many correlations with which to determine the molecular diffusion coefficients. The Knudsen diffusion coefficients can be determined as

$$D_{AK} = 97r_p\sqrt{\frac{T_2}{M_A}} \qquad (5.134)$$

The feed-side mass transfer resistance is neglected.
The energy balance gives

$$J_A\lambda_A + J_B\lambda_B = (J_A C_{pA} + J_B C_{pB})(T_M - T) - k_1\frac{dT}{dz} \qquad (5.135)$$

It is solved to give

$$T_2 = T_1 - \left(J_A \lambda_A + J_B \lambda_B\right) \frac{\delta_M}{k_1} \frac{1 - e^{-\theta}}{\theta} \tag{5.136}$$

Equations 5.129, 5.130, 5.133, 5.135 and 5.136 can be solved to get the values of J_A, J_B, p_{AV}, p_{BV} and T_2. However, it was felt that it is difficult to solve these equations. These equations were further simplified by making more assumptions, but this limited the scope of the model.

For low permeate rates, $\theta = 1$. Equation 5.136 is simplified as

$$T_2 = T_1 - \left(J_A \lambda_A + J_B \lambda_B\right) \frac{\delta_M}{k_1} \tag{5.137}$$

The permeability of A and B is defined as $L_A = \psi_A D^0_{AM} \beta_A$ and $L_B = \psi_B D^0_{BM} \beta_B$. p_{AM} and p_{BM} were eliminated to give

$$J_B = \left(\frac{1}{\dfrac{\delta_M}{L_B} + \dfrac{1}{p^0_B k_{BV}}}\right)\left(1 - \frac{P_V}{p^0_B} \frac{J_B}{J_A + J_B}\right) \tag{5.138}$$

The mass transfer coefficients in the membrane support layer are

$$k_{AV} = \frac{D_{AV}}{\delta_S R T_2} \quad \text{and} \quad k_{BV} = \frac{D_{BV}}{\delta_S R T_2}$$

5.7 More Than One Mechanism of Heat Transfer: Flat-Plate Solar Collector

Flat-plate solar collectors are used to transform solar radiation into thermal energy. They comprise a non-concentrating non-tracking type (i.e. the solar radiation falls on a flat surface which remains stationary). The top cover consists of one (or, more frequently, two) glass covers. A schematic diagram of the solar collector is given in Figure 5.11a. The solar radiation, after passing through the glass covers, falls on an absorber plate where it is absorbed. The sides and bottom of the collector are exposed to the ambient temperature. The mechanism of heat transfer is radiation and convection. The radiation cannot be neglected as it is the main mechanism of the solar collector. Figure 5.11b shows the various heat transfer resistances in a flat-plate solar collector as described below (Hseih 1986).

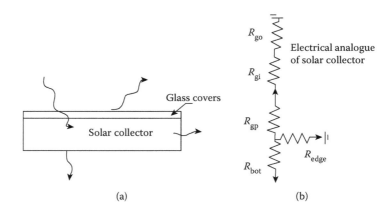

FIGURE 5.11
(a) Flat-plate solar collector and (b) electrical analogue of heat transfer resistances.

Let the temperature of ambient conditions, absorber plate, inside glass cover, outside glass cover and sky be T_a, T_p, T_{gi}, T_{go} and T_s, respectively. At steady state, the heat lost to the ambient air from the top cover is the sum of that by convection, h_{cgi}, and radiation, h_{cgi}.

$$q = A\left(h_{cgo} + h_{rgo}\right)\left(T_{g1} - T_a\right) \tag{5.139}$$

The radiative heat transfer between the top glass cover and the ambient air is given by

$$h_{rgo} = \frac{\sigma\left(T_{go}^4 - T_s^4\right)}{\dfrac{1}{\varepsilon_{go}}} = \varepsilon_{go}\sigma\left(T_{go} + T_s\right)\left(T_{go}^2 + T_s^2\right)\frac{\left(T_{go} - T_s\right)}{\left(T_{go} - T_a\right)} \tag{5.140}$$

The area of the cover is cancelled because it is present on both sides. The radiative heat transfer takes place between the glass top cover and sky, and the convective heat transfer takes place between the top glass cover and the surrounding air. The infrared emissivities of the top and lower glass covers and absorber plate are ε_{go}, ε_{gi} and ε_p, respectively.

For convective heat transfer, a suitable empirical correlation may be used. Kumar and Mullick (2010) obtained the following relationship for h_{cgo}:

$$h_{cgo} = 6.9 + 3.87 V_w \quad \text{for} \quad V_W \le 1.12 \text{ m} \cdot \text{s}^{-1} \tag{5.141}$$

The heat loss from the top cover is the sum of the radiative and convective terms:

$$q = A\left(h_{cgo} + h_{rgo}\right)\left(T_{go} - T_a\right) = \frac{\left(T_{go} - T_a\right)}{R_{go}} \tag{5.142}$$

Therefore, the heat transfer resistance between the top glass cover and ambient condition, R_{go}, is

$$R_{go} = \frac{1}{A\left(h_{cgo} + h_{rgo}\right)} \tag{5.143}$$

The heat transfer resistances between the inner and outer glass plates, R_{gi}, and the absorber plate and inner glass plate, R_{gp}, are

$$R_{gi} = \frac{1}{A\left(h_{cgi} + h_{rgi}\right)} \tag{5.144}$$

$$R_p = \frac{1}{A\left(h_{cp} + h_{rp}\right)} \tag{5.145}$$

The convective heat transfer coefficients between the inner and outer glass plates and between the absorber plate and inner glass plate are h_{cgi} and h_{cp}, respectively. The h_{rgi} and h_{rp} are heat transfer coefficients between the inner and outer glass plates and between the absorber plate and inner glass plate, respectively. The radiative heat transfer coefficients are

$$h_{rgi} = \frac{\sigma\left(T_{go} + T_{gi}\right)\left(T_{go}^2 + T_{gi}^2\right)}{\dfrac{1}{\varepsilon_{go}} + \dfrac{1}{\varepsilon_{gi}} - 1} \tag{5.146}$$

$$h_{rp} = \frac{\sigma\left(T_p + T_{gi}\right)\left(T_p^2 + T_{gi}^2\right)}{\dfrac{1}{\varepsilon_p} + \dfrac{1}{\varepsilon_{gi}} - 1} \tag{5.147}$$

The convective heat transfer coefficients can be determined using the following equations (Hollands et al. 1976) or any other suitable equations:

$$\frac{h_{cgi}L}{k_g} = 1 + 1.44\left(1 - \frac{1708}{Ra\cos\alpha}\right)\left(1 - \frac{1708(\sin 1.8s)^{1.6}}{Ra\cos\alpha}\right)$$
$$+ \left[\left(\frac{Ra\cos\alpha}{5830}\right)^{1/3} - 1\right]; \ Ra\cos\alpha > 5830$$

$$\frac{h_{cgi}L}{k_g} = 1 + 1.44\left(1 - \frac{1708}{Ra\cos\alpha}\right)\left(1 - \frac{1708(\sin 1.8s)^{1.6}}{Ra\cos\alpha}\right); \ Ra\cos\alpha \leq 5830$$

$$\frac{h_{cgi}L}{k_g} = 1; \ Ra \cos \alpha \le 1708 \tag{5.148}$$

where:

$$Ra = \frac{2gL^3(T_1 - T_2)\rho^2}{\mu^2(T_1 + T_2)}\left(\frac{C_p\mu}{k}\right)$$

L = distance between two plates
s = tilt angle

Heat transfer resistance between the absorber plate and ambient is in series and hence is added to give the following:

$$R_{top} = R_{go} + R_{gi} + R_p \tag{5.149}$$

Therefore, the heat loss from the top, q_{top}, is given by

$$q_{top} = U_{top}A(T_p - T_a) = \frac{(T_p - T_a)}{R_{top}} \tag{5.150}$$

The energy loss from the bottom is due to two heat transfer resistances. The first is due to the conduction in the insulation, and the second is due to a combination of conduction, convection and radiation in the bottom plate. Due to the low temperature of the bottom, the radiation is neglected. The heat transfer resistance in the bottom is therefore

$$R_{bot} = \frac{\delta_{ins}}{k_{ins}A} + \frac{1}{h_{cb}} = U_{bot}A \tag{5.151}$$

The bottom plate faces downwards. The resistance to heat transfer is small in comparison to the convective term, and it is often neglected.

The losses from the edges are quite small and may be neglected. A relationship for the heat transfer coefficient for natural convection at a vertical wall may also be used. A value of overall heat transfer coefficient of 0.5 $W \cdot m^{-2}$ K can be multiplied to the perimeter of the solar collector to get the edge losses, q_{edge}. In terms of the solar collector area, A,

$$U_{edge}A = 0.5A_{Peri} = \frac{1}{R_{edge}} \tag{5.152}$$

Now, we have all the required equations to estimate the heat resistance at the top, bottom and edge of the solar collector. All these resistances are acting in parallel and hence can be added in an analogous way as the electrical resistances are added to give the overall resistance, R_{ov}.

$$\frac{1}{R_{ov}} = \left(\frac{1}{R_{top}} + \frac{1}{R_{bot}} + \frac{1}{R_{edge}} \right) \tag{5.153}$$

In terms of heat transfer coefficients, we get

$$U_{ov} + U_{top} + U_{bot} + U_{edge} \tag{5.154}$$

Energy balance gives the model equation for solar collector:

$$\rho_1 C_p \frac{dT_p}{dt} = Q.A - U_{ov} A \left(T_p - T_a \right) \tag{5.155}$$

It may be noted that the plate temperature, T_p; ambient temperature, T_a; and solar radiation, Q, are functions of time. The equation may be integrated numerically using quasi–steady state.

5.8 Introducing Other Effects in Laws of Conservation

Equations obtained using the laws of conservation of momentum, mass and energy are used in a large number of situations. However, many processes involve phenomena not considered while deriving these equations. Writing shell balances for every problem is an attractive option if one realises that these equations can be used if appropriate terms are added to these. The most common situation is encountered when a reaction is carried out. In the case of reaction, the equation of motion should include the volume change if it exists. The equation of continuity for chemical species should consider the rate of reaction. The equation of change of thermal energy should include the heat effects (i.e. the heat of reaction). A few examples for consideration of reactions have been presented in Chapter 4. Here, the effects of reaction and electrokinetic potential on the transfer process are discussed.

5.8.1 Reactions

In Chapter 3, a few simple problems involving reaction and diffusion were discussed. In the case of isothermal reactions, the equation of continuity and the equation of change of momentum are solved independently. Application of laws of conservation in the presence of reaction should be dealt with differently. The equation of continuity for a species has a generation term which consists of the rate of reaction. If the reaction is exothermic or endothermic, then the rate of reaction is a function of temperature. Similarly, the equation of change of thermal energy has a generation term which involves the heat of reaction. But the heat generated depends upon the rate of reaction. Both of these equations are, therefore, coupled and hence are to be solved simultaneously. The hydrodynamics of the reactor depend upon the type of the reactor and the flow rates. These aspects are discussed in more detail in Chapter 6.

5.8.2 Electrokinetic Phenomena: Flow in Microchannels

Laws of conservation can also be used to describe the flow and transfer processes under the influence of electrical charge. Let us consider the flow of a fluid in a microchannel. It can be considered as a flow between two parallel plates. Electrokinetic phenomena can play an important role if the distance between the two plates is less and the ionic concentration is small. If E_y is the induced electric field, also called the 'electrokinetic potential', between the plates, ρ_e is the net charge density per unit volume, and the net body force is $E_y\rho_e$. The equation of change for the momentum includes additional body forces similar to the generation terms in the equation of change of thermal energy or the equation of continuity for each chemical species. For steady-state, 1D flow of an incompressible flow of Newtonian fluid, the equation of change for the momentum can be written as (Ren and Li 2005)

$$\frac{\partial p}{\partial z} - \mu \frac{\partial^2 v_z}{\partial z^2} + E_y\rho_e = 0 \tag{5.156}$$

The inertial term, $v_z \dfrac{\partial v_z}{\partial z}$, is zero as given by the equation of continuity. The electrokinetic potential can be determined in terms of the electric current. Transport of the net charge due to flow of the liquid is known as the streaming current, I_s. It can be written as

$$I_s = 2\int_0^H v_z\rho_e \, dx \tag{5.157}$$

A current in the direction opposite to the fluid is induced due to the flow. It is called the 'conduction current', I_c, and is given by

$$I_c = 2E_y\lambda H \tag{5.158}$$

At steady state, the net current should be zero:

$$I_c + I_s = 0 \tag{5.159}$$

Substitution of these values in Equation 5.156 gives

$$\frac{\partial^2 v_z}{\partial z^2} = \frac{\partial p}{\partial z} + \frac{1}{\mu} \frac{\rho_c}{\lambda H} \int_0^H v_z \rho_e \, dx \qquad (5.160)$$

The boundary conditions are as follows:

At the centre (i.e. at $x = 0$): $\dfrac{dv_z}{dx} = 0$

At the wall (i.e. $x = H$): $v_z = 0$

Ren and Li (2005) provided differential equations to determine ρ_e as a function of x, which is solved first before numerically solving Equation 5.160.

5.9 Summary

The multiphase systems involve several phases. The laws of conservation of momentum, mass and energy are applicable in a continuous phase. It cannot be applied across the interface. The involvement of interface requires the definition of transfer area and the processes across the interface. These are specified by assuming that both phases are at equilibrium at the interface.

In general, a dispersed phase and a continuous phase are present in multiphase systems. The dispersed phase may be stationary as in the case of packed beds. It may be moving (e.g. bubble columns and fluidised beds). A few models to describe the heat transfer in fluidised beds were discussed. These were based on simple models (e.g. film theory and surface renewal model) discussed in Chapters 3 and 4. Complexities to the model were added by considering another gas film or by combining different concepts, thus giving importance to more than one mechanism (e.g. fluid convection and particle convection). The model equations can be obtained by using the equations based on laws of conservation of momentum, mass and energy. The development of various models is the result of attempts to understand various mechanisms and their contributions to the process. Many models provided analytical solutions. But consideration of phenomena in more detail requires the application of numerical techniques to solve the model equations.

The models based on surface renewal and the boundary layer to explain the heat transfer in bubble columns exhibit the simplicity and importance of these approaches. Certain phenomena related to the dispersed phase are governed by the processes occurring in a region adjacent to the interface of the dispersed phase. The drag force is due to the friction at the surface. It affects the velocity of the dispersed phase (e.g. bubbles, drops and particles). The mass transfer takes place across the interface, and hence the region of interest is near the interface. Processes such as drying require simultaneous solution of the equation of change. The membrane separation processes can be modelled by resistance in series, while the solar flat-plate collector is an example of resistance in parallel.

References

Adams, R.L., and Welty, J.R. 1979. A gas convection model of heat transfer in large particle fluidized beds. *AIChE J.* 25(3): 395–404.

Al-Bosoul, M.A. 2002. Bed-to-surface heat transfer in circulating fluidized beds. *Heat Mass Trans.* 38: 295–299.

Baskakov, A.P., and Suprun, V.M. 1972. Determination of the convective component of the heat transfer coefficient to a gas in a fluidized bed. *Int. Chem. Eng.* 12: 324–326.

Bird, R.B., Stewart, W.E., and Lightfoot, E.N. 1960. *Transport Phenomena*. New York: John Wiley.

Botterill, J.S.M. 1975. *Fluid-Bed Heat Transfer*. New York: Academic Press.

Calderbank, P.H., and Moo-Young, M.B. 1961. The continuous phase heat and mass-transfer properties of dispersions. *Chem. Eng. Sci.* 16: 39–54.

Cova, D.R. 1966. Catalyst suspension in gas-agitated tubular reactors. *Ind. Eng. Chem. Proc. Des. Dev.* 5(1): 20–25.

Deckwer, W.D. 1980. On the mechanism of heat transfer in bubble column reactors. *Chem. Eng. Sci.* 35: 1341–1346.

Denloye, A.O.O., and Botterill, J.S.M. 1978. Bed-to-surface heat transfer in a fluidized bed of large particles, *Powder Technol.* 19: 197–203.

Dixon, A.G., Arias, J., and Willey, J. 2003. Wall-to-liquid mass transfer in fixed beds at low flow rates. *Chem. Eng. Sci.* 58: 1847–1857.

Dow, W.M., and Jacob, M. 1951. *Heat transfer* between a vertical tube and a fluidized air-solid flow. *Chem. Eng. Prog. Symp. Ser.* 47: 637–648.

Farkas, E.J., and Leblond, P.F. 1969. Solids concentration profile in the bubble column slurry reactor. *Can. J. Chem. Eng.* 47: 215–218.

Ganzha, V.L., Upadhyay, S.N., and Saxena, S.C. 1982. A mechanistic theory for heat transfer between fluidized bed of large particle and immersed surface. *Int. J Heat Mass Transfer.* 25(10): 1531–1540.

Garner, F.H., and Suckling, R.D. 1958. Mass transfer from a soluble solid sphere. *AIChE J.* 4(1): 114–124.

Gunn, D.J., and Khalid, M. 1975. Thermal dispersion and wall heat transfer in packed beds. *Chem. Eng. Sci.* 30: 261.

Hinze, J.O. 1959. *Turbulence*. New York: McGraw-Hill.

Hollands, K.G.T., Unny, T.E., Raithby, G.D., and Konicek, L. 1976. Free convective heat transfer across inclined air layers. *J. Heat Trans.* 98: 189–193.

Hseih, J.S. 1986. *Solar Energy Engineering*. Englewood Cliffs, NJ: Prentice Hall.

Huang, D., and Levy, E. 2004. Heat transfer to fine powders in a bubbling fluidized bed with sound assistance. *AIChE J.* 50(2): 302–310.

Imafuku, K., Wang, T.Y., Koide, K., and Kubota, H. 1968. The behaviour of suspended solid particles in the bubble column. *J. Chem. Eng. Jpn.* 1(2): 153–158.

Joshi, J.B., Sharma, M.M., Shah, Y.T., Singh, C.P.P., Ally, M., and Klinzing, G.E. 1980. Heat transfer in multiphase contactors. *Chem. Eng. Commun.* 6: 257–271.

Kanevce, G.H., Kanevce, L.P., Mitrevski, V.B., Duhkravich, G.S., and Orlande, H.R.B. 2002. Inverse approaches to drying of thin bodies with significant shrinkage effects. *J. Heat Transfer.* 129: 379–385.

Kast, W. 1962. Analyse des warmeubergangs in blasensaulen. *Int. J. Heat Mass Transfer.* 5: 329–336.

Kato, Y., Morooka, S., Kago, T., Saruwatari, T., and Yang, S. 1985. Axial hold-up distribution of gas and solid particles in three-phase fluidized bed for gas-liquid(slurry)-solid systems. *J. Chem. Eng. Jpn.* 18(4): 308–312.

Kojima, H., Anjyo, H., and Mochizaki, Y. 1986. Axial mixing in bubble column with suspended particles. *J. Chem. Eng. Jpn.* 19(3): 232–236.

Kojima, H., and Asano, K. 1981. Hydrodynamic characteristics of a suspension-bubble column. *Int. Chem. Eng.* 21(3): 473–481.

Kumar, S., and Mullick, S.C. 2010. Wind heat transfer coefficient in solar collectors in outdoor conditions. *Solar Energy.* 84: 956–963.

Laguerre, O., Amara, S.B., and Flick, D. 2006. Heat transfer between wall and packed bed crossed by low velocity airflow. *Appl. Therm. Eng.* 26: 1951–1960.

Lamont, J.C., and Scott, D.S. 1970. An eddy cell model of mass transfer into surface of a turbulent liquid. *AIChE J.* 16(1970): 513–519.

Leva, M., Weintrub, M., and Grummer, M. 1949. Heat transmission through fluidized beds of fine particles. *Chem. Eng. Progr.* 45(9): 563–572.

Levenspiel, O., and Walton, J.S. 1954. Wall-bed heat transfer in fluidized systems. *Chem. Eng. Progr. Symp. Ser.* 9(50): 1–7.

Lewis, D.A., Field, R.W., Xavier, A.M., and Edwards, D. 1982. Heat transfer in bubble columns. *Trans. Instn. Chem. Engrs.* 60: 40–47.

Linek, V., Kordac, M., and Moucha, T. 2005. Mechanism of mass transfer from bubbles in dispersions Part II: Mass transfer coefficients in stirred gas–liquid reactor and bubble column. *Chem. Eng. Proc.* 44: 121–130.

Liu, M.G., Dickson, J.M., and Cote, P. 1996. Simulation of pervaporation system in the industrial scale for water treatment part i: Extended resistance-in series model. *J. Membr. Sci.* 111: 227–241.

Mickley, H.S., and Trilling, C.A. 1949. Heat transfer characteristics of fluidized beds. *Ind. Eng. Chem.* 41(6): 1135–1147.

Mickley, H.S., and Fairbanks, D.F. 1955. Mechanism of heat transfer to fluidized bed. *AIChE J.* 1(3): 374–384.

Moreira, M.F.P., Frerreire, M.C., and Freire, J.T. 2006. Evaluation of pseudohomogeneous models for heat transfer in packed beds with gas flow and gas–liquid cocurrent downflow and upflow. *Chem. Eng. Sci.* 61: 2056–2068.

Murray, P., and Fan, L.S. 1989. Axial solid distribution in slurry bubble column. *Ind. Eng. Chem.* 28: 1697–1703.

Patel, R.D., and Simpson, J.M. 1977. Heat transfer in aggregative and particulate liquid-fluidized beds. *Chem. Eng. Sci.* 32(1): 67–74.

Ren, C.L., and Li, D. 2005. Improved understanding of the effect of electrical double layer on pressure-driven flow in microchannels. *Anal. Chim. Acta.* 531: 15–23.

Saxena, S.C. 1989. Heat transfer between immersed surfaces and gas-fluidized beds. In *Advances in Heat Transfer*, vol. 19, ed. J.P. Harttnet and T.F. Irvine, 97–190. San Diego, CA: Academic Press.

Smith, D.N., and Ruether, J.A. 1985. Dispersed dynamics in a slurry bubble column. *Chem. Eng. Sci.* 40(5): 741–754.

Suganama, T., and Yamanishi, T. 1966. Behaviour of solid particles in bubble columns. *Kogaku Kogaku.* 30: 1136–1140.

Verma, A.K. 1989. Heat transfer mechanism in bubble columns. *Chem. Eng. J.* 42: 205–208.

Wallis, G.B. 1974. The terminal speed of single drops or bubbles in an infinite medium. *Int. J. Multiphase Flow.* 1: 491–511.

Wasan, T., and Ahluwalia, M.S. 1969. Consecutive film and surface renewal for heat or mass transfer from a wall. *Chem. Eng. Sci.* 24: 1535–1542.

Washmund, B., and Smith, J.W. 1967. Wall to fluid heat transfer in liquid fluidized beds: Part 2. *Can. J. Chem. Eng.* 45(3): 156–165.

Yamanaka, Y., Sekizawa, T., and Kubota, H. 1970. Age distribution of suspended solid particles in a bubble column. *J. Chem. Eng. Jpn.* 3(2): 264–266.

Zabrodsky, S.S., Epanov, Yu.G., Galershtein, D.M., Saxena, S.C., and Kolar, A.K. 1981. Heat transfer in a large particle fluidized bed with immersed in-line and staggered bundles of horizontal smooth tubes. *Int. J. Heat Mass Transfer.* 24(4): 571–579.

Zehner, P. 1986. Momentum, mass and heat transfer in bubble columns. Part 2. Axial blending and heat transfer. *Int. Chem. Eng.* 26(1): 29–35.

Zhang, K. 2002. Axial solid concentration distribution in tapered and cylindrical bubble columns. *Chem. Eng. J.* 86: 229–307.

Zhi, Q., and Kai, G. 2009. Modeling and kinetic study on absorption of CO_2 by aqueous solutions of N-methyldiethanolamine in a modified wetted wall column. *Chin. J. Chem. Eng.* 17(4): 571–579.

Ziegler, E.N., and Brazelton, W.T. 1964. Mechanism of Heat Transfer to a Fixed Surface in a Fluidized Bed. *Ind. Eng. Chem. Fundamentals.* 3(2): 94–98.

6

Multiphase Systems with Reaction

Multiphase reactors (e.g. packed beds, emulsion phase reactors, slurry bubble columns and two- and three-phase fluidised beds) are used to carry out various chemical and biochemical reactions. These reactors also find use in pollution control equipment. The models for multiphase systems without chemical reactions can be developed using laws of conservation.

In the absence of chemical reactions, models are developed to understand the hydrodynamics and transfer processes in a process. The hydrodynamics may include pressure drop, motion of dispersed systems, heat transfer and mass transfer either between dispersed phase and the continuous phase or between the external surface and the bulk. It is easy to decide whether the region around the interface is important or the region away from the interface is important from the knowledge of location of the reaction. A few of these models based on the laws of conservation for hydrodynamics and transfer processes in multiphase systems are presented in Chapter 5. In the absence of reaction, analytical solutions to several models are available. It is extremely difficult to obtain an analytical solution when the differential equations are coupled. The existence of analytical solutions depends upon the degree of simplification, geometry and boundary conditions.

In the presence of chemical reactions, model equations are obtained by introducing generation terms. The model equations are non-linear for most of the reactions. A general analytical solution applicable to several reactions is not possible. Analytical solutions to the reactors carrying out first-order and pseudo-first-order reactions are available. These analytical solutions are the basis for developing measurement techniques to determine the mass transfer coefficient at the interface in dispersed systems.

The models for multiphase reactors require kinetic rate expressions. The models reported in the literature have been discussed to understand a systematic procedure to obtain model equations.

In a multiphase system, the reaction takes place in one or more of the phases. The rate of reaction depends upon the concentration of the reactants and the temperature in the phase in which the reaction takes place. The temperature depends upon the heat of the reaction, which depends upon the rate of reaction, the term present in the equation of continuity for the species. One of the reactants may be present in another phase, in which case the reactant has to diffuse in the first phase, cross the interface and diffuse in the second phase. In the case of solid-catalysed reactions, the reactants diffuse to the surface of the catalyst where the ions are absorbed. After the reaction, the products diffuse back to the bulk. Thus, the diffusion in different phases, the equilibrium at the interface and adsorption at the solids are considered in models for multiphase systems. Models for two- and three-phase reactors are chosen not only to illustrate various types of simplification during development of a model but also to understand the methodology to consider various aspects of kinetics in the model equations based on laws of conservation of momentum, mass and energy.

Due to a wide variety of rate expressions, the use of numerical techniques to obtain a solution of model equations is unavoidable. Since the algorithm or computer programme to solve a reaction can be used to solve model equations for another reaction by changing the rate expression, the development of a numerical technique to solve model solutions is less time consuming if the model has already been developed for one reaction.

6.1 Development of a Model for Multiphase Reactors: Common Assumptions and Methodology

No two models for multiphase reactors look the same. This is due to differences in the chemical reactions, types of the contacting devices, regions of operation of these devices and degrees of complexity or simplicity considered. Therefore, the model development requires an understanding of hydrodynamics, transfer processes and reaction kinetics. Several models for multiphase reactors are available in the literature. They have been developed by expert investigators. A methodology to develop a model for multiphase reactors can be obtained from these models.

Commonly used assumptions to simplify the model equations for multiphase reactors are made with an aim to eliminate some of the terms in equation of change for momentum, mass and thermal energy. These assumptions are discussed in the following section.

6.1.1 One-Dimensional and Two-Dimensional Models

The models can be classified based on an assumption to simplify the process. In general, packed beds, fluidised beds, bubble columns, agitated vessels etc. are cylindrical. The spatial variation of any property or parameter may be in the axial and radial directions. The axial variation of the parameter will always exist because of the depletion of reactant due to reaction. The radial variation may not be desirable. If the diameter of the packed bed is small, the fluid velocity may not be flat. It produces spatial variation of the convective transport of the reactant and products to and from the solids. The rate of reaction also exhibits spatial variation. The catalyst deactivation may show radial variation, which is not desirable.

Proper design of a distributor at the fluid inlet may result in an almost flat velocity profile except very near to the wall. In the case of isothermal reactors, in the absence of radial variation of temperature and convective transport, the parameters will vary in the axial direction only. Models developed for such cases are called 'one-dimensional (1D) models'. If the radial variation of the parameters is also considered, the model is known as a 'two-dimensional (2D) model'.

6.1.2 Homogeneous and Heterogeneous Models

In Chapter 5, the assumption of continuum was introduced. In solid-catalysed chemical reactions, the reaction takes part at the surface of the solid catalyst. However, the region of interest is not only around the catalyst, where the reaction takes place, but also in the bulk which controls the transport of reactants and products. The assumption of continuum will be oversimplification. Such models are known as homogeneous models. However, it is not possible to study the interaction between the two phases using homogenous models. To understand the inter-phase interaction, it is necessary to consider both phases separately. The models having different model equations for different phases are called heterogeneous models.

6.1.3 Two- and Three-Fluid Models

When two phases are considered separately, the laws of conservation have to be written for every phase. One set of equations is obtained by applying laws of conservation for the continuous phase. But in cases of the dispersed phase, each individual drop, bubble or solid is a phase. Because the number of dispersed units is quite large, writing different laws of conservation of each unit and then solving them are out of the question except when one decides to use computational fluid dynamics (CFD) software. Many investigators used an entirely different approach and developed two- or three-fluid models in cases of two and three-phase reactors, respectively.

To understand these models, let us consider a bubble column operated in batch mode (Figure 6.1). The radial distribution of bubbles is not important. Imagine that all the bubbles are on one side of the column. They are so close that they may be treated as only gas phase (Figure 6.1b). This is the basis of the two-fluid model. In a homogeneous model, gas–liquid dispersion is assumed to behave as a single fluid. Although the reaction takes place at the interface, defining the interfacial area per unit volume of the dispersion is enough to avoid several details regarding bubble behaviour. However, it is now required to use the properties such as viscosity and thermal conductivity of the dispersion, and not that of the fluid. Since the relationships used for the estimation of properties for the dispersion using the properties of its constituents do not involve differential equations, homogeneous models are simpler than heterogeneous models for multiphase systems.

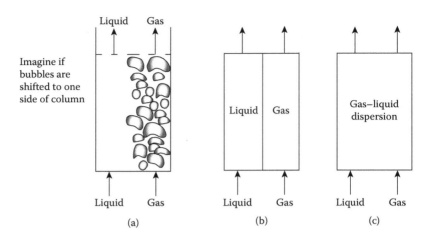

FIGURE 6.1
Bubble column: (a) an imaginary arrangement of bubbles in a bubble column, (b) a two-fluid model and (c) a homogeneous model.

In the heterogeneous models, the liquid and gases are considered separate phases. All the bubbles are assumed to be a single phase, even though they are separate identities (Figure 6.1). This model sketch is slightly misleading. The model sketch looks as if one phase is on one side and the other phase is on the other side, whereas both the phases are assumed to be present at all locations. This type of model sketch helps in writing model equations. Several characteristics of the dispersed system are retained by the simplification given in the model sketch. A few of these are as follows:

1. Axial variations of gas holdup can be considered.

2. The laws of conservation can be written by considering axial variations of concentration and temperature.

3. The bubble velocity can be incorporated in the model because it is related to superficial gas velocity by the relation $U_b = \dfrac{v_{zg}}{\varepsilon_g}$.

4. The transfer processes between the two phases can also be considered.

5. The assumption of plug flow or an axial dispersion model can be made if there is net fluid flow in the axial direction (e.g. the gas velocity is in the upward direction).

6. If one of the fluids is in batch mode, then a well-mixed condition can also be considered.

7. The bubble coalescence and bubble breakup can be incorporated into the model because these depend mainly on the axial position.

In general, axial variations of any phenomena occurring in not only the continuous phase but also the dispersed phase can be modelled. The only limitation of the two- or three-phase fluid models is that no radial variation of properties can be included in the model. Fortunately, in most of the multiphase reactors, such a detail is not expected from a model. The rigorous models providing axial as well as radial variations can be developed using CFD.

6.1.4 Thermodynamics, Kinetics, Hydrodynamics and Reactor Model

The modelling of multiphase reactors involves obtaining equations to describe the spatial and temporal variations of concentration of each species and temperature profiles. Although the velocity profiles can also be determined, it is generally observed that due to the presence of mass transfer resistances, the processes become important in the region adjacent to the interface. Therefore, the velocity profile in the liquid or gas phase does not dominate the chemical conversion. The fluid velocity influences the convective transport of the reactants and products indirectly. The model equations obtained using laws of conservation of momentum, mass and thermal energy are termed 'reactor models'.

The transport of molecules or ions across the interface is governed by the thermodynamic properties of the mixture and called 'thermodynamic models'. These may be in the form of equilibrium relations (e.g. gas–liquid equilibrium, liquid–liquid equilibrium and adsorption isotherms).

The mass and heat transfer rates depend on the hydrodynamics of the reactor. The holdup and the average velocity of the phases affect the transfer coefficients. These are obtained using either the simple models discussed earlier or empirical correlations.

The phase in which the reaction takes place uses the equation of continuity for a particular species involving the reaction rate as generation term. Several types of kinetic expressions for various chemical reactions are available. It may be a simple kinetic rate expression or a complex kinetic rate expression.

A set of equations obtained from these three models forms a complete set of model equations. More equations related to heat exchange can be added to this set of equations. The model equations obtained from the consideration of hydrodynamics, thermodynamics and kinetics are added to model equations for reactor models. All these aspects are usually studied separately and are sub-models of a reactor model. Different combinations of these generate a variety of models that look very different from each other but still follow a common approach.

The models for multiphase reactors make use of the assumptions discussed above. The critical assumption can only be determined after sensitivity analysis, which will be discussed in Chapter 8.

6.1.5 Methodology for Model Development for Multiphase Systems

A methodology for development of a model consists of the following steps:

1. Identify the following:
   a. The number of phases
   b. The presence of various chemical species in each phase
   c. The phase in which each reaction takes place

   Different reactions may take place in different or the same phases. The above information is helpful in understanding what is happening in each phase.

2. Identify the number of interfaces, and get equations describing mass and heat transfer across the interfaces. Theories of mass and heat transfer coefficients or even empirical correlations may be used. The mass transfer rate requires a mass transfer coefficient and concentration at the interface. Therefore, thermodynamic relationships (e.g. phase equilibria) are obtained. A few models which do not assume equilibrium at the interface relate the concentration in the two phases at the interface in terms of mass transfer rates, and these are called 'rate-based models'. Because experimental data are always more reliable than the values predicted from a theory, the role of empirical correlations in a model cannot be ignored. Probably, it will never be eliminated completely. Thus, the processes across various interfaces are specified.

3. Collect the rate equations to describe the kinetics. This involves the concentration of all the reactants.

4. Consider various options for the assumptions. The choice of assumption depends upon the experience. If the experimental data are to be validated, one can start with a large number of assumptions to give simple models. If the model fails to validate the data, some of the assumptions may be dropped and a new set of model equations is solved. If the model is to be used to simulate a reactor for which no data are available, then it is better to start with a smaller number of assumptions. The model will be more rigorous, but the confidence in predicted values will be higher. In such cases, the use of a theory may be preferred over the use of empirical correlations. Even in this case, it is preferred to validate the model for some other system.

5. Apply laws of conservation of mass and energy. These and the equations collected in previous steps are the model equations.

6. Solve the model equations, and validate the model. It is now ready for simulation.

The methodology discussed above is suggestive only. The application of the concepts discussed above will now be illustrated with the help of models for packed beds, trickle beds, fluidised beds, bubble columns and slurry reactors and bioreactors.

6.2 Packed Bed Reactors

A packed bed consists of stationary solids. One or more fluids may flow through it. A suitable model can be developed only when the interaction between the fluid and solid is clearly understood. The solids in the packed bed reactors can have any of the following roles:

1. The solids are used as a catalyst. The molecules of the reactants reach to the surface of the catalyst where the reaction takes place. The mechanism of movement of the reactants may be diffusion or convection. The products are formed at the surface of the catalyst. The surface is not only the outer surface but also the surface available in the pores. The products again go back to the bulk fluid by diffusion or convection.

2. The solids are used as reactants, such as in the manufacture of bleaching powder. In this process, the chlorine gas is passed through the calcium carbonate. Chlorine gas flowing through the bed comes in contact with the calcium carbonate and reacts to form bleaching powder.

The model for a packed bed reactor must consider the role of the solids in the bed.

6.2.1 Isothermal Bioreactor

Chemical reactions which involve a negligible amount of heat of reactions do not require any heat exchanger to keep the temperature constant. The temperature in the reactor can be assumed to be constant throughout the reactor. Such reactors are called 'isothermal reactors'. From a modelling point of view, the isothermal model does not consist of energy balance.

Krause and Feig (2011) carried out enzymatic hydrolysis of hexanoic acid, octanoic acid and decanoic acid with methanol in a packed bed reactor. The entire process consisted of a mixing tank used for circulating the contents. The mass transfer in pipes was also considered. However, the model equations for a packed bed reactor without circulation are presented here.

Let us consider the reaction of methyl hexanoate with water, producing hexanoic acid and methanol:

$$CH_3(CH_2)_4 COOCH_3 + H_2O \rightleftharpoons CH_3(CH_2)_4 COOH + CH_3OH$$

The biphasic mixture consisted of organic and aqueous phases. Methyl ester and the acid were present only in the organic phase. Water and methanol were present in both phases. Because the reactants hexanoic acid and water are present in the organic phase only, the reaction takes place at the solid–organic interface (Figure 6.2). The bed thus contains solid, organic and aqueous phases. The law of conservation of mass transfer is applied in organic and aqueous phases. Though the solid is porous, the diffusion in the pores is not considered.

In the aqueous phase, only methanol and water are present. The conservation provides two equations, one each for methanol and water. Subscript i denotes different phases.

$$\left(\frac{\varepsilon_{aq}}{\varepsilon_{aq} + \varepsilon_{orq}}\right)\varepsilon \frac{\partial C_i^{aq}}{\partial t} = \left(\frac{\varepsilon_{aq}}{\varepsilon_{aq} + \varepsilon_{orq}}\right)v_z^{aq}\frac{\partial C_i^{aq}}{\partial z} - k_L a\left(C_i^{aq*} - C_i^{aq}\right) \qquad (6.1)$$

In the organic phase, all the four chemical species are available, hence four equations, one for each chemical species.

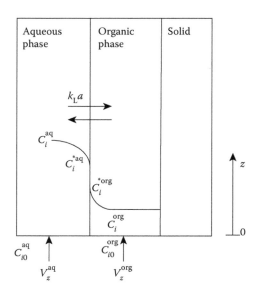

FIGURE 6.2
The model sketch for an isothermal bioreactor with inert solids.

$$\left(\frac{\varepsilon_{org}}{\varepsilon_{aq}+\varepsilon_{orq}}\right)\varepsilon\frac{\partial C_i^{org}}{\partial t}=\left(\frac{\varepsilon_{org}}{\varepsilon_{aq}+\varepsilon_{orq}}\right)v_z^{org}\frac{\partial C_i^{org}}{\partial z}+k_La\left(C_i^{org*}-C_i^{org}\right)-\rho_B\left(-r_i\right) \quad (6.2)$$

where ε_{org} and ε_{aq} are the holdup of the organic and aqueous phases, respectively; and ε is void fraction in the reactor. The rate of reaction should be estimated for concentration at the surface of the solid. The diffusion should also be considered. The boundary conditions in terms of the initial concentration C_{i0}^{aq} and C_{i0}^{org} are as follows:

At $z = 0$: $C_i^{aq} = C_{i0}^{aq}$ and $C_i^{org} = C_{i0}^{org}$ at all t

At $t = 0$: $C_i^{aq} = C_{i0}^{aq}$ and $C_i^{org} = C_{i0}^{org}$ at all z

Equations 6.1 and 6.2 are only reactor models. The kinetic models are required to replace the reaction rate. The mass transfer coefficient depends upon the hydrodynamics in the packed bed. Any empirical correlation or a theoretical model to predict the mass transfer rate at the interface can be used. The model equations can be solved numerically.

In the presence of the heat of reaction, there is a spatial distribution of temperature. The consideration of different phases in Figure 6.2 will not describe the model sketch completely. The following types of steady-state models for packed bed reactors may provide appropriate model equations.

6.2.2 The Solids as Reactant-Kinetic Models

Many chemical reactions (e.g. the combustion of solid fuels and calcinations of solids) involve solids as reactants. Several kinetic models have been described in the literature (Levenspiel 1999; Missen et al. 1999). Such reactions have at least two reactants: one is present in solid, and another is present in fluid. The products may be fluid, solid or both.

A (fluid) + bB (solid) → P (fluid and/or solid)

If the product is solid, then it remains in the solid and is called 'ash'. If product contains fluid, then it diffuses out of the solid. The equation of continuity should be written for each chemical species. The model equations can be developed after due consideration of diffusion and reaction. For the species in fluid form, the mass transfer flux term should be included. The reaction can be classified into one of three categories:

1. The reaction is slow, and it takes place throughout the particle. The size of the particle is constant. This model has been named the 'progressive conversion model' (Figure 6.3). Equation of continuity is applied in the particle. The choice of the coordinate system depends upon the shape of the particles. For slow reactions, the gas

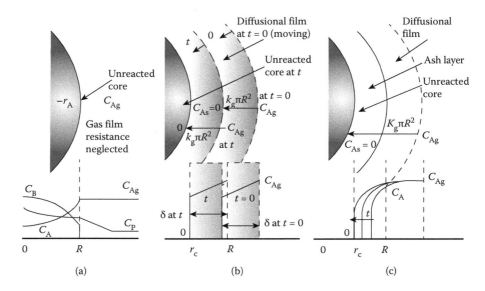

FIGURE 6.3
Fluid–solid reaction with the solid as a reactant: (a) progressive conversion model, (b) shrinking particle model and (c) ash-diffusion controlled shrinking core model for spherical particle. (Adopted from Levenspiel, O., *Chemical Reaction Engineering*, 3rd ed, John Wiley, New York, 1999.)

film resistance is usually neglected, and the equation of continuity for spherical particles for species A is written as follows (Levenspiel 1999). It is valid within the particle.

$$\varepsilon \frac{\partial C_A}{\partial t} = \frac{D_{As}}{R^2} \frac{1}{r^2} \frac{\partial}{\partial r} r^2 \frac{\partial C_A}{\partial r} - \rho_s \left(-r_A\right) \tag{6.3}$$

Here, D_{As} is the diffusivity of A in solid. Similar equations are written for every species. The boundary conditions for Equation 6.3 are as follows:

$$t = 0, C_A = C_{Ag} \text{ at all } r$$

$$\text{and } r = 0, \frac{\partial C_A}{\partial r} = 0$$

2. The reaction is fast. As soon as the fluid comes in contact with the unreacted portion of the particle, the reaction takes place, leaving ash behind. If the ash layer is soft, it is removed from the particle as in the case of fluidised beds in which the ash layer is removed due to attrition. As a result, the size of the particle decreases with time.

This model is known as the 'shrinking particle model'. Let r_a be the radius of the unreacted core (or particle) at any time. Two simple cases were mentioned by Levenspiel (1999). If the gas film is controlling, then the equation of continuity is obtained by equating the mass transfer from the bulk fluid to the solid surface to consumption of A by reaction at the surface.

$$\rho_B \left(4\pi r_a^2 \right) \frac{dr_a}{dt} = b k_g \left(r_a \right) C_{Ag} \tag{6.4}$$

with initial condition: at $t = 0$, $r = R$. The mass transfer coefficient, $k_g(r_a)$, at the surface of the particle is a function of particle size and can be estimated using any of the suitable models discussed earlier.

If the surface reaction is controlling, then the equation of continuity can be written as

$$\frac{\rho_B r_a^3}{R^2} \frac{dr_a}{dt} = b(-r_A) \tag{6.5}$$

with the following boundary condition: at $t = 0$, $r_a = R$.

3. The reaction is fast. The ash layer is not soft and remains attached to the solid particle. As time progresses, the unreacted core shrinks and the thickness of the ash layer increases. The ash layer introduces a resistance to mass transfer.

Three types of simple kinetic models for the reaction rate are possible (Figure 6.4). If the gas film is controlling, then the conservation of mass is represented by an equation similar to Equation 6.4, with r_a is replaced by R.

If the ash film is controlling, then the conservation of mass results in the following equation (Levenspiel 1999):

$$\frac{\rho_B r_a^3}{R^2} \frac{dr_a}{dt} = 4\pi r^2 D_{eff} \frac{dC_A}{dt} \tag{6.6}$$

where D_{eff} is the effective diffusivity of A in the ash layer. The boundary conditions are $t = 0$, $r = R$ and $C_A = C_{Ag}$.

If the chemical reaction is controlling, then the conservation of mass is given by Equation 6.5. If all the resistances are significant, then they are added to obtain

$$\frac{dr_a}{dt} = \frac{b C_A \rho_B}{\dfrac{r_a^2}{R^2 k_g} + \dfrac{(R - r_a) r_a}{R D_e} + \dfrac{1}{k'}} \tag{6.7}$$

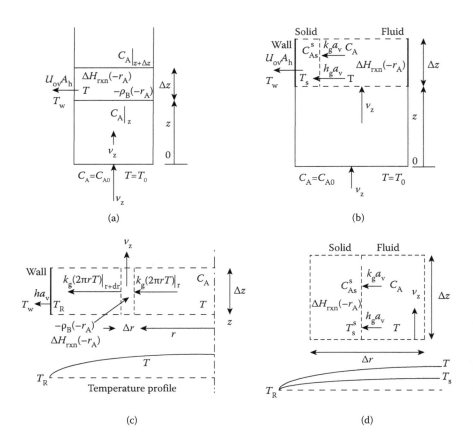

FIGURE 6.4
Steady-state models for packed bed reactors: (a) 1D homogeneous model, (b) 1D heterogeneous model, (c) 2D pseudo-homogeneous model and (d) 2D heterogeneous model.

The reactor models can be chosen depending upon the type of the reactor. The solution of equations can be obtained by substituting the rate expression and solving the model equations. The reactor models for packed towers are discussed in the following section.

6.2.3 Catalytic Packed Bed Reactors: Reactor Models

In catalytic reactors, the solids used as catalysts are porous or supported on a porous substance. The reaction takes place in the solids which are in the dispersed phase. The simplest reactor model can be developed by making use of the assumption of continuum. The concentration and temperature vary only in the axial direction and are constant at any radial position. Such models are known as '1D models'.

6.2.3.1 1D Pseudo-Homogeneous Model

Although the solids are stationary and the fluid is flowing, the properties of fluid and solid at any radial position are constant at steady state. Let us assume that the axial flow is plug flow. It means that the dispersion term is neglected. The heat and mass transfer resistances are neglected. The equation of continuity for a species A in the reaction A→B is (Figure 6.4a)

$$v_z \frac{dC_A}{dz} = -\rho_B (-r_A) \tag{6.8}$$

where ρ_B is the density of bulk. Similarly, the equation of change of thermal energy after neglecting the dispersion term is

$$v_z \rho_g C_{Pg} \frac{dT}{dz} = (-\Delta H) \rho_B r_A - 4 \frac{U_{ov}}{D_t} (T - T_w) \tag{6.9}$$

where U_{ov} is the overall heat transfer coefficient for the heat exchanger. The area of circular jacket or tube per unit volume is $\pi d_t / (\pi d_t^2 / 4) = 4/d_t$. The boundary conditions are as follows:

At $z = 0$: $C_A = C_{A0}$ and $T = T_0$

Equations 6.8 and 6.9 are ordinary differential equations and can be solved using the fourth-order Runga–Kutta method or ordinary differential equation solvers, which are available in various types of software. The model equations are general and can be solved after substituting the expression for the reaction rate and heat of reaction.

In many applications, more than one reaction takes place. For example, Fischer–Tropsch synthesis is carried out in a packed bed reactor using iron or cobalt catalyst (Jess and Kern 2009). There are three reactions. The main reaction is represented by

$$CO + 2H_2 \rightarrow (-CH_2-) + H_2O + \Delta H_{FT}$$

The other two reactions are methane formation and water-shift reaction:

$$CO + 3H_2 \rightarrow CH_4 + H_2O + \Delta H_{MF}$$

$$CO + H_2O \rightarrow CO_2 + H_2 + \Delta H_{WS}$$

While writing the equation of continuity for hydrogen, the reaction rate, $-r_A$, in Equation 6.8 will be the rate of generation from all three chemical reactions:

$$r_A = \sum (r_{H_2}) = r_{H_2,FT} + r_{H_2,MF} - r_{H_2,WS} \tag{6.10}$$

where $r_{H_2,FT}$, $r_{H_2,MF}$ and $r_{H_2,WS}$ are the reaction rates due to Fischer–Tropsch synthesis, methanol formation and water-shift reaction, respectively. The heat of reaction in Equation 6.9 will be the sum of the heat of reaction of all chemical reactions.

$$\Delta H.r_A = \sum \left(\Delta H.r_{H_2} \right) = \Delta H_{FT}.r_{H_2,FT} + \Delta H_{MF}.r_{H_2,MF} + \Delta H_{WS}.r_{H_2,WS} \quad (6.11)$$

While writing conservation of mass, as in Equation 6.10, the minus sign is used when the reactant is produced. The minus sign for the heat of reaction while writing conservation of thermal energy as in Equation 6.11 will be for an endothermic reaction.

6.2.3.2 1D Heterogeneous Model

The 1D pseudo-homogeneous model predicts a constant radial concentration and temperature profile. The model does not give acceptable results in the case of rapid reaction. If the reaction is taking place at the solid surface (i.e. the solids are catalysts), it becomes necessary to consider the conditions at the particles and fluid separately (Figure 6.4b). The equation of continuity for species A in the fluid phase is

$$v_z \frac{dC_A}{dz} = k_g a_v \left(C_A - C_{As}^s \right) \quad (6.12)$$

Because the reaction is not taking place in the fluid, the generation term is absent. The mass transfer resistance at the catalyst surface is considered.

The equation of change of thermal energy is

$$v_z \rho_g C_{pg} \frac{dT}{dz} = h a_v \left(T_s^s - T \right) - 4 \frac{U_{ov}}{D_t} \left(T - T_w \right) \quad (6.13)$$

The boundary conditions are as follows:

At $z = 0$: $C_A = C_{A0}$ and $T = T_0$

The radial variations of concentration and temperature are not considered. An equation of change of thermal energy balance at the solid surface results in the following algebraic equation. All of the mass of A consumed in the reaction is equal to the mass transferred from the fluid to the solid.

$$k_g a_v \left(C_A - C_{As}^s \right) = \rho_B r_A \quad (6.14)$$

The energy generated by the reaction is transferred to the fluid. The conservation of energy gives the following equation:

$$ha_v\left(T_s^s - T\right) = (-\Delta H)\rho_B r_A \tag{6.15}$$

Equations 6.12 and 6.13 can be solved using the fourth-order Runga–Kutta method. Equations 6.14 and 6.15 are evaluated after every step. In case of multiple reactions, the rate of reaction and heat of reaction terms are due to all chemical reactions.

In case of exothermic reactions, there is a possibility of a hot spot. These cannot be predicted if the spatial distribution of the temperature is not considered. The model should consider mass and heat transfer resistance in axial as well as radial directions. The models can be classified as homogeneous and heterogeneous reactor models.

6.2.3.3 2D Pseudo-Homogeneous Model

In pseudo-homogeneous models, the fluid and solid are considered a single phase. The concentration and temperature profiles vary in axial as well as radial directions. Let us assume that the axial flow is the plug flow (i.e. the dispersion in the axial direction is neglected in comparison to the convective transport of mass and thermal energy). Also, in absence of radial velocity, the only mechanisms of mass and heat transfer in the radial direction are diffusion and conduction, respectively (Figure 6.4c). Therefore, the equation of continuity for a species A in a cylindrical coordinate is (Lerou and Forment 1982).

$$\varepsilon D_e\left(\frac{\partial^2 C_A}{\partial r^2} + \frac{1}{r}\frac{\partial C_A}{\partial r}\right) - v_z\frac{\partial C_A}{\partial z} = \rho_B r_A \tag{6.16}$$

The diffusion in the radial direction is multiplied by the porosity because the diffusion is taking place in the fluid phase and not in the solid phase.

The equation of the change of thermal energy is

$$k_g\left(\frac{\partial^2 T}{\partial r^2} + \frac{1}{r}\frac{\partial T}{\partial r}\right) - v_z\rho_g C_{pg}\frac{\partial T}{\partial z} = (-\Delta H)\rho_B r_A \tag{6.17}$$

The boundary conditions are as follows:

At $z = 0$: $C_A = C_{A0}$ and $T = T_0$ for $0 \le r \le R$

For all z; at $r = 0$: $\dfrac{\partial C_A}{\partial r} = 0$, $\dfrac{\partial T}{\partial r} = 0$

At $r = R$: $\dfrac{\partial C}{\partial r} = 0$, $k_g\dfrac{\partial T}{\partial r} = -h(T_R - T_w)$

The derivative of concentration at the wall of the vessel is zero because there is no mass transfer across the wall.

The 2D model may also consider axial dispersion, which has been neglected in Equations 6.16 and 6.17. It will be discussed in Section 6.2.3.5 while discussing dynamic models.

6.2.3.4 2D Heterogeneous Model

Highly exothermic reactions result in radial temperature profiles in reactors. It becomes necessary to consider the two phases separately. The conservation of mass and energy equations are written for fluid and solid phases separately.

Equation of continuity for a chemical species in the fluid is

$$v_z \frac{\partial C_A}{\partial z} = \varepsilon D_e \left(\frac{\partial^2 C_A}{\partial r^2} + \frac{1}{r} \frac{\partial C_A}{\partial r} \right) - k_g a_v \left(C_A - C_{As}^s \right) \tag{6.18}$$

where C_{As}^s is the concentration of A at the surface of the solid. Equation 6.18 will be written for each chemical species. Because the reaction is not taking place in the fluid phase, the generation term is absent (Figure 6.4d). Mass transfer resistance in the fluid phase can be evaluated using any appropriate empirical correlation for the packed bed. Alternatively, any theory such as surface renewal and boundary layer can be applied.

The equation of change of thermal energy in the fluid gives

$$v_z \rho_g C_{pg} \frac{\partial T}{\partial z} - k_{eg} \left(\frac{\partial^2 T}{\partial r^2} + \frac{1}{r} \frac{\partial T}{\partial r} \right) = h_g a_v \left(T_s^s - T \right) \tag{6.19}$$

where T_s^s is the temperature at the surface of the solid.

The equations for the solid phase can be written by equating the rate of mass transfer and the reaction at the solid surface.

$$k_g a_v \left(C - C_s^s \right) = \rho_B \eta_e r_A \tag{6.20}$$

Here the effectiveness factor, η_e, takes into account the conversion in pores of the solids. Another equation to estimate the value of η_e can be obtained by writing the equation of continuity for all the species within the solid.

Let k_e^g and k_e^s are effective thermal conductivity of the gas and solid respectively. The equation of change of thermal energy in the solid, assumed to be a spherical particle, provides

$$h_g a_v \left(T_s^s - T \right) = (-\Delta H) \rho_B r_A + k_e^s \left(\frac{\partial^2 T_s}{\partial r^2} + \frac{1}{r} \frac{\partial T_s}{\partial r} \right) \tag{6.21}$$

The boundary conditions are as follows:

At $z = 0$: $C_A = C_{A0}$ and $T = T_0$ for $0 \leq r \leq R$

At all axial positions at the centre of the reactor, all the gradients are zero:

$$r = 0, \quad \frac{\partial C_A}{\partial r} = 0, \quad \frac{\partial T}{\partial r} = \frac{\partial T_s}{\partial r} = 0, \quad (6.22)$$

And at the reactor wall:

$$r = R, \quad \frac{\partial C_A}{\partial r} = 0, \quad \frac{\partial T}{\partial r} = -\frac{h_w^g}{k_e^g}(T - T_W), \quad (6.23)$$

At the surface of the particle, the temperature gradient is given by

$$\frac{\partial T}{\partial r} = -\frac{h_e^s}{k_e^s}(T_S - T_W) \quad (6.24)$$

Here h_e^s and h_w^g are heat transfer coefficient on fluid side at the surface of the solid and reactor wall respectively.

6.2.3.5 Unsteady-State or Dynamic Models

The steady-state models do not predict the behaviour of a packed bed reactor with time. The dynamic models can also be developed as 1D or 2D models. A general 2D dynamic model with assumption of plug flow with dispersion in the axial direction and only dispersion in the radial direction contains the unsteady-state term and dispersion terms in model equations as discussed above.

The model equations in the packed bed reactors are relatively well defined and can be classified as one of the above. The difference between various models is the sub-model used to describe the behaviour of the individual sorbent pellets or particles in the bed (Efthimiadis and Sotirchos 1993).

For example, a 2D pseudo-homogeneous with axial dispersion plug flow and radial dispersion was mentioned by Koning et al. (2006). The reaction in the solids was taken into account by the effectiveness factor.

The conservation of mass gives

$$\varepsilon \frac{\partial C_i}{\partial t} + v_z \frac{\partial C_i}{\partial z} = D_{ez} \frac{\partial^2 C_i}{\partial z^2} + \varepsilon D_{er}\left(\frac{\partial^2 C_i}{\partial r^2} + \frac{1}{r}\frac{\partial C_i}{\partial r}\right) - \sum n_i r_i \eta_e \quad (6.25)$$

where n_i is the stoichiometry coefficient for the ith component.

The conservation of thermal energy gives

$$\left[\varepsilon \rho_g C_{pg} + (1-\varepsilon)\rho_s C_{ps}\right]\frac{\partial T}{\partial t} + v_z \rho_g C_{pg}\frac{\partial T}{\partial z}$$

$$= k_{ga}\left(\frac{\partial^2 T}{\partial z^2}\right) + k_{gr}\left(\frac{\partial^2 T}{\partial r^2} + \frac{1}{r}\frac{\partial T}{\partial r}\right) + \sum \Delta H_i r_A \eta_e \tag{6.26}$$

Equations 6.25 and 6.26 are applicable for multiple reactions.

The unsteady terms can be added to other model equations also to get corresponding dynamic models. 1D dynamic models can be obtained by dropping all the terms containing the variable r from Equations 6.25 and 6.26.

Fluidised bed, trickle bed, slurry reactor and slurry bubble columns are used to carry out gas–liquid reactions catalysed by solids. They can be treated in a similar manner; however, the difference is due to the flow field in these different types of reactors (Levenspiel 1999). The kinetic models are the same, and the mechanism of mass transfer may be the same but the transfer coefficients are different. These differences are illustrated through some of the models reported in the literature.

6.3 Trickle Bed Reactors

When gas and liquid flow concurrently in a packed bed reactor and the liquid flow rate is small, the reactor is called a 'trickle bed reactors'. While only two phases were present in the packed bed reactors, there are three phases in trickle bed reactors. If the solids are inert, then the reaction will take place in the fluid phase only. The model equations will be similar to the equations obtained for flow of biphasic mixture in the packed bed. If the reaction is catalytic (i.e. the solids act as a catalyst), then the reaction will take place in the solid phase.

In a trickle bed reactor, four different flow regimes are reported in the literature (Attou and Ferschneider 2000; Gunjal et al. 2005). At very low gas and liquid velocities, the liquid flows down, forming a film over a solid surface. This regime is called a 'film flow'. Both liquid and gas phases are continuous in this regime. The solid surface may be completely or partially wetted. At low liquid flow rates and high gas flow, a part of the liquid is present in the form of droplets. The liquid is entrained by the gas. This regime is called a trickle flow regime. At a very high gas flow rate, when all the liquid is in the form of droplets, a spray flow regime is observed. At high liquid and gas, flow rates are so high that two distinct gas and liquid-rich pulses are observed. This regime is called a 'pulse flow regime'. At a very high liquid flow rate, the gas is present in the form of bubbles, and the regime is called a bubbly flow regime. The operation of trickle bed reactors in trickle and pulse flow regimes is desirable.

The transport processes depend upon the manner in which the fluid phases contact each other and form the interface. The contacting method determines the processes near the interface and affects the transfer rates across the interface. Various theories to estimate the transfer processes and the conditions under which these are applicable are discussed in Chapter 4. It is essential to identify different flow regimes so that a suitable correlation or theory can be used to estimate various parameters in the reactor model. These parameters are usually mass and heat transfer coefficients, radial and axial dispersion of fluids, holdups and fractional areas of the solids wetted by the liquid.

The development of a reactor model for a trickle bed is discussed below. It is a 1D three-phase heterogeneous model.

These parameters are usually mass and heat transfer coefficients, radial and axial dispersion of fluids, holdups and fractional areas of the solids wetted by the liquid.

$$CO + 2H_2 \rightleftarrows CH_3OH$$

$$H_2O + CO \rightleftarrows CO_2 + H_2$$

The plug flow for the gas and plug flow with axial dispersion for the liquid were assumed. The surface of the spherical catalyst particle was partially wetted (i.e. a part of the solid was exposed to gas directly). The remaining part was in contact with the liquid. The sketch of the model is shown in Figure 6.5a–c. Because the particle is partially wet, let f be the fraction of

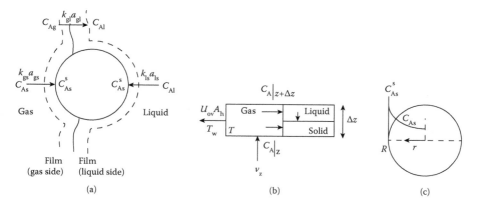

FIGURE 6.5
Model sketch for trickle bed reactor: (a) mass transfer for partially wetted particle, (b) mass transfer and heat transfer in an axial elemental volume of the reactor and (c) concentration profile in the particle.

the catalyst surface exposed to the gas. These assumptions indicate that the model is a 1D three-phase heterogeneous model. The conservation of mass in the gas phase gives

$$\rho_g v_{zg} \frac{\partial C_{Ag}}{\partial z} = -k_{gl}a_{gl}\left(\frac{C_{Ag}}{H_A} - C_{Al}\right) - (1-f)k_{gs}a_{gs}\left(\frac{C_{Ag}}{H_A} - C_{As}^s\right) \qquad (6.27)$$

where C_{Al} and C_{Ag} are the concentration of species A in liquid and gas respectively, k_{gl} and k_{gs} are mass transfer coefficient from gas to liquid and gas to solid respectively. The corresponding interfacial area are a_{gl} and a_{gs} (Figure 6.5). Wang et al. (2005) used different mass transfer coefficients for each species. It is generally believed that the mass transfer coefficient depends upon hydrodynamics. Therefore, it should be the same for each species. The values of k_{gl} for all the species used by them were close. Therefore, one value for all the species was used. Equations similar to Equation 6.27 for other species (reactants and products) are written. Equation 6.27 takes into account two interfaces: the gas–liquid and gas–solid present in the reactor.

Let k_{ls} be mass transfer coefficient from liquid to solid. The conservation of mass in the liquid phase gives

$$\rho_l v_{zl} \frac{\partial C_{Al}}{\partial z} = \varepsilon_l D_A \frac{\partial^2 C_{Al}}{\partial^2 z} + k_{gl}a_{gl}\left(\frac{C_{Ag}}{H_A} - C_{Al}\right) - fk_{ls}a_{ls}\left(C_{Al} - C_{As}^s\right) \qquad (6.28)$$

Similarly, equations for other species are also written. The boundary conditions are as follows:

At $z = 0$: $\rho_l v_{zl} C_{Al}\big|_{z=0} = \varepsilon_l D_A \dfrac{\partial C_{Al}}{\partial z}\bigg|_{z=0}$

At $z = L$: $\dfrac{\partial C_{Al}}{\partial z}\bigg|_{z=L} = 0$

All the reactions take place in the solid phase. It is assumed that the particles are spherical. The reactant diffuses into the solid. The conservation of mass in the solid phase gives

$$D_{Ae}\left(\frac{\partial^2 C_{As}}{\partial r^2} + \frac{2}{r}\frac{\partial C_{As}}{\partial r}\right) = n_A \rho_B(-r_A) \qquad (6.29)$$

While writing conservation of mass equations, the sign of the reaction rate should be kept in mind (e.g. for water, the sign on the left-hand side will change because water is produced).

The boundary conditions are as follows:

$$\text{At all } z \text{ and at } r = R: D_A \left.\frac{\partial C_{As}}{\partial r}\right|_{r=R} = (1-f)k_{gs}\left(\frac{C_{Ag}}{H_A} - C_{As}^s\big|_{r=R}\right) - fk_{ls}\left(C_{Al} - C_{As}^s\big|_{r=R}\right)$$

$$\text{At all } z \text{ and at } r = 0: \ D_A \left.\frac{\partial C_{As}}{\partial r}\right|_{r=0} = 0$$

For the reactor, cylindrical coordinates were used. For exothermic reactions, it is required to remove the heat from the reactor; therefore, the geometry of the reactor is important. For the particles, the model equations were written in spherical coordinates. The equation of change of thermal energy per unit volume is

$$v_z C_1 C_{pl} \frac{\partial T}{\partial z} = -4\frac{U_{ov}}{D_t}(T - T_w) + \sum \frac{3(1-\varepsilon_g)}{R} D_A \left.\frac{\partial C_{As}}{\partial r}\right|_{r=R} \cdot \Delta H_{rxn}\rho_p \quad (6.30)$$

The boundary condition is as follows:

At $z = 0$: $T = T_0$

The model equations can be solved numerically.

6.4 Slurry Reactors

Many gas–liquid reactions catalysed by solid particles are carried out in slurry reactors. The solid particles are small enough so that they can be suspended in the liquid. The suspensions are called a slurry. The gases are sparged into the slurry. One common gas–liquid contactor is a slurry bubble column in which the solids are suspended due to bubble-induced liquid motion. The solids can also be kept in suspension using mechanical stirrers.

In general, the reactions carried out in a slurry reactor are of the following form:

$$A \text{ (gas)} + B \text{ (liquid)} \xrightarrow{\text{At solid surface}} C \text{ (liquid)}$$

The reactant A is absorbed by the liquid at the gas–liquid interface. It is then transported towards the catalyst surface by diffusion and convection. The second reactant B present in the liquid is also transported to the catalyst surface.

The reaction takes place at the catalyst surface. The product C is again transported to the bulk of the liquid.

6.4.1 Mechanically Agitated Slurry Reactors

The solids are maintained under suspension in mechanically agitated vessels by the rotating impeller at a speed above a critical speed required to suspend the solids. This value depends upon the type of the impeller, the geometrical parameters, the particle size and the properties of solid and fluid.

In mechanically stirred slurry reactors, the contents of the vessel can be assumed to be well mixed. The particles may be porous. The reaction takes place in the solid or at the surface of the solid. The reactant present in the gas phase enters the liquid phase. The reactant diffuses into the solid through the liquid–solid interface (Figure 6.6). It then diffuses into the pores of the solid. The reaction rate depends upon local concentration of the reactants in the pores as the reaction takes place in them. One of the approaches to model this situation is to use a sub-model which relates the concentration of the reactant at the surface of the solid and the effectiveness factor for the reaction rate within the pores of the solids.

The effectiveness factor can be defined as the ratio of the reaction rate in the presence of mass transfer to the reaction rate without mass transfer. The observed reaction rate in any spatial position can be written in terms of either concentration of reactant at the catalyst surface or concentration in the bulk. For example, the observed reaction rate in terms of the concentration in the gas phase, p_{Ag}, for first-order reaction kinetics for the following reaction is given by Levenspiel (1999):

$$A \text{ (gas)} + bB \text{ (liquid)} \longrightarrow \text{Products}$$

The observed rate of reaction, r_{Aobs}, is

$$r_{A,obs} = \frac{1}{\dfrac{1}{k_{Ag}a_i} + \dfrac{H_A}{k_{Al}a_i} + \dfrac{H_A}{k_{As}a_s} + \dfrac{H_A}{k\overline{C_B}\eta f}} p_{Ag} \tag{6.31}$$

$$r_{B,obs} = \frac{1}{\dfrac{1}{k_{Bs}a_s} + \dfrac{1}{k_B\overline{C_A}\eta_B f}} C_{Bl} \tag{6.32}$$

Here k_{Ag}, k_{Al} and k_{As} are mass transfer coefficient for A in gas, liquid and solid respectively, f is solid loading and η is effectiveness factor. The average concentration of A and B in solid are $\overline{C_A}$ and $\overline{C_B}$ respectively. The observed reaction rate for B does not have mass transfer terms at the gas–liquid interface because B is already present in the liquid phase. Equations 6.31 and 6.32 can be solved only numerically.

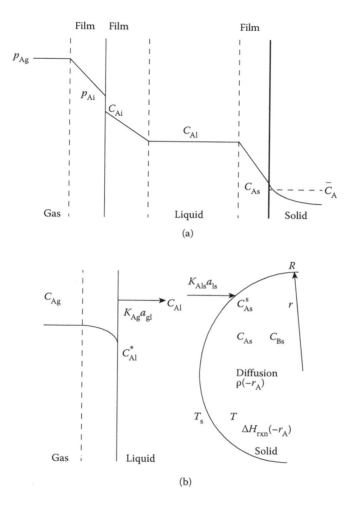

FIGURE 6.6
Model sketch for (a) a mechanically agitated slurry reactor and (b) a slurry bubble column with a spherical particle.

The reactor model can now be written in terms of the observed rates. For mechanically agitated reactors, the gas flow is well mixed. For all types of liquid flow rates, the reactor model is written as follows (Levenspiel 1999):

$$F_{A0}X_A = -r_{A,obs}V_r \qquad (6.33)$$

6.4.2 Slurry Bubble Columns

A dynamic model for the reaction above can be developed by writing the equation of continuity for each of the reactants. The concentration varies only in the axial direction. The radial variation of concentration is absent because the diffusion of A in the liquid starts from the gas–liquid interface, which is present at all almost radial positions. The same is true for the concentration of A and B at the liquid–solid interface. Therefore, a 1D model may be used. A model developed by Toledo et al. (2001) is presented here. Let the gas and liquid enter at the bottom at velocities v_{zg} and v_{zl}, respectively. Let us assume plug flow of the liquid and gas. Evaporation of the liquid is neglected. The role of the evaporation in the law of conservation of mass is to introduce a generation term which makes the chemical species disappear from one phase and appear in the second phase. This type of generation term is used whenever any phase change occurs.

Let ε_g and ε_l be the gas and liquid holdup, respectively. In the gas phase, only A is present; therefore, writing an unsteady 1D equation of continuity for A:

$$\varepsilon_g \frac{\partial C_{Ag}}{\partial t} + v_{zg} \frac{\partial C_{Ag}}{\partial z} = D_{Ag} \frac{\partial^2 C_{Ag}}{\partial z^2} - K_{Agl}a_{gl}\left(C_A^* - C_{Al}\right) \tag{6.34}$$

where D_{Ag} is the diffusion coefficient of A in the gas phase; K_{Agl} is the mass transfer coefficient at the gas–liquid interface; C_{Ag} and C_{Al} are concentrations of A in bulk of gas and liquid phases, respectively; C_A^* is the equilibrium concentration of A at the interface (i.e. the solubility of A in the liquid); and gas holdup is ε_g.

If the dimensionless axial position $Z = z/L$ is used, the following equation is obtained:

$$\varepsilon_g \frac{\partial C_{Ag}}{\partial t} + \frac{v_{zg}}{L} \frac{\partial C_{Ag}}{\partial z} = \frac{D_{Ag}}{L^2} \frac{\partial^2 C_{Ag}}{\partial z^2} - K_{Agl}a_{gl}\left(C_A^* - C_{Al}\right) \tag{6.35}$$

The boundary conditions are (Figure 6.6b) as follows:

At the distributor plate (i.e. $Z = 0$): $\left.\dfrac{D_{Ag}}{L}\dfrac{\partial C_{Ag}}{\partial z}\right|_{Z=0} = v_{zg}\left(C_{Ag} - C_{Ag0}\right)$ for all $t > 0$

At the top (i.e. $Z = 1$): $\left.\dfrac{\partial C_{Ag}}{\partial z}\right|_{Z=1} = 0$ for all $t > 0$

C_{Ag0} is the concentration of A in the gas at the inlet. The liquid phase contains both A and B; hence, equation of continuity will be written for A and B. The last term in Equation 6.30 is due to the disappearance of A from the gas phase as a result of mass transfer from the gas to liquid phase.

The equation of continuity for A in the liquid phase is

$$\varepsilon_l \frac{\partial C_{Al}}{\partial t} + \frac{v_{zl}}{L} \frac{\partial C_{Al}}{\partial z} = \frac{D_{Al}}{L^2} \frac{\partial^2 C_{Al}}{\partial z^2} + K_{Agl}a_{gl}\left(C_A^* - C_{Al}\right) - K_{Als}a_{ls}\left(C_{Al} - C_{As}^s\right) \tag{6.36}$$

where C_{As}^s is the concentration of A at the surface of the solid. Equation 6.31 contains two generation terms. Here K_{Agl} and K_{Als} are mass transfer coefficient for A in gas–liquid and liquid–solid surface respectively. The first term is the mass of A, which has entered into the liquid phase from the gas phase. The second term is due to the disappearance of A from the liquid to the solid phase. Recall that the reaction is not taking place in either the gas or liquid phases. Hence, there are no generation terms due to the chemical reaction in Equations 6.35 and 6.36.

The following boundary conditions are used:

At the distributor plate (i.e. $Z = 0$): $\dfrac{D_{Al}}{L} \dfrac{\partial C_{Al}}{\partial z}\bigg|_{Z=0} = v_{zl}\left(C_{Al} - C_{Al0}\right)$ for all $t > 0$

At the top (i.e. $Z = 1$): $\dfrac{\partial C_{Al}}{\partial z}\bigg|_{Z=1} = 0$ for all $t > 0$

where C_{Al0} is the concentration of A in the liquid phase at the inlet.

The equation of continuity for B in the liquid phase can similarly be written as

$$\varepsilon_l \frac{\partial C_{Bl}}{\partial t} + \frac{v_{zl}}{L} \frac{\partial C_{Bl}}{\partial z} = \frac{D_{Bl}}{L^2} \frac{\partial^2 C_{Bl}}{\partial z^2} - K_{Bls} a_{ls} \left(C_{Bl} - C_{Bs}^s\right) \tag{6.37}$$

Toledo et al. (2001) used the same diffusivity for A and B in liquid phases (i.e. $D_{Al} = D_{Bl}$). Different diffusivities have been used in the Equations 6.36 and 6.37 for the clarification of the model equations.

The boundary conditions are as follows:

At the distributor plate (i.e. $Z = 0$): $\dfrac{D_{Bl}}{L} \dfrac{\partial C_{Bl}}{\partial z}\bigg|_{Z=0} = v_{zl}\left(C_{Bl} - C_{Bl0}\right)$ for all $t > 0$.

At the top (i.e. $Z = 1$): $\dfrac{\partial C_{Bl}}{\partial z}\bigg|_{Z=1} = 0$ for all $t > 0$

where C_{Bl0} is the concentration of B in the liquid at the inlet.

Equations 6.35 through 6.37 are in rectilinear coordinates. The thickness of the diffusional boundary layer is so small that the curvature of the gas–liquid interface may be neglected. Even at the liquid–solid phase, the curvature was neglected for the diffusional boundary layer in the liquid by Toledo et al. (2001). Further, the solids particles are small and were assumed to be spherical. Hence, the equation of continuity for A and B in the solid phase was written in spherical coordinates.

The equations of continuity for A and B in the solid phase in terms of the dimensionless radius of the particle are as follows:

$$\varepsilon_s \frac{\partial C_{As}}{\partial t} = \frac{D_{As}}{R^2} \frac{1}{r^2} \frac{\partial}{\partial r} r^2 \frac{\partial C_{As}}{\partial r} - \rho_s\left(-r_A\right) \tag{6.38}$$

$$\varepsilon_s \frac{\partial C_{Bs}}{\partial t} = \frac{D_{Bs}}{R^2} \frac{1}{r^2} \frac{\partial}{\partial r} r^2 \frac{\partial C_{Bs}}{\partial r} - \rho_s(-r_B) \tag{6.39}$$

The first terms on the right-hand side of both equations are, respectively, the diffusion of A and B in the solid. There is no convection in the solid. The reaction taking place in the solid phase results in the disappearance term. The boundary conditions are as follows:

At the centre of the particle (i.e. $r = 0$): $\left. \frac{\partial C_{As}}{\partial r} \right|_{r=0} = 0$, $\left. \frac{\partial C_{Bs}}{\partial r} \right|_{r=0} = 0$ for all $t > 0$.

At the surface of the particle (i.e. $r = R$), the mass of A and B entered into solid due to diffusion from the liquid phase, respectively, is equal to the diffusion in the solid at the surface.

$$\left. \frac{D_{As}}{R} \frac{\partial C_{As}}{\partial r} \right|_{r=R} = K_{Als} a_{ls} \left(C_{Al} - C_{As}|_{r=R} \right)$$

$$\left. \frac{D_{Bs}}{R} \frac{\partial C_{Bs}}{\partial r} \right|_{r=R} = K_{Bls} a_{ls} \left(C_{Bl} - C_{Bs}|_{r=R} \right)$$

If the heat of the reaction is small to the extent that it can be neglected, then the reaction is isothermal and the equation of change of thermal energy is not required. If the heat of the reaction is significant, then the equation of change of thermal energy is to be used. Because the reaction is taking place in the solids, the equation of change of thermal energy is written only for the solid phase. This is generated due to the heat of reaction, $-\Delta H_R$.

$$\rho_s C_{ps} \frac{\partial T_s}{\partial t} = \frac{k_s}{R^2} \frac{1}{r^2} \frac{\partial}{\partial r} r^2 \frac{\partial T_s}{\partial r} - \rho_s(-r_A)(-\Delta H_R) \tag{6.40}$$

The boundary conditions are as follows:

At $r = 1$: $\left. \frac{k_s}{R} \frac{\partial T_s}{\partial r} \right|_{r=R} = h_s \left(T - T_s|_{r=R} \right)$

At $r = 0$: $\left. \frac{\partial T_s}{\partial r} \right|_{r=0} = 0$

The heat generated (or adsorbed) due to reaction affects the temperature of the gas–liquid dispersion. Let us assume that as a consequence of good mixing, the temperatures of gas and liquid are equal. No heat is generated in the fluid as there is no reaction in the fluid. The equation of change of thermal energy can hence be written for the gas–liquid dispersion.

$$\left(\varepsilon_g\rho_g C_{pg} + \varepsilon_l\rho_l C_{pl}\right)\frac{\partial T}{\partial t} + \frac{\left(\varepsilon_g\rho_g C_{pg}\upsilon_{zg} + \varepsilon_l\rho_l C_{pl}\upsilon_{zl}\right)}{L}\frac{\partial T}{\partial z}$$

$$= \frac{\left(\varepsilon_g k_g + \varepsilon_l k_l\right)}{L^2}\frac{\partial^2 T}{\partial z^2} + h_s a_{ls}\left(T_s\big|_{r=1} - T\right) - \frac{4U_{ov}}{D_t}(T - T_c) \tag{6.41}$$

With the following boundary conditions:

At the distributor plate (i.e. $Z = 0$):

$$\frac{\left(\varepsilon_g k_g + \varepsilon_l k_l\right)}{L}\frac{\partial T}{\partial z}\bigg|_{Z=0} = \left(\varepsilon_g\rho_g C_{pg}\upsilon_{zg} + \varepsilon_l\rho_l C_{pl}\upsilon_{zl}\right)(T - T_0)$$

At the top (i.e. $Z = 1$): $\dfrac{\partial T}{\partial z}\bigg|_{Z=1} = 0$

The last term in Equation 6.41 is due to heat removal by the cooling fluid used in the jacketed heat exchanger (Toledo et al. 2001). It will change if the type of heat exchanger is changed.

Toledo et al. (2001) also presented a simplified model in which the internal resistance in the particle was not considered.

6.5 Fluidised Bed Reactors

Solids are kept in suspension due to motion of a fluid in fluidised bed. The fluidised beds exhibit good heat and mass transfer characteristics. Hydrodynamics in fluidised beds influence its performance. The hydro-dynamics is characterised by several factors. These are minimum velocity to fluidise the bed, bed expansion behaviour, flow regimes etc. Commonly used equations are discussed below.

6.5.1 Minimum Fluidisation Velocity

The minimum fluidisation velocity, U_{mf}, can be estimated using the Ergun equation. It was derived based on the understanding that at a certain velocity at which onset of fluidisation is observed, the total weight of the particle is supported by the drag force on the particles. The Ergun equation is

$$\left[\frac{1.75(1-\varepsilon_{mf})\rho_g U_{mf}^2}{\varepsilon_{mf}^3\phi_s d_p}\right] + \left[\frac{150(1-\varepsilon_{mf})^2\mu_g U_{mf}}{\varepsilon_{mf}^3\left(\phi_s d_p\right)^2}\right] = \frac{\Delta P_{mf}}{H_{mf}} \tag{6.42}$$

The pressure drop, ΔP_{mf}, in the bed is given by

$$\Delta P_{mf} = H_{mf} g \left(1 - \varepsilon_{mf}\right)\left(\rho_s - \rho_g\right) \tag{6.43}$$

The Reynolds number, Re, in Equation 6.42 is defined as

$$Re = \frac{d_p U \rho_g}{\mu_g} \tag{6.44}$$

6.5.2 Bed Expansion

With increasing superficial gas velocity, the bed expands, changing the porosity of the bed. The bed expansion depends upon the regime of fluidisation. The bed expansion in particulate fluidised beds is usually described by the following equation (Richardson and Zaki 1954). The porosity of the bed, ε, at any superficial fluid velocity, U, is expressed in terms of porosity at minimum fluidisation velocity, ε_{mf}:

$$\frac{U}{U_{mf}} = \left(\frac{\varepsilon_{mf}}{\varepsilon}\right)^n \tag{6.45}$$

where Richardson and Zaki's exponent, n, is given by

$$n = 4.65 + 19.5 \frac{d_p}{D_c} \text{ for Re} < 0.2 \tag{6.46}$$

$$n = \left(4.35 + 17.5 \frac{d_p}{D_c}\right) Re^{-0.03} \text{ for } 0.2 < Re < 1 \tag{6.47}$$

$$n = \left(4.45 + 18 \frac{d_p}{D_c}\right) Re^{-0.1} \text{ for } 1 < Re < 200 \tag{6.48}$$

$$n = 4.45 \, Re^{-0.1} \text{ for } 200 < Re < 500 \tag{6.49}$$

$$n = 2.39 \text{ for Re} > 500 \tag{6.50}$$

In a bubbling fluidised bed, the bed expansion is given by (Geldart 2004)

$$\frac{H}{H_{mf}} = \frac{\bar{U}_{bs}}{\bar{U}_{bs} - Y\left(U - U_{mf}\right)} \tag{6.51}$$

where

$$Y = 1.64 \, Ar^{0.2635} \tag{6.52}$$

Archimedes number, Ar, is given by

$$Ar = \frac{\rho_g \left(\rho_p - \rho_g\right) g d_p^3}{\mu_g^2}$$ (6.53)

Geldart (2004) pointed out that for group B powders, the bubbles grow with the distance above the distributor plate and with increasing $(U - U_{mf})$. The equivalent bubble diameter, d_{bV}, based on the volume is given by

$$d_{bV} = \frac{0.54}{g^{0.2}} (U - U_{mf})^{0.4} \left(H + 4N^{-0.5}\right)^{0.8}$$ (6.54)

Here, N is the number of holes in the distributor. The average bubble velocity, $\bar{U}_{bs}$, is given by

$$\bar{U}_{bs} = \phi \left(g d_{bV}\right)^{0.5}$$ (6.55)

where, for class B powders:

$$\phi = 0.64 \text{ for } D \le 0.1 \text{ m}$$ (6.56)

$$\phi = 1.6 D^{0.4} \text{ for } 0.1 < D < 1 \text{ m}$$ (6.57)

$$\phi = 1.6 \text{ for } D \ge 1 \text{ m}$$ (6.58)

For other class of powders suitable equations may be used to predict the bed expansion.

6.5.3 Flow Regimes

The fluidised beds exhibit several flow regimes, as shown in Figure 5.5. The aggregative fluidisation exhibits different flow regimes as the superficial gas velocity increases. The fluid velocity at which the transition from one flow regime to another flow regime takes place should be known. Slugging should be avoided to achieve a good heat transfer coefficient. Constantineau et al. (2007) mentioned three criteria for slugging to occur:

1. The bed height must be lower than a minimum bed height, H_L, at which the bubble coalescence is complete and stable slug spacing is achieved. The value of, H_L, is given by

$$H_L = 1.3 D^{0.175}$$ (6.59)

2. The superficial gas velocity, U, must be less than the minimum slug-ging velocity, U_{ms}, which, for a shallow bed, is given by

$$U_{ms} = U_{mf} + 0.07\sqrt{gD} + 0.16(L - L_{mf})^2 \qquad (6.60)$$

3. The maximum stable bubble diameter must be smaller than the col-umn diameter. It may be calculated as a function of bed height and superficial gas velocity.

As the slugging adversely affects the transfer processes, operation of a fluidised bed in the slugging regime is usually undesirable.

6.5.4 Two-Phase Fluidised Bed Reactors

Two-phase fluidised beds consist of solid as one phase and liquid or gas as the second phase. Due to good mixing of the fluid in fluidised beds, these are used as isothermal reactors. Therefore, spatial variations of temperature may be neglected (Forment and Bischoff 1979). Equations for conservation of thermal energy are not included in the reactor model for steady-state pro-cesses. However, temporal variation of temperature exists under unsteady state. Therefore, dynamic models consider the equation of change of thermal energy.

The interaction of the two phases depends upon the flow regimes observed in the fluidised beds. The following flow regimes in fluidised beds are observed.

1. *Particulate fluidisation*: The particles move as individuals. If the reac-tion is taking place at the outer surface of the solid, the assumption of continuum can be made. The fluid flows in the axial direction. The model equations will be similar to those discussed in the packed beds. For example, let us assume that the flow of the fluid is plug flow with axial dispersion. There is no radial dispersion. The reac-tion takes place at the surface of the particle. The dynamic reactor model will be of the 1D pseudo-homogeneous type. The conserva-tion of mass and energy results in the following equation:

$$\varepsilon \frac{\partial C_i}{\partial t} + v_z \frac{\partial C_i}{\partial z} = D_{ez} \frac{\partial^2 C_i}{\partial z^2} - n_i r_i \qquad (6.61)$$

Equation 6.61 is quite similar to Equation 6.21 for the packed beds. The boundary conditions are also the same as those for Equation 6.21. The difference is in the values of the dispersion coefficient. In the packed bed, the particles are stationary. Therefore, with increasing fluid velocity, the mixing induced by the flow field around the par-ticles increases. It results in enhanced dispersion coefficient. In the fluidised bed, as the fluid velocity increases, the bed expands and the inter-particle distance increases. The flow field around the particle becomes less vigorous; hence, the dispersion coefficient decreases. It means that although the reactor model for fixed and fluidised beds looks similar, it is the hydrodynamics of the reactor which results in different behaviour exhibited by packed and fluidised beds.

2. *Aggregative fluidisation:* This is usually observed in gas fluidised beds. In an aggregative fluidised bed, some of the gas flows in the interstitial spaces between the particles and the rest flow in the form of bubbles in the bubbling fluidised bed or in the form of slugs in the slugging fluidised beds. The former, with solid particles, forms emulsion and can be assumed to behave as a single phase. It, however, excludes the gas present in bubbles. The situations have been modelled by considering one phase as emulsion and the other phase as gas. This assumption leads to models for fluidised beds that are called 'two-fluid models' (Figure 6.7). The inter-phase transfer and interaction at any height of the model are modelled by making suitable assumptions. The assumption of continuum is applied to both fluids separately.

3. *Turbulent fluidised beds:* When superficial gas velocity is increased further, the bubbles disappear. Regions of gas and emulsion can still be seen. These regions are dynamic. The bubbles do not have a definite shape. The emulsion is also not continuous and is present in the form of packets.

4. *Circulating fluidised beds:* At very high gas velocities, the particles are collected at the top and then are circulated so that they remain in the bed. These are called circulating fluidised beds or fast fluidised beds.

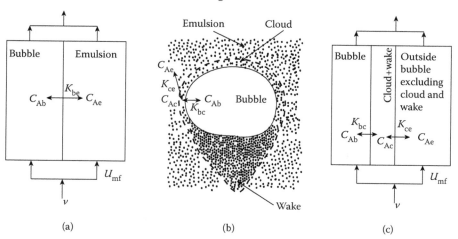

FIGURE 6.7
Models for bubbling fluidised beds: (a) two-fluid model, (b) bubble–wake–cloud around a bubble and (c) consideration of wake and cloud by Kunni and Levenspiel. (Adopted from Levenspiel, O., *Chemical Reaction Engineering*, 3rd ed, John Wiley, New York, 1999.)

5. *Spouted beds*: When one attempts to fluidise large particles, they recirculate in the bed in a particular manner. Three zones are formed: spout at the centre, annulus and the fountain. The spouted beds are not only used for drying but also may be attractive for applications such as pyrolysis of biomass and waste plastics.

6.5.5 Fluidised Bed Bioreactors

Bioreactors using immobilised enzymes are similar to catalysed chemical reactors. The growth of microorganisms or the dynamic nature of biofilm thickness is absent. Emery and Cardos (1978) proposed a steady-state 1D homogeneous model for fluidised beds using an immobilised enzyme as the catalyst for conversion of starch to glucose. It is a liquid fluidised bed and exhibits particulate fluidisation. The reactor model is

$$v_z \frac{\partial C_A}{\partial z} = \hat{D} \frac{\partial^2 C_A}{\partial z^2} - r_A \tag{6.62}$$

The reaction rate is as follows: Michaelis–Menton kinetics is given below in terms of mass units.

$$-r_A = \frac{162 k_2' E C_s}{\varepsilon V_R (K_M' + C_s)} \tag{6.63}$$

Here, E is the enzyme activity and V_R is the reactor volume. Substituting the rate constant in the reactor model and using dimensionless parameters, Equation 6.59 reduces to

$$\frac{\hat{D}}{v_z L} \frac{\partial^2 C_A}{\partial Z^2} - \frac{\partial C_A}{\partial Z} - \frac{\gamma}{Q} C_A = 0 \tag{6.64}$$

where

$\gamma = \dfrac{162 k_M' E}{K_M'}$ without any mass transfer resistance

$\gamma = \dfrac{1}{\left(\dfrac{K_M'}{162 k_2' E} + \dfrac{d_p}{6(1-\varepsilon) k_1 V_C} \right)}$ after consideration of mass transfer resistances

Q = the volumetric flow rate
V_C = the volume of the immobilised enzyme

The heat generation is negligible, and fluidised beds behave as isothermal reactors. Therefore, the conservation equation for heat was not required. Equation 6.64 is the same as Equation 4.133 with a first-order concentration term for which the analytical solution is given by Equation 4.129. However, while using Equation 4.133 to predict an axial concentration profile for the substrate, a constant value of γ was considered. At a given flow rate, the mass transfer coefficient and the porosity of the bed remains constant. As the flow rate of the fluid is increased, these parameters and the dispersion coefficient also change.

6.5.6 Two-Fluid Models

In the bubbling fluidised beds, a part of the gas passes through interstitial space between the particles. The amount is independent of superficial gas velocity and is equal to the total gas flow at incipient fluidisation. The remaining gas flows in the form of bubbles (Figure 6.7). In a two-fluid model, it is assumed that there are two phases. The emulsion phase contains all the particles and the gas passing through it. The bubble phase, also called the 'dispersed phase', contains only gas. It may contain a few particles also. Thus, the two fluids are not gas and solid but emulsion and bubble. They may also be called 'continuous phase', 'dispersed phase' or heavy and light phases. The laws of conservation of mass, momentum and heat are written for each phase separately.

Total superficial gas velocity is given in terms of superficial velocity for each phase.

$$U = \sum_j \varepsilon_j U_j \tag{6.65}$$

Similarly, the total flow rates and flow rate for each species are given by

$$F = \sum_j F_j \tag{6.66}$$

$$F_j = \varepsilon_j U_j A C_{Aj} = \varepsilon_j U_j A \sum_i C_{Aji} \tag{6.67}$$

$$F_{ji} = \varepsilon_j U_j A C_{Aji} \tag{6.68}$$

Constantineau et al. (2007) developed a steady-state 1D two-fluid model to predict conversion in a two-phase fluidised bed reactor. The model was applicable to a fluidised bed operating in bubbling as well as slugging regimes. The conservation of mass for species can be written for both phases as

$$\frac{d}{dz} F_{ji} + (-1)^j k_{12} \varepsilon_1 \left\{ \sum_j (-1)^j \frac{F_{ji}}{\varepsilon_j U_j} \right\} + k \phi_j b_i \frac{F_{j,rxn}}{U_j}$$

$$- (-1)^j k \phi_2 \Delta b_i \frac{F_{2,rxn}}{U_2} \frac{F_{1i}}{U_1} = 0, \ j = 1,2 \tag{6.69}$$

where $j = 1$ for the dispersed phase or bubble phase and $j = 2$ for the continuous phase or emulsion phase. The second term corresponds to the mass transfer between the two phases and the generation of species i due to reaction.

It may be noted that $k_{1-2} = -k_{2-1}$. The third term is gas produced due to reaction. Because the solids are present in emulsion as well as in the bubble phase, the reaction takes place in both phases. The excess gas produced goes to the bubble phase. The flow rate of gas in the emulsion phase remains constant. The fourth term takes into account the transfer of extra gas due to a change of volume during the reaction. The difference in gas total stoichiometric coefficients due to chemical reaction is b_i. The volume fraction of solids in phase j is ϕ_j.

The boundary conditions are as follows:

At $z = 0$: $F_{ji} = \varepsilon_j U_j A C_{Aji}\big|_{z=0}, j = 1,2$

The two differential equations given by Equation 6.6 can be solved numerically provided the transport properties and hydrodynamic parameters (e.g. mass transfer coefficient, phase holdups and interfacial area) are known. The holdup of the bubble phase is given as

$$\varepsilon_1 = \frac{U - U_{mf}}{U_1} \tag{6.70}$$

The differential equation is to be integrated from $z = 0$ to $z = L$. The expanded bed height, L, can be obtained from

$$L - L_{mf} = \int_0^L \varepsilon_1 \, dz \tag{6.71}$$

The following correlation (Sit and Grace 1981) for mass transfer coefficient, k_{12}, was used by Constantineau et al. (2007):

$$k_{12} = \frac{U_{mf}}{3} + 2\left(\frac{2D_A \varepsilon_{mf} U_{b\infty}}{\pi d_b}\right)^{1/2} \tag{6.72}$$

The interfacial area for spherical particles is given by

$$a_i = \frac{6}{d_b} \tag{6.73}$$

For the bubbling fluidised bed:

$$U_{b\infty} = 0.71\sqrt{g d_b} \tag{6.74}$$

The minimum fluidisation velocity, U_{mf}, can be estimated using Ergun's equation. The bubble velocity, U_b, is estimated from the following correlation (Thompson et al. 1999):

$$U_b = (U - U_{mf})\left(1 + \frac{0.71}{U}\sqrt{g d_b}\right) = U_1 \tag{6.75}$$

All other parameters can now be estimated. For a given reaction, the kinetic parameters are known. The model presented by Constantineau et al. (2007) is applicable to a slugging fluidised bed, provided the hydrodynamic parameters for a slugging fluidised bed are used.

The two-fluid model may fit the reactor data but with meaningless negative values of some of the parameters (Levenspiel 1999). As a result, the use of such a model as a generalised model for simulation of a fluidised bed reactor is of limited use. A modified model effort considered three phases: the bubble phase, the wake phase and the emulsion phase (Figure 6.7b and Figure 6.7c). The wake is a region dragged behind the bubble. It contains solids which are exchanged with the emulsion phase due to vortex shedding. The exchanges between the bubble and wake phases and the wake and emulsion phases take place. The wake volume is also required. A region adjacent to the bubble and surrounding it is called a 'cloud'. The transfer between cloud and bubble is also considered.

6.5.7 Three-Phase Fluidised Bed Bioreactors

The treatment of microbial reactions is somewhat different than that of enzyme-catalysed reactions. The microorganisms are supported on the particle in the form of a biofilm. The biochemical reactions take place in the biofilm. It also results in the growth of the biofilm. The biofilm is not porous; hence, the only mechanism of mass transfer is the diffusion. When the particles are suspended, the mass transfer between the liquid takes place through a fictitious liquid film (film model), diffuses into the biofilm and is consumed (Figure 6.8). It is generally assumed that the support is non-porous and is totally inert. Mowla and Ahmadi (2007) studied biodegradation of hydrocarbon-polluted water in a three-phase fluidised bed bioreactor. It was assumed that the oxygen penetrated the biofilm completely so that the reaction was not oxygen limited.

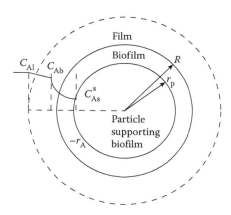

FIGURE 6.8
Mass transfer in a biofilm.

The reaction rate under such conditions was described by the Monod kinetics. Let us assume that the reaction rate in the biofilm is very slow; therefore, the reaction may be approximated as a first-order reaction.

$$-r_A = \frac{\mu_{max} X C_A}{Y(K_c + C_A)} = \frac{\mu_{max} \rho_{bf} C_A}{Y K_c} \qquad (6.76)$$

where X is the biomass concentration, Y is the yield coefficient and μ_{max} is the maximum specific growth rate. The density of the dry biofilm, ρ_{bf}, is given by

$$\rho_{bf} = \frac{W}{N_p (4/3) \pi (R^3 - r_p^3)} \qquad (6.77)$$

The equation of continuity for a chemical species in the biofilm involves diffusion and reaction only.

$$D_A \left(\frac{\partial^2 C_A}{\partial r^2} + \frac{2}{r} \frac{\partial C_A}{\partial r} \right) = -r_A \qquad (6.78)$$

The boundary conditions are as follows:

At $r = r_p$: $\dfrac{\partial C_A}{\partial r} = 0$

At $r = R$: $C_A = C_A^s$

The concentration profile in the biofilm is given by

$$C_A = \frac{C_A^s R}{r} \frac{\lambda r_p \cosh \lambda(r - r_p) + \text{Sinh}\lambda(r - r_p)}{\lambda r_p \cosh \lambda(R - r_p) + \text{Sinh}\lambda(R - r_p)} \qquad (6.79)$$

where $\lambda = \sqrt{\dfrac{\rho_{bf} k}{D_A}}$ and k is the reaction rate constant. The effectiveness factor, η, is defined as

$$\eta = \frac{N_p 4\pi R^2 \left(-D_A \dfrac{dC_A}{dr}\bigg|_{r=R} \right)}{k C_A X_W} \qquad (6.80)$$

where X_W is the total mass of the biomass.

Equation 6.80 can be evaluated by determining differentiating Equation 6.79 to give

$$\eta = \frac{3(R/r_p)^3}{(R/r_p)^3 - 1} \frac{1}{(\lambda R)^2} \frac{(R r_p \lambda^2 - 1)\text{Sinh}\left[\lambda(R - r_p)\right] + (R - r_p)\lambda \cosh \lambda(R - r_p)}{\lambda r_p \cosh \lambda(R - r_p) + \text{Sinh}\lambda(R - r_p)} \qquad (6.81)$$

Equation 6.78 is the kinetic sub-model which is to be used in the reactor model. The reaction rate can now be expressed in terms of the concentration at the surface of the biofilm.

$$-r_A = \eta k C_A^s X_W \tag{6.82}$$

Mowla and Ahmadi (2007) used a 1D steady-state reactor model. The concentration at any axial position is C_A^b. Taking mass balance at the surface of the biofilm, we get

$$k_1 A_b \left(C_A^b - C_A^s \right) = \eta k C_A^s X_W \tag{6.83}$$

The concentration at the surface of the biofilm at any axial position can now be expressed after rearrangement of Equation 6.83.

$$C_A^s = \frac{C_A^b}{\left(1 + \dfrac{\eta k X_W}{k_1 A_b} \right)} \tag{6.84}$$

Application of the law of conservation of mass gives

$$v_z \frac{\partial C_A^b}{\partial z} = \hat{D} \frac{\partial^2 C_A^b}{\partial z^2} + \frac{\eta k X_W}{\left(1 + \dfrac{\eta k X_W}{k_1 A_b} \right)} C_A^b \tag{6.85}$$

The following boundary conditions are applied:

At $z = 0$: $v_z C_A^b \big|_{z=0} + \varepsilon_1 E_1 \dfrac{dC_A^b}{dz} = v_z C_A^b$

At $z = L$: $\dfrac{dC_A^b}{dz} = 0$

The equation of change of thermal energy is not considered as the heat of reaction involved is negligible. Equation 6.82 was solved to give the outlet concentration. Equation 6.85 is the same as Equation 4.133; hence, the solution is given by Equation 4.128.

6.5.8 Dynamic Model for Three-Phase Fluidised Bed Bioreactors

The three-phase fluidised bed involves all three phases. The particles are kept in suspension by the liquid motion in the upward direction. The gas is also introduced in the column, but the fluid motion induced by the gas phase is insufficient to keep the solids in suspension. Many solid-catalysed gas–liquid reactions are carried out in the three-phase fluidised beds.

Aerobic bioreactors require a supply of oxygen, the introduction of nutrients and microorganisms or enzymes supported on catalysts. Thus, the presence of gas, liquid and solid phases is quite common. While high mechanical stirring is undesirable, the use of packed beds results in microbial fouling. Due to these reasons, it is desirable to maintain the microorganism or immobilised enzymes in a suspended state. Three-phase fluidised beds and slurry bubble columns are the types of bioreactors which meet this requirement.

The bioreactor models differ from chemical reactors due to differences in the nature of kinetic models and presence of biofilms. The thickness of the biofilms changes with time; hence, the density of the particles on which the microorganisms are supported also changes. The bed expansion behaviour and hence the dispersion coefficient also exhibit temporal changes. The mass transfer resistance in the biofilm is dynamic in nature. The bioprocesses require a large residence time due to slow reaction rates. However, the fluidised bed requires the liquid to flow; hence, the residence time is less. The residence time of the liquid is enhanced by recirculating the liquid. All these make a case for using dynamic models in fluidised bed bioreactors.

A general dynamic 1D heterogeneous model was given by Fuentes et al. (2009). They presented a three-fluid model for the bioreactor involving 24 species (Figure 6.9). The general model is being presented here without going into the details of the biochemical reactions. It may be used in other cases also. The equation of change of momentum for each of the fluids requires a few additional terms in the Navier–Stokes equations presented in Chapter 4.

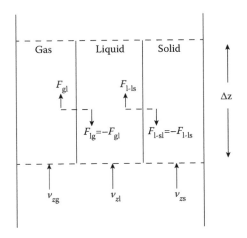

FIGURE 6.9
Model sketch for three-phase fluidised beds.

$$\begin{bmatrix} \text{Accumulation} \\ \text{rate of momentum} \end{bmatrix} = \begin{bmatrix} \text{Convective} \\ \text{momentum flux} \end{bmatrix} + \begin{bmatrix} \text{Molecular flux} \\ \text{of momentum} \end{bmatrix} + \begin{bmatrix} \text{Other forces} \end{bmatrix}$$

$$(6.86)$$

The other forces are forces due to interactions with another force. The forces which act on each phase are gravitational, pressure and interaction between two phases. These forces are responsible for the variation of velocity within a phase. The gravitational force is given by

$$F_{gi} = -\varepsilon_i \rho_i g \tag{6.87}$$

where $i = 1$ for gas, 2 for liquid and 3 for solid. The consistence equation should also hold:

$$\sum \varepsilon_i = 1 \tag{6.88}$$

The pressure force is due to hydrostatic head and is given by

$$F_{pi} = \varepsilon_i \left(\sum_i \varepsilon_i \rho_i \right) g \tag{6.89}$$

There are three phases; hence, there will be three binary interaction terms between gas and liquid, between liquid and solid, and between gas and solid. These forces are buoyancy and drag forces. The former is due to the relative motion of the phases, and the latter is due to density difference. The interaction forces between the gas and liquid phases are given by

$$F_{I,lg} = -F_{I,gl} = \frac{36 \xi_g \mu_1 \left(v_{zg} - v_{zl} \right)}{d_b^2} \tag{6.90}$$

The interaction forces between the gas and liquid phases are given by

$$F_{I,sl} = -F_{I,ls} = \xi_s \left(\rho_s - \rho_1 \right) g \left(\frac{v_{zg} - v_{zs}}{U_{b\infty}} \right)^{4.8/n} \xi_1^{-3.8} \tag{6.91}$$

Here, ξ_g and ξ_s are the fractions of gas and solid in gas–liquid and liquid–solid dispersed systems. The equation of change for the momentum can now be written in terms of the holdups:

$$\frac{\partial}{\partial t}\left(\varepsilon_j \rho_j v_{zj}\right) + \frac{\partial}{\partial z}\left(\varepsilon_j \rho_j v_{zj}^2\right) = \frac{\partial}{\partial z}\left(\mu_j \frac{\partial}{\partial z}\left(\varepsilon_j \rho_j v_{zj}\right)\right) + F_{ik} + F_{pj} + G_{gj} \tag{6.92}$$

where $j = 1,2,3$.

The second term represents convection. The equation of continuity for each of the gas ($j = 1$), liquid ($j = 2$) and solid ($j = 3$) phases was given as

$$\frac{\partial}{\partial t}\left(\varepsilon_j \rho_j\right) + \frac{\partial}{\partial z}\left(\varepsilon_j \rho_j v_{zj}\right) = \frac{\partial}{\partial z}\left(\hat{D}_j \frac{\partial}{\partial z}\left(\varepsilon_j \rho_j\right)\right) + \sum T_{ij} \tag{6.93}$$

where $j = 1,2,3$, $\sum T_{jj} = 0$.

The last term, $\sum T_{ij}$, accounts for mass transfer from one phase to another across the interface. For completely wet solids, there is no interface between gas and solid; therefore, only one term for mass transfer across the interface will be present. This term will consist of the mass transfer coefficient between gas and liquid. The liquid forms two interfaces, one with the gas and another with the solids. Therefore, two mass transfer terms will be present.

The equation of continuity for each species in the gas, liquid and solid phases can be written as

$$\frac{\partial}{\partial t}\left(\varepsilon_j C_{ij}\right) + \frac{\partial}{\partial z}\left(\varepsilon_j C_{ij} v_{zj}\right) = \frac{\partial}{\partial z}\left(D_{ij} \frac{\partial}{\partial z}\left(\varepsilon_j C_{ij}\right)\right) + \sum_k R_{ijk} + \sum T_{ij}; \ j = 1,2,3 \tag{6.94}$$

The term $\sum_k R_{ijk}$ indicates that there are k generation terms for ith species corresponding to k reactions taking place in phase j. These equations are written for each of the species. Equation of change of thermal energy is not required for biological systems. Fuentes et al. (2009) solved the set of Equations 6.92 through 6.94 numerically for 24 species and 18 variables.

The effect of variation of the thickness of the biofilm on the density of the particle and bed expansion was also taken into account.

The diameter of the particle with biofilm thickness is

$$d_{bp} = d_p + 2\delta \tag{6.95}$$

The volume fraction of the biofilm is

$$f_{bp} = \left(\frac{d_{bp}}{d_p}\right)^3 - 1 \tag{6.96}$$

The density of the particle with biofilm is

$$\rho_{bp} = \frac{\rho_p + f_{bp}\rho_{bf}}{1 - f_{bp}} \tag{6.97}$$

The fraction of the biofilm is governed by the bioprocesses taking place in the bioreactor.

6.6 Summary

A model to predict the performance of a multiphase reactor can be developed using the laws of conservation of mass, momentum and thermal energy. However, because these models are applicable in a single phase, the equation of continuity for a chemical species contains terms corresponding to the generation of chemical species due to chemical reaction and terms to take into account for generation or disappearance due to mass transfer across the boundary.

Most of the reactors – such as packed beds, fluidised beds and slurry bubble columns – have cylindrical geometry. Therefore, these equations are written in cylindrical coordinates. The variables may be considered to vary in axial (longitudinal) or axial and radial directions resulting in 1D or 2D models. In packed beds, the bed may be modelled as a pseudo-homogeneous or heterogeneous process.

The model for reactors involves four sets of equations. The reactor model is obtained using the laws of conservation. The kinetic models are based on the reactions taking place in different phases. It is necessary to know the phase in which a reaction takes place. The thermodynamic model provides a set of equations to estimate the concentration at the interface. The hydrodynamic models are a set of empirical and theoretical equations to describe variables such as holdup and mass and heat transfer coefficients.

The role of the particle can change the model equations. Different model equations and boundary conditions are obtained if the solids are used as inert, reactant or catalyst solids. Slow and fast reaction rates affect the controlling steps in the kinetic sub-model.

The trickle bed reactors and fluidised bed reactors exhibit different flow regimes. The model equations for the same reactor operating in different flow regimes will be different due to differences in the assumptions made. The bioreactors may not consider the equation of change of thermal energy.

The fluidised beds are modelled as heterogeneous reactors. The reactor models are often similar to the model equations for the packed beds. But the hydrodynamics of fluidised beds are quite different than that for other reactors. The hydrodynamics also differs from each other for different regimes of fluidisation (e.g. particulate, aggregative, slugging and circulating fluidised beds).

Two-fluid models and three-phase models are available in the literature. A general methodology is to treat the assumption of continuum for different phases separately. The laws of conservation are written for each species in different phases.

The biochemical reactions are quite commonly carried out in three-phase fluidised beds. The bioprocesses taking place in the biofilm can be used as sub-models.

In general, a model for a multiphase reactor can be made of a few sub-models for the reactor, thermodynamics, kinetics and hydrodynamics.

References

Attou, A., and Ferschneider, G. 2000. A two-fluid hydrodynamic model for the transition between trickle and pulse flow in a cocurrent gas-liquid packed-bed reactor. *Chem. Eng. Sci.* 55: 491–511.

Constantineau, J.P., Grace, J.R., Lim, C.J., and Richards, G.G. 2007. Generalized bubbling–slugging fluidized bed reactor model. *Chem. Eng. Sci.* 62: 70–81.

Efthimiadis, E.A., and Sotirchos, S.V. 1993. Experimental validation of a mathematical model for fixed-bed desulfurization. *AIChE J.* 39(1): 99–110.

Emery, A.N., and Cardoso, J.P. 1978. Parameter evaluation and performance studies in a fluidized-bed immobilized enzyme reactor, *Biotechnol. and Bioeng.* 20: 1903–1929.

Forment, G.F., and Bischoff, K.B. 1979. *Chemical Reactor Analysis and Design*. New York: John Wiley.

Fuentes, M., Mussati, M.C., Scenna, N.J., and Aguirre, P.A. 2009. Global modeling and simulation of a three-phase fluidized bed bioreactor. *Comput. Chem. Eng.* 33: 359–370.

Geldart, D. 2004. Expansion of gas fluidized beds. *Ind. Eng. Chem. Res.* 43: 5802–5809.

Gunjal, P.R., Ranade, V.V., and Chaudhari, R.V. 2005. Hydrodynamics of trickle-bed reactors: Experiments and CFD modeling. *Ind. Eng. Chem. Res.* 44: 6278–6294.

Jess, A., and Kern, C. 2009. Modeling of multi-tubular reactors for Fischer-Tropsch synthesis. *Chem. Eng. Technol.* 32(8): 1164–1175.

Koning, G.W., Kronberg, A.E., and Swaaij, W.P.M. 2006. Improved one-dimensional model of a tubular packed bed reactor. *Chem. Eng. Sci.* 61: 3167–3175.

Krause, P., and Fieg, G. 2011. Experiment based model development for the enzymatic hydrolysis in a packed-bed reactor with biphasic reactant flow. *Chem. Eng. Sci.* 66: 4838–4850.

Lerou, J.J., and Froment, G.F. 1982. Fixed-bed reactor design and simulation. In *Computer-Aided Process Design*, ed. M.E. Leesley, 367–383. Houston, TX: Gulf.

Levenspiel, O. 1999. *Chemical Reaction Engineering*, 3rd ed. New York: John Wiley.

Missen, R.W., Mims, C.A., and Saville, B.A. 1999. *Introduction to Chemical Reaction Engineering and Kinetics*. New York: John Wiley.

Mowla, D., and Ahmadi, M. 2007. Theoretical and experimental investigation of biodegradation of hydrocarbon polluted water in a three phase fluidized-bed bioreactor with PVC biofilm support. *Biochem. Eng. J.* 36(2): 147–156.

Richardson, J.F., and Zaki, W.N. 1954. Sedimentation and fluidisation: Part 1. *Chem. Eng. Res. Des.* 32: 35–53.

Sit, S.P., and Grace, J.R. 1981. Effect of bubble interaction on interphase mass transfer in gas fluidized beds. *Chem. Eng. Sci.* 36: 327–335.

Thompson, M.L., Bi, H.T., and Grace, J.R. 1999. A generalized bubbling/turbulent fluidized-bed reactor model. *Chem. Eng. Sci.* 54: 2175–2185.

Toledo, E.C.V., Santana, P.L., Maciel, M.R.W., and Filho, R.M. 2001. Dynamic modelling of a three-phase catalytic slurry reactor. *Chem. Eng. Sci.* 56: 6055–6061.

Wang, J., Anthony, R.G., and Akgerman, A. 2005. Mathematical simulations of the performance of trickle bed and slurry reactors for methanol synthesis. *Comput. Chem. Eng.* 29: 2474–2484.

7

Population Balance Models and Discrete-Event Models

Models based on the application of laws of conservation of momentum, mass and energy have been described in Chapters 4 to 6. These models were also applied to multiphase systems. Different phases were considered by either making the assumption of continuum or writing different equations for different phases (e.g. a two-fluid model for fluidised beds). A detailed description of phases is not considered; as a result, the models have limited predictive capabilities. The information lost due to the assumption of continuum for different phases is examined in this chapter.

Let us consider a process vessel in which a phase enters, is dispersed, and is collected from the outlet. The inlet and outlet streams may be a continuous phase (e.g. if gas is sparged in the liquid, the dispersion of gas in the form of bubbles takes place in the process vessel). When the gas comes out, it is a continuous fluid. The properties of the gas leaving the vessel can be defined without requiring the properties of individual bubbles. By assuming the gas phase to be continuous, the information about the gas bubbles in the process vessel and hence any changes in the properties of the gas due to bubble behaviour in the process vessel are lost. The properties of the product stream are not lost since, at the exit, the product stream may form a homogeneous mixture. An assumption of continuum for gas may be accepted. The same may be true for liquid also but cannot be applied to solids, because the solids leaving the process vessel may not be uniform in size unless they are melted and mixed. It is not possible to predict the properties related to non-uniformity of the leaving solid stream and the information about the dispersed phase in the process vessel. These properties can be particle size distribution and all of the properties related to it. The assumption of continuum does not fulfil the aim of modelling, if these properties are to be predicted.

The dispersed-phase units are of different sizes and move with different velocities. Both are random variables and can be described by probability density functions. These and similar random variables should be properly taken into consideration while modelling a multiphase system for the prediction of product quality.

The modelling of the multiphase systems involving the dispersed phase may consider both the continuous phase and the dispersed phase as two independent continua. These occupy the same phase but with different velocities and volume fractions for each phase. This approach is called the Eulerian–Eulerian approach. The models discussed in Chapter 6 are examples of this approach.

If the assumption of continuum is made only for the continuous phase and the dispersed phase is treated by identifying each individual separately, then it is known as the Eulerian–Lagrangian (E-L) approach. In this approach, the dispersed units (individuals) are considered as point sources.

7.1 Stochastic Models

The need to consider the random nature of the movement of the dispersed phase, such as bubbles and particles, led to the development of stochastic models. Some of the simple models were based on the assumptions that the velocity of the dispersed phase changed randomly. The time and position were taken as discrete variables. The variation of the velocity was according to specific stochastic rules used to determine the probability of change of velocity from one value to another. The nature of the simple stochastic model can be understood by the following example.

Iordache and Munteean (1981) proposed a stochastic model to predict gas holdup in a bubble column. It introduced a number density, $n_k(x,v_k,t)$, to represent the number of bubbles with velocity v_k with a position between x and $x + \Delta x$ at time t. The velocity of the bubble is influenced by either the fluid surrounding it or any other bubble approaching it. Let a bubble interact with fluid with probability $a\Delta t$ and with another bubble with probability $(1 - a\Delta t)$. Let P_{ek} be the probability that a bubble changes its velocity from v_k to v_e during time interval Δt due to bubble–bubble interaction. It is assumed that the bubble–fluid interaction does not change the velocity of the bubble. The number of bubbles with velocity v_k with a position between x and $x + \Delta x$ at time $t + \Delta t$ can be written as

$$n_k\left(x,v_k,t+\Delta t\right)\Delta x = \left[\left(1-a\Delta t\right)\sum_{e=1}^{m} P_{ek} n_e\left(x - v_k\Delta t, v_{ek}, t\right)\right]\Delta x$$

$$+ a\Delta t \sum_{e=1}^{m} n_k^0\left(x - v_k\Delta t, v_k, t\right)\Delta x \tag{7.1}$$

Equation 7.1 makes use of probabilities, which are not easy to estimate or measure experimentally. The probability, P_{ek}, was given by

$$P_{ek} = b n_k\left(x, v_e, t\right)\left(v_k - v_e\right)\Delta t, \ k,e = 1,2,....,m \ \left(v_e < v_k\right) \tag{7.2}$$

where b is the proportionality constant. With this definition of the transition probability, Equation 7.1 is reduced to the following equation in terms of continuous variables:

$$\frac{\partial n}{\partial t} + v_k \frac{\partial n}{\partial x} = -a\left(n - n^0\right) - b.m.n\left(\bar{v} - v_k\right) \tag{7.3}$$

The number density in dilute dispersions is n^0. In dilute dispersions, the bubbles are far from each other so that they do not influence the motion of any other bubble. Thus, the number density in dilute dispersions is represented by n^0. The symbol m represents the number of bubbles measured in a linear dimension. The linear concentration of bubbles and average velocity of bubbles are given by

$$n(x,t) = \int_0^\infty n(x,v,t)dv = \int_0^\infty n^0(x,v,t)dv \tag{7.4}$$

$$\bar{v}(x,t) = \frac{\int_0^\infty vn(x,v,t)dv}{\int_0^\infty n(x,v,t)dv} \tag{7.5}$$

Assuming uniform distribution in the entire column and at steady state, the left side of Equation 7.3 is zero, and hence the solution is given by

$$n(v) = \frac{n^0(v)}{1 + \frac{b}{a}m\left(\bar{v} - v\right)} \tag{7.6}$$

Iordache and Corbu (1986) used Equations 7.3 through 7.6 to describe the sedimentation also. Use of Equation 7.6 was validated for dilute solutions only.

The bubble coalescence and bubble breakup observed in bubble columns were not considered by the model. The dispersed-phase systems are more complex than the above model since the growth or shrinking of particles due to chemical reaction and mass transfer may influence the size of the particle. In brief, the model ignored the changing size of the dispersed phase during the process. In many processes (e.g. crystallisation and polymerisation), it becomes more important.

7.2 The Complex Nature of the Dispersed Phase

It is a formidable task to keep track of each dispersed-phase unit (i.e. the point source). The situation is simplified by distributing the entire phase into several groups. Let us call the entire group a population and each unit of the dispersed phase an individual. The individuals can be counted. However, the number of individuals may vary with time and may depend upon the position in the process vessel. The number of individuals may increase due to breakage and creation of new nuclei. These are analogous to a birth process in a population of living species. The number of individuals may decrease due to agglomeration of individuals, coalescence and similar processes. The individuals may grow in size (e.g. in a crystallisation process). Microbial fermentation involves the actual birth and death processes of microorganisms. If a reaction is considered at the molecular level, then the generation or depletion of chemical species may be identified as birth and death processes. All these processes are similar in nature. Due to the complexity of the problem, a statistical description of the process is a comfortable simplification since the properties of the product stream are generally described statistically. Thus, instead of tracking the size of each individual, the sizes of these point sources are described statistically using the probability distribution function. The dispersed-phase units may be characterised by many variables and have been termed 'particle vectors' (Ramakrishna 2000). These characteristics may be position, size, shape, density and sometimes the concentration profile within the particle. The choice of the variables in a particle vector depends upon the processes being affected by these variables. The term 'particle vector' was used while treating the particulate system (Ramakrishna 2000). In the case of gas–liquid systems, one may call it a 'bubble vector', or in the case of emulsions, the term 'drop vector' may be used. In general, let us call it a 'dispersed-phase vector'. Similarly, the properties of the continuous phase may be expressed as a 'continuous-phase vector'.

The interphase transfer processes can now be looked on as the local properties of the continuum phase affecting the dispersed-phase vector, and the dispersed-phase vector as affecting the surrounding continuous phase.

7.2.1 Size Variation of the Dispersed Phase

The changes in the size of the dispersed phase are due to the following factors:

1. *Mass transfer between dispersed and continuous phases*: Mass transfer between the phases changes the size of the dispersed units due to the accumulation of mass. The change of size may be termed 'swelling' or 'shrinking' phenomena. The number of the dispersed phase does not change.

2. *Chemical reaction in the dispersed phase*: If a chemical reaction takes place in the dispersed phase, it may result in the depletion or production of mass in the dispersed phase. The size of the dispersed-phase individual may decrease or increase accordingly. If the reaction takes place in the fluid phase, it changes the properties of the continuous phase and does not affect the size of the dispersed-phase unit. The number of the dispersed phase does not change due to chemical reaction.

3. *Coalesce and breakup*: The collision between the dispersed phase results in coalescence. It may be treated as both the disappearance of dispersed-phase individuals and the generation of a new, large

dispersed-phase individual. The net outcome is the change of the number of population. A dispersed-phase individual can also break into two or even more smaller units of the dispersed phase. It is equal to the disappearance of a single dispersed-phase unit and the appearance of two or more smaller units of the dispersed phase.

4. *Nucleation*: Many processes (e.g. crystallisation) involve the nucleation of new sites. It is equivalent to the generation of new dispersed-phase units increasing the total population.

The mass transfer and chemical reaction do not change the number of dispersed phase, but they change the size of the dispersed phase. Coalescence, breakup and nucleation immediately change the number of the dispersed phase. The size also changes as a consequence of agglomeration or division into smaller units.

7.2.2 Movement of the Dispersed Phase

The dispersed phase may move within the system. If the dispersed-phase units are so small that Brownian motion can be considered, then the net movement is due to concentration difference of the particles. The dispersed-phase motion is governed by diffusion and local velocity of the continuous phase and is independent of the size of dispersed-phase unit. This situation is applicable in the case of aerosols and similar systems.

Large dispersed-phase units require a force balance to estimate acceleration as a function of time. The forces acting on dispersed-phase units may be due to forces (e.g. buoyancy and drag forces). The velocity of the dispersed phase depends upon its size.

Interaction of the dispersed-phase units with a wall is not described by shear force. The dispersed-phase units may be either repelled by the wall or absorbed by the wall. In the latter case, the particles are deposited on the wall.

7.2.3 Discrete Nature of Time

All the processes taking place in the system can now be classified as the processes changing the population of the dispersed phase and the processes changing the dispersed-phase vector. The former are known as birth and death processes, and the latter is used while applying laws of conservation of momentum, mass and energy.

Birth and death processes are not continuous. They take place at a particular time; therefore, the processes cannot be described as a continuous function of time. Every time a birth or death occurs, the behaviours of the individual and hence the population change. It becomes necessary to treat the time as a discrete variable.

The problem now is how to keep track of the population size and the dispersed-phase vector. The former is formulated by population balance. The latter is estimated by application of the appropriate laws of conservation.

In situations where change in the continuous-phase vector has no effect on the dispersed-phase vector, the population balance is not required. The assumption of continuum for the dispersed phase also simplifies the model.

7.3 Population Balance Equation

The population of the dispersed phase is tracked by dividing the group of dispersed phases into various smaller groups and then counting all the dispersed-phase individuals in all the groups. The number in each such group may increase or decrease due to the coalescence and breakup and for other reasons discussed in this chapter. It may be noted that bubble coalescence combines two smaller dispersed-phase units to decrease the number in each size group, but it also increases the number in a larger size group. Similarly, the breakup of a dispersed-phase unit decreases the number in that size group, but as smaller dispersed-phase units, called 'daughters', are generated, the number in that size group increases. Let us consider that in a continuous stirred tank of volume, V, the feed containing few dispersed-phase units enters the system. The rate of change of number of dispersed-phase units of size, r, can be written as

$$
\begin{bmatrix} \text{Accumulation rate} \\ \text{of dispersed units} \\ \text{of size } 'r' \end{bmatrix} + \begin{bmatrix} \text{Rate at which the} \\ \text{dispersed units of} \\ \text{size } 'r' \text{ enters in feed} \end{bmatrix} - \begin{bmatrix} \text{Rate at which the} \\ \text{dispersed units of} \\ \text{size } 'r' \text{ leaves the system} \end{bmatrix}
$$

$$
\begin{bmatrix} \text{Rate of change of} \\ \text{size of dispersed} \\ \text{units of (size } <r - \text{size } r) \\ \text{due to growth} \end{bmatrix} = \begin{bmatrix} \text{Generation due to} \\ \text{breakage of dispersed} \\ \text{units of size } >r \end{bmatrix} - \begin{bmatrix} \text{Disappearance due to} \\ \text{breakage of dispersed} \\ \text{units of size } 'r' \end{bmatrix}
$$

$$
+ \begin{bmatrix} \text{Generation of dispersed} \\ \text{units of size } 'r', \text{ due to} \\ \text{coalescence of small units} \end{bmatrix} - \begin{bmatrix} \text{Disappearance of dispersed} \\ \text{units of size } 'r' \text{ due to} \\ \text{coalescence with other units} \end{bmatrix} \qquad (7.7)
$$

The terms on the left-hand side of Equation 7.7 correspond to accumulation, inlet–outlet streams and swelling or shrinking of the dispersed-phase units. If a bubble grows beyond the upper size limit, it is transferred to a class of larger particles. The number of dispersed-phase units decreases in this class. The terms on the right-hand side, also called the 'generation terms', correspond to breakage and coalescence phenomena which are random processes and should be treated in terms of probabilities. Let the number of dispersed-phase units per unit volume be n. The number probability density function is $f(D)$. Hence, the number of dispersed-phase units having size between D and $D + \Delta D$ is $nf(D)\, dD$.

Ramakrishna (2000) denoted the number probability density function by the 'particle vector', $f(x, D, t)$, where D is the internal vector and x is the external vector. The internal vector specifies all properties of a dispersed-phase unit that affect the number of particles. The external vector refers to the local properties related to the surrounding continuous phase. Since the values of $f(x, D, t)$ are always estimated after each time step and using the properties of the surrounding fluid, the x and t are being dropped to give a simpler look to the equation.

The dispersed-phase units of sizes larger than D may break and may produce dispersed-phase units of any two sizes. Only a fraction of such breakage will generate dispersed-phase units of size D. Let $\beta(D, D')$ be the probability distribution function for the daughter dispersed-phase units (i.e. it is the probability that a dispersed-phase unit of size D will be produced if a unit of size $D' > D$ breaks).

Not all dispersed-phase units have a chance to coalesce. This is taken into account by collision frequency, $h(D, D')$, which is the frequency of collision between the dispersed-phase units of sizes D and D'. Not all the collisions result in coalescence. Let $\lambda(D, D')$ be the collision efficiency (i.e. the fraction of the collisions that result in successful coalescence). The sizes of dispersed-phase units coalescing to form a dispersed-phase unit of size D are determined by the volume of coalescing dispersed-phase units.

Most of the models consider that during coalescence, only two dispersed-phase units are involved. In the case of agglomeration, more than one dispersed-phase unit may coalesce to form a larger unit. The former may be termed 'binary coalescence', and the latter may be called 'multi-dispersed-phase units'.

Equation 7.7 may now be written as follows (Singh et al. 2009):

$$\frac{\partial}{\partial t}\{nf(D)V\} + \{n_{out}f_{out}(D)F_{out} - n_{in}f_{in}(D)F_{in}\} + \frac{\partial}{\partial D}nf(D)\frac{\partial D}{\partial t}$$

$$= \int_D^\infty \beta(D,D')v(D')g(D')nVf(D')dD' - g(D)nf(D)V$$

$$\times \int_0^{a\sqrt{2}} \lambda\left\{\left(D^3 - D'^3\right)^{1/3}, D'\right\}h\left\{\left(D^3 - D'^3\right)^{1/3}, D'\right\}nf\left\{\left(D^3 - D'^3\right)^{1/3}\right\}nf(D)VdD'$$

$$- \int_0^\infty \lambda\{D,D'\}h\{D,D'\}nf\{D'\}VdD' \qquad (7.8)$$

Equation 7.8 is one of the general forms of the one-dimensional population balance equation. The term corresponding to swelling and shrinking of dispersed-phase units involves time derivative of D, which may be again a function of size, D, and position, z. The equation is used in many situations involving bubbles, drops or particles as the dispersed phase. The breakage kernel, or coalescence kernel, source terms can be considered sub-models for the model represented by Equation 7.8 (Figure 7.1). The sub-model describes the behaviour of the dispersed phase.

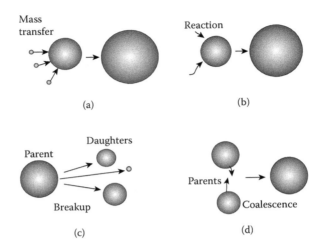

FIGURE 7.1
Factors affecting the size of individual dispersed-phase units: (a) mass transfer, (b) chemical reaction, (c) breakup of bubbles, droplets or particles and (d) coalescence of bubbles or droplets, and agglomeration of particles.

Various simplified forms of Equation 7.8 are available in the literature. Similar to a deterministic model, the population balance model may be a lumped parameter model without considering the spatial distribution of the dispersed-phase properties. One-dimensional model takes into account only the axial variation. It may consider velocity of the dispersed phase or may be assumed to move with a constant velocity. The models incorporating radial variation of the variables are complicated and require computational fluid dynamics (CFD) software. They are beyond the scope of the present discussion.

The development of various sub-models of population balance equation and its applications in chemical and biochemical processes are illustrated in Section 7.4. A few commonly used probability density functions in dispersed-phase systems also are described in Section 7.4.

7.4 Probability Distribution Functions

Equation 7.8 has terms with infinity as the integration limit. This seems unreasonable since the volume of the system is finite and hence the number of dispersed-phase units is also finite. However, if the number of dispersed-phase units is very large, then the probability density function may be approximated, being continuous. Some of the useful probability distributions functions used for smooth as well as discrete quantities are discussed throughout the remainder of this section.

7.4.1 Normal or Gaussian Distribution

The probability distribution function, $f(x)$, with mean, μ, and standard deviation, σ, is given as

$$f(x) = \frac{1}{\sigma\sqrt{2\pi}} \exp\left[-\frac{1}{2}\left(\frac{x-\mu}{\sigma} \right)^2 \right], \quad -\infty < x < \infty \tag{7.9}$$

Equation 7.9 is called 'normal distribution'. It is called 'standard normal distribution' if $\mu = 0$ and $\sigma = 1$. Several authors call Equation 7.9 'Gaussian distribution', and standard normal distribution is called normal distribution. Since this distribution produces negative values also, it cannot be used to represent the distribution of size; however, it can be used for the velocities of dispersed-phase units. The distribution is symmetric about the mean.

7.4.2 Logarithmic Normal Distribution

The probability distribution function, $f(x)$, is said to be log-normal distribution if $\log(x)$ follows normal distribution:

$$f(x) = \frac{1}{\sqrt{2\pi}\zeta x} \exp\left[-\frac{1}{2}\left\{ \frac{\ln(x) - \lambda}{\zeta} \right\}^2 \right], \quad 0 < x < \infty \tag{7.10}$$

The mean and standard deviations are λ and ζ, respectively. The values of x are always positive and can be used to describe the size of the dispersed-phase units. The mean and standard deviation of log-normal distribution are related to those of Gaussian distribution in the following way:

$$\zeta^2 = \ln\left(1 + \frac{\sigma^2}{\mu^2} \right) \tag{7.11}$$

and

$$\lambda = \ln\mu - \frac{1}{2}\zeta^2 \tag{7.12}$$

This distribution is skewed.

7.4.3 Poisson Distribution

The probability distribution function, $f(x)$, for Poisson distribution is given as

$$f(x) = \lambda^x \frac{e^{-\lambda}}{x!} \tag{7.13}$$

Both the mean and variance are λ.

Poisson distribution is valid under the following assumptions:

1. An event occurs at random (i.e. anytime and anywhere in the space).
2. The event in any time interval is independent of the event in any other non-overlapping time interval.
3. The probability of occurrence of an event in a small time interval, Δt, is proportional to the time interval.

For $\lambda > 9$, Poisson distribution may be approximated by the standard normal distribution with mean $= \lambda$ and variance $= \lambda$. Since in most applications, this condition is satisfied, the use of standard normal distribution is quite common.

7.4.4 Gamma Distribution

If the occurrence of an event follows Poisson distribution, then the time until the kth occurrence of an event is described by Gamma distribution, as given here:

$$f(x) = \frac{v(vx)^{k-1}}{(k-1)!}e^{-vx} \tag{7.14}$$

The mean time and standard deviations are k/v and k/V^2, respectively. When used as a general-purpose probability distribution, it is given as

$$f(x) = \frac{v(vx)^{k-1}}{\Gamma k}e^{-vx} \tag{7.15}$$

7.4.5 Beta Distribution

Beta distribution is suitable to describe the probability distribution of random numbers whose values are bounded between a and b. It is given by

$$f(x) = \frac{(x-a)^{q-1}(b-x)^{r-1}}{B(q,r)(b-a)^{q+r-1}}, a \le x \le b \tag{7.16}$$

$$= 0 \quad \text{elsewhere}$$

Here,

$$B(q,r) = \frac{\Gamma q \Gamma r}{\Gamma(q+r)}$$

The mean and variance are given as

$$\mu = a + \frac{q}{q+r}(b-a) \tag{7.17}$$

and

$$\sigma^2 = \frac{qr}{(q+r)^2(q+r+1)}(b-a)^2 \tag{7.18}$$

The distribution is useful when describing a size distribution that has lower and upper limits.

7.4.6 Exponential Distribution

For events occurring according to Poisson distribution, the 'time till the first event occurs' follows exponential distribution as given here:

$$f(x) = \frac{1}{b}e^{-x/b} \tag{7.19}$$

The mean and standard deviations are b and b^2, respectively.

7.5 Population Balance Models: Simulation Methodology

The population balance equation requires the information about the probability density function in Equation 7.8. The increase in the population is considered the birth of new dispersed-phase units. The decrease in the population is considered the death of dispersed-phase units. The mechanisms of birth and death are different for bubbles, drops and solid particles. A clear understanding of the mechanisms for coalescence, breakup, agglomeration etc. is necessary to obtain frequencies and probabilities. After defining the terms in Equation 7.8 completely, the simulation studies can be carried out. The flow field, temperature and concentration field for the continuous phase affect the size distribution of the dispersed phase. The effects of mass transfer and chemical reactions on the size distribution of the dispersed phase should be modelled. The velocity, coalescence and breakup of dispersed phase also affect the population density. The simulation methodology used by Darmana et al. (2005) is discussed to illustrate the effects of all factors on the evolution of population density. A discrete bubble model was used to study the gas–liquid reaction in a 3D bubble column. The E-L model presented by them combined CFD simulations also. This discussion will be limited to the method of considering hydrodynamics, mass transfer and reaction kinetics which can be developed for other situations also. Darmana et al. (2005) used four different time scales to combine all factors. These time scales, in increasing order, were for chemical reactions, bubble collisions, liquid–gas mass transfer and flow field of the liquid.

1. The mass and momentum balance for each bubble was solved first. The mass transfer rate was estimated using a two-film model between the liquid and the bubble. A correlation for the Sherwood number and enhancement factor due to chemical reaction was used. The Reynolds number requires bubble velocity. At this step, the fluid velocities and local concentrations of various chemical species are required. A weighted average was taken of the values of various variables at the vortices of a rectangular cell in which the bubble lies. The change in the size of the bubble was estimated using the mass transfer rate. The swelling or shrinking phenomena were thus considered.

2. In the next step, issues related to bubble collisions with neighbouring bubbles were resolved. The force balance on each bubble was applied to determine the acceleration of the bubble. The bubble velocity at the end of the time scale was estimated. The liquid flow field was computed.

3. The information after this step was used to determine the change in the concentration of each species due to chemical reaction. The sequential method of combining all the aspects of the model may be applied in many other situations.

7.5.1 Discrete Form of the Population Balance Equation

Equation 7.8 uses continuous probability density functions. Since it is an integro-differential equation, it has to be solved numerically. The numerical techniques to solve the population balance equation require discretisation of the variable.

The dispersed phase can be divided into several classes according to their sizes, d_i, with each of the class containing N_i dispersed-phase units. The distribution in classes can be on the basis of volume or mass. The most obvious choice is to use multiple small dispersed units, but this has many disadvantages, including a large number of classes increasing the computational effort. One of the approaches is to use a geometric grid. In a frequently used approach, the dispersed phase is distributed into classes such that the volume of each class is exactly twice as large as the volume of the preceding class. Thus, if the volume of the smallest class is v_1, the volume of a bubble belonging to the ith class is

$$v_i = 2v_{i-1} = 2^{i-1} v_1 = \frac{\pi}{6} d_i^3 \tag{7.20}$$

The regions are separated by Δd_i. The discrete number density in each, n_i, is given by

$$n_i = \frac{N_i}{\Delta d_i} \tag{7.21}$$

The geometric discretisation rule reduces the required number of classes to a small number; 6–10 classes are sufficient to provide useful results. Kiparissides et al. (2004) described the fractional geometric discretisation rule as follows:

$$v_i = 2^{1/q} \, v_{i-1} \tag{7.22}$$

where q is an integer. For $q = 1$, Equation 7.22 reduces the geometric discretisation rule given by Equation 7.20. The value of q can be taken between 2 and 4 (Kiparissides et al. 2004). The boundaries of the classes have been termed as 'pivots'. A dispersed phase may have any value of the volume, but it is assigned to two adjacent pivots by the following rule:

$$v = xv_i + (1-x)\, v_{i+1} \tag{7.23}$$

where x is the fraction of the volume assigned to class i. Kumar and Ramakrishna (1996) gave the following rule for assigning a fractional volume to discrete volumes in the case of coalescence:

$$n_{coal} = \frac{v_{i+1} - v}{v_{i+1} - v_i} \text{ for } v_i < v < v_{i+1}$$

and

$$n_{coal} = \frac{v - v_{i-1}}{v_i - v_{i-1}} \text{ for } v_{i-1} < v < v_i \tag{7.24}$$

In the case of breakup, the following is used:

$$n_{break} = \int_{v_i}^{v_{i+1}} \frac{v_{i+1} - v}{v_{i+1} - v_i} k(v, v_k) u(v_k) dV + \int_{v_{i-1}}^{v_i} \frac{v - v_{i-1}}{v_i - v_{i-1}} k(v, v_k) u(v_k) dV \tag{7.25}$$

Every dispersed-phase unit produces at least two – or, in general, several – daughter dispersed phases following the size distribution given by $k(v,v_k)$. Here, $u(v_k)$ is the number of daughter dispersed-phase units of volume v_k. In discrete form, the following population balance equation was used by Kumar and Ramakrishna (1996):

$$\frac{dN_i(t)}{dt} = \sum_{k=1}^{M} n_b g(v_k) N_k(t) + \sum_{\substack{j,k \\ v_{i-1} \le v_j + v_k \le v_{i+1}}}^{j \ge k} n_c \beta(v_j, v_k) N_j(t) N_k(t) - g(v_i) N_i(t)$$

$$- N_i(t) \sum_{k=1}^{M} \beta(v_i, v_k) N_k(t) \tag{7.26}$$

The discrete form of the population balance equation depends on the variable (e.g. mass or volume) and the discretisation rule. Other discretisation methods are also reported in the literature (Kiparissides et al. 2004).

7.5.2 Bubble Coalescence and Breakup

The bubble coalescence in a gas–liquid system depends upon the flow field. Coalescence is dominated by turbulent eddies at intermediate superficial gas velocity. The coalescence due to wake entrainment becomes important at high superficial gas velocity. The coalescence due to differences in bubble rise velocities is also considerable, especially when the bubble rise velocity is sensitive to the bubble size.

The coalescence rate, $c\ (d_i, d_j)$, can be written as the product of the collision frequency, $\omega\ (d_i, d_j)$, and the coalescence efficiency or coalescence probability, $P_c\ (d_i, d_j)$, for dispersed-phase units of sizes d_i and d_j.

$$c\ (d_i, d_j) = \omega\ (d_i, d_j)\ P_c\ (d_i, d_j) \tag{7.27}$$

The coalescence rate is also called the 'coalescence kernel'.

7.5.2.1 Bubble Coalescence due to Turbulent Eddies

The bubble collision frequency due to turbulent eddies may be considered analogous to the gas kinetic theory. However, due to the considerable volume of the bubbles, the free space for movement of the bubbles is reduced, hence, the bubble collision frequency increases. Wang et al. (2005) gave the following expression for collision frequency:

$$\omega\left(d_i, d_j\right) = \frac{\pi}{4} \frac{\varepsilon_{g,max}}{\left(\varepsilon_{g,max} - \varepsilon_g\right)} \Gamma_{ij} \sqrt{2} \varepsilon_e^{1/3} \left(d_i + d_j\right)^2 \left(d_i^{2/3} + d_j^{2/3}\right)^2 \tag{7.28}$$

where l_{bij} is the mean distance between bubbles of size d_i and d_j, and l_b' is the bubble turbulent path length. Γ_{ij} is a function of the ratio of these two, as given here:

$$\Gamma_{ij} = \exp\left[-\left(\frac{l_{bij}}{l_b'}\right)^6\right] \tag{7.29}$$

It was assumed that l_b' is the length that an eddy of size d_i, moved during its lifetime. It was found that l_b' is approximately equal to 0.89 d_i. The average value of l_b' for bubbles of size d_i and d_j is now

$$\overline{l_b'} = 0.89\sqrt{d_i^2 + d_j^2} \tag{7.30}$$

The value of l_{bij} is given by

$$l_{bij} = \frac{M_b}{\left(n_i + n_j\right)^{1/3}} \tag{7.31}$$

The constant M_b is determined from bubble size distribution. As the number or size of the bubbles increases, l_{bij} decreases.

Wang et al. (2005) used the following expression, given by Luo (1993), for the coalescence probability in case of bubble coalescence dominated by turbulent eddies:

$$P_c\left(d_i,d_j\right)=\exp\left[-M_1\frac{\sqrt{0.75\left(1+\xi_{ij}^2\right)\left(1+\xi_{ij}^3\right)}}{\left(\frac{\rho_g}{\rho_l}+0.5\right)\left(1+\xi_{ij}\right)^3}\sqrt{W_{ij}}\right] \tag{7.32}$$

Here, $\xi_{ij}=d_i/d_j$. The Weber number, W_{ij}, is defined as $W_{ij}=\rho_l d_i\left(\overline{u_{ij}}\right)^2/\sigma$, where $\overline{u_{ij}}$ is the geometric average of the mean turbulent velocity of a bubble $\left(\text{i.e. } \overline{u_{ij}}=\sqrt{\left(\overline{u_i}\right)^2+\left(\overline{u_j}\right)^2}\right)$. The mean turbulent velocity of a bubble of size $\overline{u_i}$ is evaluated as $\overline{u_i}=\sqrt{2}\left(\varepsilon d_i\right)^{1/3}$. The value of the modelling parameter M_1 was taken as equal to 1.

7.5.2.2 Bubble Coalescence for Small Bubbles

Small bubbles are entrained into the wake of large moving bubbles. Bubble collision frequency due to wake entrainment was given by

$$\omega\left(d_i,d_j\right)=M_2 d_i^2 u_{\text{slip},i} \tag{7.33}$$

The constant M_2 is an adjustable parameter. Wang et al. (2005) used a value of 15.4. In a bubble column, the slip velocity can be estimated as

$$u_{\text{slip},i}=0.71\sqrt{gd_i} \tag{7.34}$$

The following correlation for the coalescence probability was used.

$$P_c\left(d_i,d_j\right)=\exp\left[-M_3\frac{\rho_l^{1/2}\varepsilon^{1/3}}{\sigma^{1/2}}\left(2d_{ij}\right)^{5/6}\right] \tag{7.35}$$

Wang et al. (2005) used $M_3=0.46$.

Since small bubbles have such small wakes that they did not have any significant wake entrainment, the bubble coalescence rate, $c\ (d_i,\ d_j)$, was modified as

$$c\ (d_i,\ d_j) = \theta\omega\ (d_i,\ d_j)\ P_c\ (d_i,\ d_j) \tag{7.36}$$

where $\theta=\dfrac{\left(d_j-d_c/2\right)^6}{\left(d_j-d_c/2\right)^6+\left(d_c/2\right)^6}$ for $d_j\geq d_c/2$ and $\theta=0$ for $d_j<d_c/2$.

The value of d_c was calculated by the relationship given by Ishii and Zuber (1979).

$$d_c = 4\sqrt{\frac{\sigma}{g(\rho_l - \rho_g)}}$$ (7.37)

7.5.2.3 Bubble Coalescence due to the Relative Velocity of Bubbles

The bubble collision frequency due to the relative velocity of bubbles is given in terms of the bubble size and bubble rise velocity in the size group of bubbles.

$$\omega(d_i, d_j) = \frac{\pi}{4}(d_i + d_j)^2 |u_{bi} - u_{bj}|$$ (7.38)

The coalescence probability due to the relative velocity of bubbles is, unfortunately, not available. Therefore, Equations 7.35, 7.40 or 7.41 may be used.

7.5.2.4 Film Drainage and Bubble Coalescence

Chen et al. (2005) used the following collision frequency given by Saffman and Turner (1956):

$$\omega(d_i, d_j) = \frac{\pi}{4} n_i n_j \varepsilon_e^{1/3} (d_i + d_j)^2 \left(d_i^{2/3} + d_j^{2/3}\right)^2$$ (7.39)

Attempts have been made to model bubble coalescence. It is considered that as two bubbles come closer to each other, the liquid film between the bubbles drains out of the film (Figure 7.2). The film thickness decreases to a critical film thickness when the bubbles merge to complete the coalescence process.

Prince and Blanch (1990) proposed the following equation for the coalescence efficiency:

$$P_c(d_i, d_j) = \exp\left[-\frac{\dfrac{(0.5 d_{ij})^3 \rho_l}{16\sigma}\varepsilon^{2/3}}{(0.5 d_{ij})^{2/3}}\ln\left(\frac{h_{0,ij}}{h_{f,ij}}\right)\right]$$ (7.40)

where $h_{0,ij}$ and $h_{f,ij}$ are the initial and critical film thickness of the film, respectively.

Chesters (1991) proposed the following equation for the coalescence efficiency:

$$P_c(d_i, d_j) = \exp\left[-M_6\left(\frac{We_{ij}}{2}\right)^{1/2}\right]$$ (7.41)

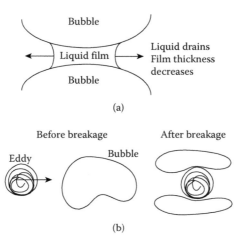

FIGURE 7.2
Mechanisms of (a) bubble coalescence and (b) bubble breakup.

where

$$W_{eij} = \frac{\rho_1 (\varepsilon d_{ij})^{2/3} d_{ij}}{2\sigma}$$

and

$$d_{ij} = \frac{2 d_i d_j}{(d_i + d_j)}$$

The time required to drain the liquid from the film between the bubbles is given by

$$\tau_{ij} = \sqrt{\frac{(0.5 d_{ij})^3 \rho_1}{16\sigma}} \ln\left(\frac{h_{0,ij}}{h_{f,ij}}\right) \tag{7.42}$$

Pohorecki et al. (2001) used $h_{0,ij} = 5 \times 10^{-4}$ m and $h_{f,ij} = 1 \times 10^{-8}$ m. The coalescence efficiency was given by

$$P_c(d_i, d_j) = \exp\left(\frac{\tau_{ij}}{\tau_{ij}^c}\right) \tag{7.43}$$

The contact time is given by

$$\tau_{ij}^c = \frac{(0.5 d_{ij})^{2/3}}{\varepsilon^{1/3}} \tag{7.44}$$

7.5.3 Bubble Breakup

The bubble breakup may be due to eddy collision or due to the instability of large bubbles. The latter occurs at high values of gas holdup. A detailed description of the breakup process is not well understood. It is common to assume that after breakup, the dispersed phase produces two small dispersed-phase units of equal size. The bubble diameter was correlated by log-normal distribution.

The breakup efficiency is given by

$$P_b\left(d_i, d_j\right) = \exp\left(-\frac{u_{ci}^2}{u_{te}^2}\right) \tag{7.45}$$

The critical velocity, u_{ci}, to break the bubble is given by

$$u_{ci} = \left(\frac{We_c \sigma}{d_i \rho_1}\right)^{0.5} \tag{7.46}$$

And turbulent velocity, u_{te}, is given by

$$u_{te} = 1.4\varepsilon^{1/3} d_e^{1/3} \tag{7.47}$$

The eddy size, d_e, is estimated assuming isotropic turbulence:

$$d_e = \left(\frac{v_1^3}{\varepsilon}\right)^{0.25} \tag{7.48}$$

Colella et al. (1999) studied the breakage and coalescence efficiency of the bubbles in bubble columns. The model considered no change in the size of the bubbles due to mass transfer and hydrostatic pressure. The bubbles' shape was oblate ellipsoid. The main mechanism was the wake entrainment. Bubble interactions were assumed to be binary for bubble coalescence and breakup. The drainage of the fluid between two bubbles and the wake entrainment were considered by applying Bernoulli's theorem. The model was validated with the experimental data. The relationships were discrete. The coalescence efficiency was given as

$$P_c\left(d_i, d_j\right) = 1 - M_4 \frac{\pi}{16} \frac{\tau |u_{bi} - u_{bj}|}{v_j} \frac{R_{ij}^4}{h_{0,ij}^2} \tag{7.49}$$

Here, τ is the time taken to drain the film, and M_4 is the adjustable model parameter. The radius of the film between the bubbles and the bubble, R_{ij}, was taken as

$$R_{ij} \propto \left(\frac{d_{ij}}{2}\right)^3 \tag{7.50}$$

The initial thickness of the film between the interacting bubbles of ith and jth class, $h_{0,ij}$, is given by

$$h_{0,ij} \propto \frac{3\mu_l \left(u_{b\infty,i} - 0.5 |u_{bi} - u_{bj}| \right)}{2\sigma} R_{ij} \tag{7.51}$$

The breakage efficiency, $P_b\,(d_i,\,d_j)$, was given by

$$P_b\left(d_i,d_j\right) = M_5 \sqrt{\frac{\rho_g}{\rho_l}} u_{w,i} \left(s_b^2 - 1\right)^2 \tag{7.52}$$

where $u_{w,i}$ is the velocity of the wake behind the leading bubble and M_5 is the adjustable model parameter.

The collision frequency was given as

$$\omega\left(d_i,d_j\right) = n_i n_j |u_{bi} - u_{bj}| \frac{V_{w,i}}{l_{bij}} \tag{7.53}$$

The volume affected by the wake of bubbles belonging to the ith class, $V_{w,i}$, and relative velocity $|u_{bi} - u_{bj}|$ were determined by the method described by Nevers and Wut (1971).

7.5.4 Monte Carlo Simulation

Renganathan and Krishnaiah (2004) used the Monte Carlo technique to estimate pressure drop, bed voidage and minimum fluidisation velocity under steady-state conditions in a liquid–solid inverse fluidised bed. Unsteady-state bed expansion for a step change in liquid flow rate was also studied. The simulation involved two steps. At first, the initial configuration of the bed was obtained by randomly choosing the initial positions of the bed after determining the number of particles present in the bed. The energy of the particles is calculated. In the next stage, particles are placed in their new positions by moving them a fixed distance. The energy of the particle is calculated. The net change in the energy was given as

$$\Delta E = (F_B - F_D)\,\Delta z \tag{7.54}$$

where Δz is the change of the position of the particle in the vertical direction. F_B is the difference of buoyancy and gravity forces, and F_D is the drag force acting on the particle. The drag force may be estimated using any suitable relationships available in the literature. If the energy is decreased (i.e. ΔE is negative), then the move is accepted. If the energy increases, then the probability of accepting the move is given by

$$p_a = \exp\left(-\frac{\Delta E}{KE} \right) \tag{7.55}$$

The kinetic energy of the particle is estimated as

$$KE = \frac{1}{2} m \left(\frac{v_z}{\varepsilon'} \right)^2 \tag{7.56}$$

where m is the mass of the particle and v_z is the superficial velocity of the liquid.

The local void fraction, ε', was calculated in terms of the ratio of the sum of intersected particle areas' total cross-sectional area over a horizontal plane passing through the centre of the particle.

$$\varepsilon' = 1 - \frac{\sum A_i}{A} \tag{7.57}$$

The intersected area of a particle was calculated in terms of the distance between the particle and the plane of intersection, z_I:

$$A_i = \pi \left(\frac{d_p^2}{4} - z_I^2 \right) \tag{7.58}$$

The simulation is repeated until the minimum energy was achieved. At this, the pressure drop was estimated as

$$\Delta P = \frac{N F_D}{A} \tag{7.59}$$

7.5.5 Stochastic Simulation

In continuous emulsion polymerisation, oscillations in monomer conversion are observed. The process was modelled for continuous emulsion polymerisation of vinyl acetate using a stochastic model by Yano et al. (2007). It was assumed that the total number of particles remains constant with time. The model considered the change in size of the particle due to its growth and aggregation. The particles could leave the reactor according to a residence time distribution, similar to the tank in the series model. As soon as a particle leaves the reactor, another particle with minimum size appears in the reactor due to nucleation. After each time step, the particles staying in the reactor grow by a volume, ΔV. The diameter of the spherical particle is given by

$$\Delta d_p = \left(\frac{6}{\pi} \Delta V + d_p^3 \right)^{1/3} - d_p \tag{7.60}$$

When two-particle collision was considered, then the resulting size was the sum of the size of both particles:

$$d_p = d_{p1} + d_{p2} \tag{7.61}$$

The particles leaving the reactor were replaced by particles produced by nucleation. The simulations were carried out to study the dynamic behaviour of particle size distribution (PSD). The model predicted that PSD oscillated with time. The dynamic behaviour of PSD was affected by the colloidal stability of the latex particles and the feed flow rate. This simple model gave only qualitative results. The model may be improved by removing the assumption that the number of particles remains constant.

7.5.6 Crystallisation

The role of mass transfer between the continuous and dispersed phases dominates the population balance. During crystallisation, the size of the particle increases due to mass transfer between solid (dispersed phase) and solution (continuous phase). Caillet et al. (2007) used the population balance model to describe the crystallisation of monohydrate citric acid. In this process, neither agglomeration nor breakage of the crystals (particles) takes place. Therefore, there is no source term in the population balance equation. The size of the crystal changes due to growth of the particle as a result of mass transfer. The population balance gives

$$\frac{\partial n(D,t)}{\partial t} + \frac{\partial}{\partial t}\left[G(D,t)n(D,t)\right] = 0 \tag{7.62}$$

where the growth rate is given as

$$G(D,t) = \frac{dD}{dt} = k_{gc}\left(C - C_s\right)^{m1} \tag{7.63}$$

In Equation 7.63, k_{gc} is the growth-rate constant (i.e. the mass transfer coefficient). Concentration of the solute and solubility of the monohydrate crystals are C and C_s, respectively. The boundary conditions for Equation 7.72 are given by

$$n(D_c,t) \approx n(0,t) = \frac{R_{sn}(t)}{G(D,t)} \approx \frac{R_{sn}(t)}{G(0,t)} \tag{7.64}$$

where D_c is the critical radius of the particles (i.e. the minimum size of crystal which is visible) and $R_{2n}(t)$ is the rate of secondary nucleation and is expressed in terms of the mass transfer coefficient.

$$R_{2n}(t) = k_{2n}C_s^{m2}\left(C - C_s^s\right)^{m3} \tag{7.65}$$

where k_{2n} is the rate constant for secondary nucleation and C_s^s is the concentration of the crystals in the suspension. The exponents m^2 and m^3 are constants.

The maximum size of the crystals satisfies the following equation:

$$G(\infty,t)\,n(\infty,t) = G(D_{max},t)\,n(D_{max},t) = 0 \tag{7.66}$$

The concentration is estimated as

$$C = \frac{\rho_c \phi_p}{M_W} \int_{D_{min}}^{D_{max}} n(D,t)\,D^3\,dD \tag{7.67}$$

The ρ_c, ϕ_p and M_w are the density, shape factor and molecular weight of the crystals, respectively.

The exponents, m^1, m^2 and m^3 and rate constants were determined by comparing the simulation results with the experimental data.

Attempts have been made to model the evolution of the shape of the crystals during crystallisation. Kuvadia and Doherty (2013) developed multidimensional population balances to study the growth of a crystal. Different growth rates for different faces of a crystal were taken into account by the proper specification of growth terms, as given here:

$$\frac{\partial n}{\partial t} + (n - n_{in})\tau = -G_1\left(R_1 \frac{\partial n}{\partial h_1} + R_2 \frac{\partial n}{\partial h_2} + R_3 \frac{\partial n}{\partial h_3} + ...\right) \tag{7.68}$$

where $R_i = G_i/G_1$ is the relative growth rate of the ith face of the crystal and G_1 is the growth rate of the slowest growing face. R_i is assumed to be constant.

7.5.7 Analytical Solution of Population Balance Models

Kostoglou and Karabelas (2004) presented an analytical solution to the following dimensionless general population balance equations considering primarily breakage with spatial dependency.

$$\frac{\partial F(D,Z,t)}{\partial t} = D(D) + \nabla_r^2 F(D,Z,t)$$

$$+ \int_D^\infty p(D,D')b(D')F(D,Z,t)dD' - b(D)F(D,Z,t) \tag{7.69}$$

where the dimensionless parameters are

$$D = \frac{D}{D_0}; \; D(D) = \frac{D(D)}{D(D_0)}; \; Z = \frac{z}{L}; \; b(D) = \frac{b(D)L^2}{D(D_0)};$$

$$p(D,D') = \frac{p(D,D')}{D_0}; \text{ and } t = \frac{D(D_0)t}{L^2}$$

A diffusion term was introduced in the breakage equation. The general analytical solution of a homogeneous problem (in the absence of the particle generation term) was given by the following recursive formula:

$$n_\delta(D,t;D_0) = \delta n(D,D_0)e^{-B(D_0)t} + e^{-B(D_0)t}\sum_{i=1}^{\infty}C_i(D,D_0)\frac{t^i}{i!} \qquad (7.70)$$

where

$$C_{i+1}(D,D_0) = \int_{D_0}^{D}A(D,D')C_i(D',D_0)dD' + \left[B(D_0)-B(D)\right]$$

and

$$C_1(D,D_0) = A(D,D_0)$$

The walls were non-reflective (i.e. when the particles reaching the wall were absorbed). In the presence of an arbitrary particle generation term, analytical solutions in closed form for an orthogonal parallelepiped, cylinder and sphere were obtained. Closed-form solutions for binary random and equal-sized multiple breakages were also presented.

7.5.8 Polymerisation

Sajjadi (2009) used the population balance model to predict PSD in fully monomer-starved semi-batch emulsion polymerisation. Different balance equations were written for polymer particles with no radical, one monomeric radical, and one, two and three polymeric radicals. The models considered that the radicals were formed by the initiator decomposition in the water phase. These radicals entered into micelles and particles, propagated and terminated. Equations for each of the phenomena, considered as sub-models, were described by suitable rate equations which depended upon the degree of polymerisation. The other sub-models were for the growth of the polymer particles and diffusion-controlled termination and were the condition for monomer partitioning among the polymer phase, micelles and water phase.

Vale and McKenna (2007) presented a detailed solution scheme to solve the population balance equation for emulsion polymerisation in 0-1 and 0-1-2 systems. The former assumes that particles have zero and one radicals while the latter mechanisms assumes that particles have zero, one and two radicals.

Several polymers are produced by suspension polymerisation, and the monomer is initially dispersed in the continuous aqueous phase by using surface active agents and agitation. The polymerisation occurs in the monomer droplets that are progressively transformed into a solid polymer. The transient droplet size distribution depends upon the drop breakage and drop coalescence rates. Since the quality of the end product depends upon the PSD, the population balance model was applied to study polymerisation (Kiparissides et al. 2004).

Ethayaraja et al. (2006) developed a population balance model to explain the mechanism of nanoparticle formation in water-in-oil micro-emulsion, a self-assembled colloidal template. A similar model for nanoparticle formation was used by Provis and Vlachos (2006).

The breakage kernels proposed by several investigators do not conserve the mass of the daughter bubbles or drops. In general, a loss of mass is observed. It leads to the incorrect prediction of the evolution of the properties of the dispersed phase. Zhu et al. (2010) developed a least-square method which ensured the conservation of mass for a breakage kernel which otherwise would have resulted in a loss of mass.

7.6 Summary

Processes with random phenomena can be solved by stochastic processes which give only moments. The development of a population balance equation to describe the random behaviour of the dispersed phase and the evolution of size distribution was a major step in formulating the problem. The discretisation form of the integro-differential population balance equation is used to obtain the solution of the problem.

Use of these equations to describe the behaviour of bubble column reactors involving gas as a dispersed phase, crystallisation involving the growth of dispersed solid particles, and a polymerisation process was illustrated. Use of a population balance equation in cases of reacting systems has been presented. The other applications reported in the literature include the study of crystal morphology, evolution of nanoparticles etc.

References

Caillet, A., Othman, N.S., and Fevotte, G. 2007. Crystallization of monohydrate citric acid. 2. Modeling through population balance equations. *Cryst. Growth Des.* 7(10): 2088–2095.

Chen, P., Sanyal, J., and Dudukovic, M.P. 2005. Numerical simulation of bubble columns flows: Effect of different breakup and coalescence closures. *Chem. Eng. Sci.* 60: 1085–1101.

Chesters, A.K. 1991. The modeling of coalescence processes in fluid–liquid dispersions. *Trans. Inst. Chem. Eng.* 69: 259–270.

Colella, D., Vinci, D., Bagatin, R., Masi, M., and Abu Bakr, E. 1999. A study on coalescence and breakage mechanisms in three different bubble columns. *Chem. Eng. Sci.* 54: 4767–4777.

Darmana, D., Deen, N.G., and Kuipers, J.A.M. 2005. Detailed modeling of hydrodynamics, mass transfer and chemical reactions in a bubble column using a discrete bubble model. *Chem. Eng. Sci.* 60: 3383–3404.

Ethayaraja, M., Dutta, K., and Bandyopadhyaya, R. 2006. Mechanism of nanoparticle formation in self-assembled colloidal templates: Population balance model and Monte Carlo simulation. *J. Phys. Chem. B.* 110: 16471–16481.

Iordache, O., and Corbu, S. 1986. A stochastic approach to sedimentation. *Chem. Eng. Sci.* 41(10): 2589–2593.

Iordache, O., and Munteean, O.I. 1981. Stochastic approach to the hydrodynamics of gas-liquid dispersions. *Ind. Eng. Chem. Fund.* 20: 204–207.

Ishii, M., and Zuber, N. 1979. Drag coefficient and relative velocity in bubbly, droplet or particulate flows. *AIChE J.* 25: 843–855.

Kiparissides, C., Alexopoulos, A., Roussos, A., Dompazis, G., and Kotoulas, C. 2004. Population balance modeling of particulate polymerization processes. *Ind. Eng. Chem. Res.* 43: 7290–7302.

Kostoglou, M., and Karabelas, A.J. 2004. Analytical treatment of fragmentation-diffusion population balance. *AIChE J.* 50(8): 1746–1759.

Kumar, S., and Ramkrishna, D. 1996. On the solution of population balance equations by discretizations. I. A fixed pivot technique. *Chem. Eng. Sci.* 51(8): 1311–1332.

Kuvadia, Z.B., and Doherty, M.F. 2013. Reformulating multidimensional population balances for predicting crystal size and shape. *AIChE J.* 59(9): 3468–3474.

Luo, H. 1993. Coalescence, breakup and liquid circulation in bubble column reactors. *D. Sc. Thesis*, Norwegian Institute of Technology.

Nevers, N., and Wut, J.L. 1971. Bubble coalescence in viscous fluids. *AIChE J.* 17: 182–186.

Pohorecki, R., Moniuk, W., Bielski, P., and Zdrojkowski, A. 2001. Modelling of the coalescence-redispersion processes in bubble columns. *Chem. Eng. Sci.* 56: 6157–6164.

Prince, M.J., and Blanch, H.W. 1990. Bubble coalescence and break-up in air-sparged bubble columns. *AIChE J.* 36(10): 1485–1499.

Provis, J.L., and Vlachos, D.G. 2006. Silica nanoparticle formation in the TPAOH-TEOS-H2O system: A population balance model. *J. Phys. Chem. B.* 110: 3098–3108.

Ramakrishna, D. 2000. *Population Balances: Theory and Applications to Particulate Systems in Engineering*. San Diego, CA: Academic Press.

Renganathan, T., and Krishnaiah, K. 2004. Stochastic simulation of hydrodynamics of a liquid-solid inverse fluidized bed. *Ind. Eng. Chem. Res.* 43: 4405–4412.

Saffman, P.G., and Turner, J.S. 1956. On the collision of drops in turbulent clouds. *J. Fluid Mech.* 1: 16.

Sajjadi, S. 2009. Population balance modeling of particle size distribution in monomer-starved semibatch emulsion polymerization. *AIChE J.* 55(12): 3191–3205.

Singh, K.K., Mahajani, S.M., Shenoy, K.T., and Ghosh, S.K. 2009. Population balance modeling of liquid-liquid dispersions in homogeneous continuous-flow stirred tank. *Ind. Eng. Chem. Res.* 48: 8121–8133.

Vale, H.M., and McKenna, T.F. 2007. Solution of the population balance equation for emulsion polymerization: Zero-one and zero-one-two systems. *Ind. Eng. Chem. Res.* 46: 643–654.

Wang, T., Wang, J., and Jin, Y. 2005. Theoretical prediction of flow regime transition in bubble columns by the population balance model. *Chem. Eng. Sci.* 60: 6199–6209.

Yano, T., Endo, K., Nagamitsu, M., Ando, Y., Takigawa, T., and Ohmura, N. 2007. Stochastic modeling of dynamic behavior of particle size distribution in continuous emulsion polymerization. *J. Chem. Eng. Jpn.* 40(3): 228–234.

Zhu, Z., Dorao, C.A., and Jakobsen, H.A. 2010. Mass conservative solution of the population balance equation using the least-squares spectral element method. *Ind. Eng. Chem. Res.* 49: 6204–6214.

8

Artificial Neural Network–Based Models

The methodology for the development of models discussed in Chapters 4 through 7 of this book is based on a simplifying representation of the process of applying appropriate governing physical laws (e.g. laws of conservation of mass, momentum and heat and laws of kinetics and thermodynamics). The effect of randomness in the process on the process' performance and product quality was considered using stochastic methods. Population balance models are a special type of stochastic model. However, in some situations, either there is no clear understanding of the process or there is not sufficient time to develop a model based on first principles that require a clear understanding of the process. For example, in cases of complex chemical reactions, the number of reactions is so large that it becomes almost impossible to obtain a unique global reaction rate equation. The change of rate-determining steps with operating conditions may not be clearly understood. In the absence of a chemical reaction, the performance of process and equipment may depend upon the flow regime; this may not be characterised easily, or there may not be sufficient time to develop a sound model for these aspects.

A model for a complex process may be composed of several sub-models, each having a specific scope. In absence of even a single sub-model, the development of a model cannot be completed. Therefore, it is essential to develop a working sub-model even if the relevant mechanism in the process is not understood. Artificial neural networks (ANNs) fill this gap. Although there are several arguments in favour of using ANNs, they should always be replaced by a model based on sound knowledge of the process whenever this becomes possible.

The development of ANNs is inspired by the functioning of the brain, which contains various neurons with several interconnections. A detailed discussion of the similarity of neurons with ANNs is beyond the scope of the book.

8.1 Artificial Neural Networks

An ANN consists of several interconnected nodes (neurons) arranged in layers as discussed in Chapter 2 (Figure 2.7). The nodes are basic computational elements which perform computations mathematically, according to predefined formulae known as activation functions. Each node receives input from several nodes, makes calculations and produces output. The interconnections transmit the output to other nodes. The weights are assigned to the interconnections. Each node calculates its own output using weights associated with connections. The weights are adjusted to minimise the error between the predicted output and the experimental data points. The process of adjusting the weights is called a learning process.

All of the nodes accept a weighed set of inputs and respond with an output. Depending upon the connectivity, various types of ANNs have been proposed. However, for modelling purposes, a feedforward multi-layered network and some of its variants are frequently used in chemical and biochemical applications and will be discussed in this chapter. These are also called multilayer perceptrons (MLPs) or back propagation neural networks (BPNNs). They will now be addressed by a single term in this book: ANNs.

Feedforward BPNNs consist of nodes arranged in layers. Each node in the first layer accepts one input from the user. Thus, the number of layers in the first layer is equal to the number of inputs to the ANN model. The last layer is the output layer. Each node in the output layer corresponds to one output of the model. The output layer may even have a single node. The input and output layers are called visible layers because they communicate with the user. There are a few other layers in between the visible layers; these are called the hidden layers. Each hidden layer consists of a set of nodes. For example, Figure 2.7 shows three layers; one hidden layer consists of four nodes. The ANN model accepts three inputs. Each node in the input layer accepts one input. The model predicts two outputs, one by each of the nodes in the outer layer. All nodes in a layer are connected to all of the nodes present in the previous layer and the next layer. They are called feedforward neural networks because, while the computations use ANNs for the prediction of the output parameter, they proceed from the input layer in the forward direction towards the output layer. The back propagation indicates the training algorithm used to develop the ANN model.

8.1.1 Information Processing through Neurons

Neural networks are considered to be universal approximators (i.e. they are capable of approximating any non-linear relationship). This concept has been derived from Kolmogorov's theorem, which states that any continuous real-valued function, $f(x_1, x_2, \ldots, x_n)$, defined on $[0,1]^n$, $n \geq 2$, can be represented in the following form:

$$f(x_1, x_2, \ldots, x_n) = \sum_{j=1}^{2n+1} g_j \sum_{j=1}^{n} \phi_{ij}(x_i)$$

(8.1)

where the g_j terms are properly chosen continuous functions of one variable, and the ϕ_{ij} functions are continuous, monotonically increasing functions independent of f. The implication of the theorem is that a continuous multivariate function on a compact set can be expressed in terms of sums and the composition of a finite number of single variable functions. This fact has been the basis of the development of ANNs.

Neuron or nodes accept a weighed set of inputs from other neurons in the previous layer they respond with an output. A picture of one unit in a neural network is presented in Figure 8.1. The output from a neuron is a function of the sum of all of the inputs to this neuron (node). This function is called the activation function. In general, sigmoid functions are used as activation functions. The role of the sigmoid is to keep the output within the required limit. Some of the commonly used sigmoid functions are presented in Table 8.1.

Let us consider a single-layer network consisting of n number of nodes. Each node (e.g. the ith node) receives a weighted input, x_i, from m nodes in the previous layer with weight w_{ji}.

Let $\bar{x} = (x_1, x_2, \ldots, x_m)$ be the input vector, and let f be the activation function. The $m \times n$ weight matrix is

$$\bar{w} = \begin{bmatrix} w_{11} & w_{12} & \ldots & w_{1m} \\ w_{21} & w_{22} & \ldots & w_{2m} \\ \ldots & \ldots & \ldots & \ldots \\ w_{n1} & w_{n2} & \ldots & w_{nm} \end{bmatrix} \tag{8.2}$$

The output of a neuron is given by

$$y_i = f\left(\sum_{j=1}^{n} w_{ji} x_i + b_j \right) \tag{8.3}$$

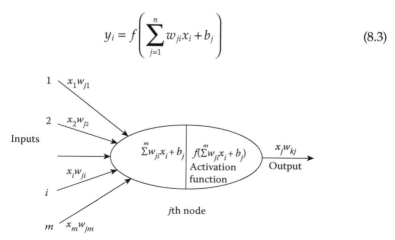

FIGURE 8.1
The information processing in a single neuron.

TABLE 8.1

Sigmoid Functions

Name	Formula	Characteristics
Log-sigmoid	$f(x) = \dfrac{1}{1+e^{-x}}$	Gives only positive values
Hyperbolic tangent sigmoid	$f(x) = \dfrac{e^{2x}-1}{e^{2x}+1}$	Gives positive as well as negative values
Arctangent function	$f(x) = \dfrac{2}{\pi}\arctan(x)$	
Saturation	$f(x) = \dfrac{x}{1+|x|}$	Gives value in the range of -1 to 1
Linear	$f(x) = x$	Equivalent to accepting the summation of all inputs as such
Hard limit	$f(x) = 1$, if $x \geq 0$ $= 0$, if $x < 0$	Used for pattern recognition

where y_i is the output of the jth neuron, f is the activation function, b_j is the bias of the jth neuron and w_{ji} is the weight associated with the interconnection between the jth input neuron and the ith neuron in the present layer. Here, n is the number of input signals to the jth neuron. The bias acts as an activation threshold.

Several such layers arranged in a series may be used. Such a type of structure is called an MLP (Figure 2.7). In its simplest form, it is called a feedforward multilayered network. In it, each node in a layer is connected with all of the nodes in the following layer. However, there are no interconnections within a single layer. Information is processed in such networks only in the forward direction.

The first layer of this network is known as the input layer, which receives information from an external source. It passes this information to the next layer for further processing. The layer following the input layer is called the hidden layer. There may be one or more hidden layers. The last layer is the output layer, which sends the results out to the user.

A neural network is composed of such layers of nodes and weighed unidirectional connections between them. In some neural networks, the number of units may be in the thousands. The output of one unit typically becomes an input for another. There may also be units with external inputs and/or outputs.

8.1.2 Radial Basis Function Networks

Radial basis function networks (RBFNs) may be considered a kind of MLP. An RBFN's architecture is much simpler as it has only one hidden layer.

TABLE 8.2

Radial Basis Functions

Name	Formula	Remarks
Gaussian	$f(x) = \exp\left(-\dfrac{(x-\mu)^2}{2\sigma^2}\right)$	
Generalised multi-quadratic and inverse multi-quadratic	$f(x) = (x^2 + \sigma^2)^\alpha$	$\alpha > 0$ (generalised multi-quadratic), $\alpha < 0$ (generalised inverse multi-quadratic), $\alpha = 1/2$ (multi-quadratic), $\alpha = -1/2$ (inverse multi-quadratic)
Thin-plate spline function	$f(x) = x^2 \ln(x)$	
Linear	$f(x) = x$	

The other layers are input and output layers which are similar to those present in MLPs. Therefore, the architecture of the RBFN can be decided by determining the number of nodes in the hidden layer only. The difference between RBFNs and MLPs is the type of activation function. The RBFN uses radially symmetric functions (e.g. a Gaussian function). These functions are called radial basis functions; some radial basis functions are given in Table 8.2. An MLP with a linear sigmoid function and one hidden layer is same as an RBFN with a linear radial basis function.

8.2 Development of ANN-Based Models

The ANN model is specified by four components: the architecture, weight matrix and bias vector, type of activation function and input–output parameters. Because ANNs use a black-box approach, they should not be used when a model based on first principles can be developed comfortably. The four components of the ANN model are described below.

8.2.1 Architecture

The number of nodes, number of layers and interconnections are called the architecture of the ANN. An ANN model is characterised by the inputs to the model, number of hidden layers, number of nodes in each layer, activation function in each layer, weight matrix and bias vector for each layer, and output of the matrix. The activation function may be different in each layer.

8.2.2 Identification of Inputs

Choosing the right set of inputs is one of the most important steps in developing an ANN-based model. The inputs are the parameters which affect the output. It has been discussed in this chapter that ANNs with a suitable architecture can correlate any set of input and output data. However, if a parameter has no relationship with the output, an ANN may not get trained successfully. Therefore, one cannot choose any arbitrary set of parameters as input and train the ANN successfully. The list of inputs should include only the relevant parameters which show a cause-and-effect relationship with the output; these are often called the target values. Even if the output is less sensitive to any parameter, the training of ANNs will be satisfactory. It may be desirable to leave out those parameters which have insignificant effects on the output or for which there are not sufficient data. It must be kept in mind that the aim of the ANN is not to exactly fit the entire data set. Such an attempt will be over-fitting the data and will lose any generalisation of the data, and the ANN model cannot be used to predict the output for unknown (i.e. unseen) data.

The parameters affecting the process output may include operating conditions, geometrical parameters and concentration. These are used as inputs. For online application of ANNs, these parameters should be easily measurable.

The data are generally normalised so that they remain in the range of 0–1. This can be done in the following manner:

$$\hat{x} = \frac{x - x_{min}}{x_{max} - x_{min}} \tag{8.4}$$

The ANN model can be used to model both the steady state and the dynamic processes. The dynamic processes are those processes which show variation with time. In process problems control the characteristics of the dynamic process such as rise time, settling time, overshoot and integrated sum of error can also be chosen as input (Loquasto and Seborg 2003).

As the ANN accepts only numerical values, the analogue form of the signals is not useful. The discrete data are also known as time-series data (Figure 8.2a). These data generally appear after a constant time interval. The input to the ANN to be developed can include information extracted from this time series. In general, the information can be extracted in the following manner:

1. *Statistical parameters*: The average and standard deviation are the first and second moments. These and higher moments can be taken as input to ANN-based models (Figure 8.2a). When there are more than one time series, cross-correlations can also be used.

2. *Auto-correlations and spectral analysis*: One of the important features of a time series is periodicity in the signal. This periodicity may be

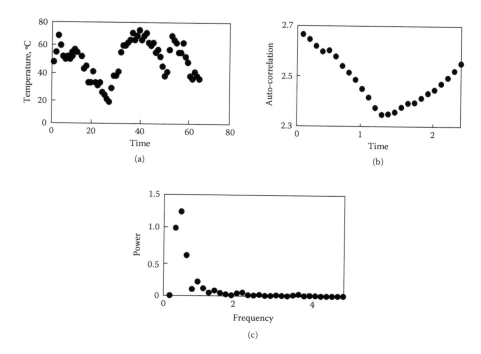

FIGURE 8.2
Time series (a) moments, (b) auto-correlation and (c) spectrum.

a result of hydrodynamics (e.g. a flow regime) or any other physical phenomena. Thus, inputs based on the periodic nature of the signal indirectly relate the input to the physical phenomena responsible for such behaviour.

The periodicity can be expressed in terms of an auto-correlation function (Figure 8.2b) or dominant frequencies in the signal obtained using Fourier transform (Figure 8.2c). However, these require additional computations. The periodicity is included in previous data at regular intervals. Partial auto-correlations and variograms can also be used (Loquasto and Seborg 2003). A variogram is the ratio of variance of i-steps to the variance of j-step data. It is a measure of the dynamic behaviour of the process.

A combination of all these parameters may also be used. For example, Xie et al. (2003) used ANNs to recognise a flow regime in a gas–liquid–pulp fibre slurry. The input to the ANN model was obtained from the pressure fluctuation signals taken at a sampling rate of 100 Hz for 2 s. An ANN with a 10-7-4 structure was trained to identify bubbly, plug flow, churn turbulent flow and slug flow regimes. The 10 inputs used were standard deviation, coefficient of skewness, coefficient of kurtosis and seven second-order auto-correlation terms of the normalised pressure signals: $T(2)$, $T(5)$, $T(10)$, $T(20)$, $T(50)$, $T(100)$ and $T(200)$. The normalised signals were obtained as

$$p^* = \frac{\left(p - \bar{p}\right)}{\sqrt{\overline{\left(p - \bar{p}\right)^2}}} \tag{8.5}$$

The second-order correlation term was obtained as

$$T(d) = p^*(n)\, p^*\, (n + d) \tag{8.6}$$

One special type of MLP is time-delay neural networks (TDNNs). These are used to accept time series (i.e. a fixed number of previous values in the time series) as input. The time delay is the time interval between two successive data in the time series. In TDNNs, the next value of the time series is correlated with its previous values. The learning procedure is similar to that of MLP. The output layer may give the estimate of the next value in the sequence, or it may extract feature from the time series.

8.2.3 Choice of the Architecture

After a thoughtful choice of inputs, the architecture of the ANN has to be chosen. The number of nodes in the input layer represents the number of inputs. The number of outputs is the number of nodes in the outlet layer. Unfortunately, there is no short-cut method to determine the number of hidden layers and the number of nodes present in them. Therefore, it is a good practice to start from the simplest possible architecture. Several rounds of trial and error are required before accepting an ANN architecture with small errors. 'Minimal error' is not a suitable criterion since it can result in a large ANN architecture that over-fits the data. In other words, in the absence of any systematic method to determine a suitable architecture or an ANN, a trial-and-error method is used.

If the number of nodes in a hidden layer is increased, then the size of the weight matrix also increases. The number of elements in all the weight matrix and bias vectors should be fewer than the number of data points; otherwise, it is equivalent to fit the data with more adjustable parameters than the data size. The maximum number of nodes in the hidden layers may be estimated using the following formula (Rene et al. 2011):

$$N_H \le \frac{N_T}{N_I + 1} \tag{8.7}$$

where N_H, N_T and N_I are the number of nodes in the hidden layer, the number of training data sets and the number of inputs to the ANN, respectively.

Loquasto and Seborg (2003) used the following three heuristic design methods to determine the number of nodes in the hidden layer. However, these ANNs were used for pattern recognition.

$$N_H = N_T \tag{8.8}$$

$$N_H = \sqrt{N_T N_I} \tag{8.9}$$

$$N_H = 1.7095 \log_2(2N_T) \tag{8.10}$$

8.2.4 Training the ANNs

ANNs learn from their set of input and output data. Here, 'learning' means that it determines the weight matrix and bias, if any; the learning depends upon the set of data. Therefore, if more data are added to the existing data, the weight matrix is refined. Therefore, it is also said that ANNs possess self-learning properties. The ANNs learn by training, which is nothing but the adjustment of the weights associated with the connections between nodes.

The next step is to choose an appropriate sigmoid activation function. Neurons give outputs through an activation function. To get the output within the desired range, sigmoid activation functions are used. 'Sigmoid' means a function that has a curve in two directions. Some of the commonly used sigmoid functions are given in Table 8.1. The types of responses for various sigmoid functions are given in Figure 8.3.

The weight matrix and bias vector for each layer are still unknown. Determination of these using the input–output data set is called 'training of the ANN'. To avoid over-fitting, the data are randomly divided into three groups. About 60% of the data are used for training the ANN with the chosen architecture. The second group of data contains about 20% data and is used to check the proper generalisation by the ANN and over-fitting, if any. The remaining 20% data are called unseen data, and the performance of the model is evaluated for this set of data.

The weight matrix and bias vector are determined by minimising the summation of errors for the entire test data set. Here, the error is the difference between the value predicted by the ANN and the test data. An ANN is said to be trained if the overall summation of error is within the permissible limit. If it is not, there may be several reasons, including:

1. The optimiser is struck. One may start from the new initial weight matrix and bios vectors.

2. The architecture is not proper. Number of hidden layers and number of nodes may not be appropriate. One can increase the number of

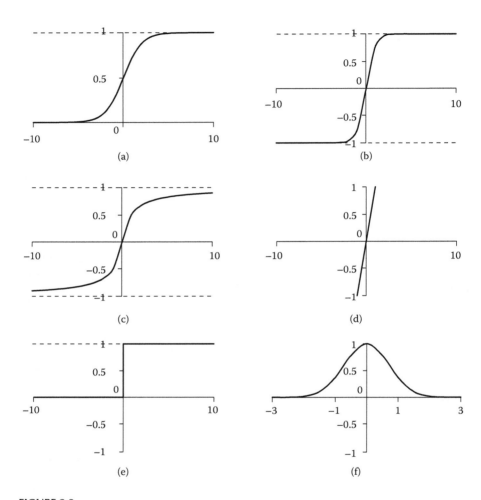

FIGURE 8.3
The output from activation functions (a) log-sigmoid, (b) hyperbolic tangent sigmoid, (c) arctangent, (d) linear, (e) hard linear and (f) Gaussian-type radial basis functions.

nodes in a hidden layer. The number of hidden layers can be increased. However, the number of hidden layers should not be larger than three, and the maximum number of nodes in hidden layers is given in Equations 8.7 through 8.10.

3. The training may be unsuccessful due to wrong choice of the inputs. If there is no relationship between the inputs and output, the ANN cannot forcibly correlate them.

There are several training algorithms for minimisation of the overall error. These are discussed later in Section 8.2.6.

Experimental data are required to train an ANN. The data for the training of an ANN can be obtained from the following sources:

1. *Experimental data available from the literature*: Several investigators carry out experiments under different operating conditions. Even then, the process may be poorly understood and no generalised correlation is available. These data may be collected and used. Unfortunately, the published data may not have some of the variables which seem to be an important input. A large amount of published data may have to be left.

2. *Unpublished data*: A research group may have a lot of unpublished data for various reasons. These data contain variables which are not published at that time.

3. *Old data from an existing process plant*: All process plants record data on a regular basis. These may be used, if available. Features corresponding to failure and faults can be extracted.

4. *Experimental data*: Experiments can be planned and required data can be obtained. However, this step itself is a time-consuming process. For process control applications, the data are collected online and used.

5. *Simulated data:* Various process simulators (e.g. ASPEN plus, HYSIS and ProSim, ChemCAD) can be used to simulate a process and obtain the necessary data for the training of the ANN. Simulation thus plays a role in ANN model development also.

8.2.5 Performance of ANN Model

After training the ANN successfully, all the weight matrices and bias vectors are known. It is ready to predict the values of the output for a given input. The outputs for the unseen data set are predicted, and goodness of the fit is checked. If the ANN fails to predict the unseen data satisfactorily, then the proposed architecture of ANN is rejected and a new one is tried.

8.2.6 Learning Methods

Learning is the process of adjustment of weights of the ANN to give a minimum summation of square of errors.

$$e = \sum_{i=1}^{N_T} \left(y_i^2 - y_{i\,\text{ANN}}^2 \right) \tag{8.11}$$

Here, y_i is the experimental data, $y_{i,\text{ANN}}$ is the data predicted by the ANN model and N_T is the total number of data points. There are several training methods reported in the literature. These are based on various optimisation techniques. The learning techniques require an initial estimate of the weight matrices, which is done by randomly assigning all the weights in the neural network. The error is determined, and the weights are adjusted according to a well-defined scheme. Back propagation algorithms and its variations are some of these. The weight correction starts in the direction of output nodes to the inputs. The direction of the correction of weights is based on the gradient of the error. Back propagation with adaptive learning rate, α, and momentum, μ, is given as follows:

$$\overline{w}_{k+1} = \overline{w}_k - \alpha_k g_k + \mu(\overline{w}_{k+1} - \overline{w}_k) \tag{8.12}$$

Back propagation without momentum can also be used. Earlier learning rules had no momentum, and the learning rate was also constant. If the learning rate is small, it takes more iterations to train the ANN. If the learning rate is large, then the training becomes unstable. However, the optimisation technique of Levenberg and Marquardt can be applied as it is more efficient than the back propagation methods.

$$\overline{w}_{k+1} = \overline{w}_k - (J^T J + \mu I)(J^T e) \tag{8.13}$$

where e is the network error, and J and J^T are the Jacobian matrix of the first derivative of error with respect to weights and transpose of the Jacobian matrix.

8.2.7 Over-Fitting and Under-Fitting

As the number of nodes in a hidden layer increases, the ANN fits the test data very well. It fits even the noise in the data. The ANN loses its generalisation capability. The error in predicting new data increases. It is undesirable because the ANN is to be used to predict the unseen data. The condition in which the ANN reproduces all the test data very precisely but fails to predict values under new situations is called 'over-fitting'. When the data are over-fitted, they pass through many points, although in between the points the predicted values show wild fluctuations. If an over-fitted ANN is used to predict output from unseen data, its performance is unsatisfactory (Figure 8.4a).

It is now desired to identify over-fitting and avoid it. The entire data are divided into training set, test set and unseen data set. The error as defined above goes on decreasing with the training epochs and may lead to over-fitting. To check the existence of over-fitting, the ANN is used to predict the values for the test data also. The summation of the square of errors, called the test error, decreases first, and it may start increasing after many training epochs. This is the point at which over-fitting is about to occur (Figure 8.4b). The training can be stopped at this point to avoid over-fitting (Kasabov 1998).

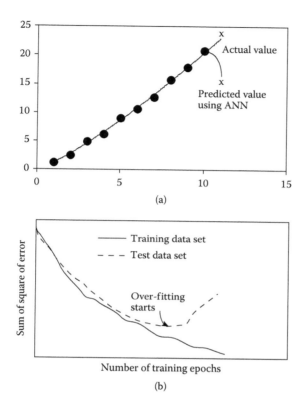

FIGURE 8.4
(a) Over-fitting and (b) avoiding over-fitting during training.

The entire process has to be repeated for various ANN architectures. The architecture giving the least summation of square of errors is chosen.

If the number of neurons is too few, then the ANN does not fit the test data. This condition is termed 'under-fitting'. The fitting can be improved by

1. *Training again with a new set of initial weight matrix*: Sometimes the optimisation scheme stuck, and by starting the training process again, the ANN can be trained.

2. *Increasing the number of neurons in the hidden layer*: The number of neurons may be increased to decrease the error. However, it may not work because increasing the number of neurons may deviate from the optimum, and the error might decrease also. The aim is to find the number of neurons with minimum error.

3. *Including a hidden layer can be considered*: Normally the training starts with one hidden layer. If the trained ANN does not provide

small error satisfactorily, then the number of hidden layers may be increased. However, at the most three hidden layers should be used. In most of the cases, one or two hidden layers are sufficient.

4. *Defining the inputs again*: If the change in the architecture does not reduce the error to a low level, then it may be inferred that the choice of the input variables is not proper. Either some important variable is missing or at least one of the variables is unnecessarily included. It may be concluded that an ANN can be trained properly only when the correct choice of inputs parameters is made.

8.3 Applications of ANNs in Chemical Engineering

ANNs have been applied in several applications in chemical, biochemical and environmental engineering problems. A few applications are discussed in this section.

8.3.1 ANN-Based Correlations

The ability to correlate the non-linear relationship between input and output has been utilised to correlate parameters in terms of easily measurable variables. The work includes correlating gas-holdup in gas–liquid systems, mass transfer coefficient in multiphase systems and estimation of thermodynamic properties. Only a few of these have been discussed to give an idea of the variety of applications.

An ANN model was used to obtain particle-fluid mass transfer coefficient in a fast fluidised bed (Zamankhan et al. 1997). The inputs to the ANN were the mean solid fraction at the upper entrainment section, superficial gas velocity, particle diameter and through-flow of solids. There were nine neurons in one hidden layer. The ANN was trained by the data obtained by evaporation of the naphthalene balls in glass fluidised beds. Overall mass transfer coefficient in a stirred tank reactor was correlated using an ANN model by Yang et al. (1999). The ANN was trained using the values reported in the literature.

ANN has been used to fit the data in a multidimensional space. For example, Pollock and Eldridge (2000) used ANN to relate height equivalent to a theoretical plate (HETP) in terms of density, viscosity, diffusivity and velocity of gas and liquid phases and relative volatility.

An ANN model was used to predict the instantaneous heat transfer rates in bubble columns (Tsutsumi et al. 2001). The architecture was 6-10-1. The inputs were the previous three values of passage time across the heat transfer probe for liquid as well as for bubbles. It had only one hidden layer containing 10 nodes. The model predicted the time series (instantaneous heat transfer rates) and captured the characteristics of heat transfer rates. The model showed the relationship between the hydrodynamic parameters and the heat transfer rate.

Jamialahmadi et al. (2001) used the ANN model with the single hidden layer to predict the bubble size in a bubble column. The inputs to the model were Bonds number (Bo), Galaleo number (Ga) and Froude number (Fr).

Overall gas-holdup in bubble columns is an important, easily measure-able hydrodynamic parameter required for the scale-up of bubble columns. However, there is no single correlation that can predict it over a wide range of operating conditions, physical properties of the gas and liquid, and column dimensions. Shaikh and Al-Dahhan (2003) proposed an ANN model with a 4-11-1 architecture (i.e. there were four inputs, and overall gas-holdup was a single output). It had one hidden layer consisting of 11 nodes. The inputs cho-sen were four dimensionless groups: the Reynolds number of gas (Re_g), Froude number of gas (Fr_g), ratio of Evotos number to Morton number (Eo/Mo) and gas–liquid density ratio (ρ_g/ρ_l). These dimensionless numbers were identified based on an understanding of the factors affecting the overall gas-holdup and grouping them to minimise the number of inputs so that the ANN model has a simple structure.

An ANN model with architecture of 4-5-1 to predict equilibrium loading of solutes for the ternary adsorption system was developed (Jha and Madras 2005). Binary data to train the ANN were developed. The ternary system consisted of 2,6- and 2,7-dimethylnaphthalene (DMN) isomers dissolved in $ScCO_2$ and adsorbed onto NaY-type zeolite. The inputs were the concen-tration of 2,6-DMN and 2,7-DMN in $ScCO_2$ of the ternary system, and the stationary phase loading of 2,6-DMN and 2.7-DMN in binary systems (pure solute-$ScCO_2$–zeolite).

A capillary tube is widely used in small-scale vapour compression refriger-ation and air-conditioning systems. An ANN-based generalised correlation for a refrigerant flow rate through adiabatic capillary tubes was developed by Zhang (2005). The data for refrigerants R12, R22, R134a, R290, R407C, R410A and R600a were used to train the ANN. The ANN used seven inputs in the form of dimensionless numbers. It had a single hidden layer and one output. The inputs were sub-cooling temperature, condensing temperature and inner diameter. The ANN correlation predicted reasonable trends for the mass flow rate. The correlation correctly predicted the mass flow rate at the zero sub-cooling.

Solid crystalline compounds stabilised by the inclusion of gas molecules inside cavities formed by water molecules through hydrogen bonding are known as gas hydrates. Gas hydrate inhibitors are injected in the pipeline to check gas hydrate formation. An ANN-based model with architecture of 2-7-1 was developed to predict the hydrate suppression temperature using a refractive index of the aqueous phase and the molecular weight of the inhibi-tor as input (Mohammadi and Richon 2006).

The biological treatment of wastewater is not clearly understood from the modelling point of view. Molga et al. (2006) used ANN to model the behaviour of activated sludge. The inputs to the model were various components present, COD and BOD. The model predicted the amount of components removed.

The crystal growth rate, nucleation rate and agglomeration coefficient for an anti-solvent crystallisation system consisting of ciprofloxacin hydrochloride, H_2O and ethanol were correlated to super-saturation (Yang and Wei 2006). Temperature, volume fraction, mass of solid particles per unit volume of slurry and agitation rate were used as inputs in ANNs with architecture of 5-10-1. The ANNs were trained with experimental data.

An ANN with architecture of 2-4-1 was used to predict the flue gas sulphuric acid dew point in stack and heat recovery systems (Nezhad and Aminian 2010). The inputs to the ANN model were the concentrations of SO_3 and water vapour.

Rene et al. (2011) developed ANN models with simple architectures of 3-5-1 and 2-5-1 to predict the removal efficiency with inlet concentration of styrene concentration, pressure drop and unit flow in biofilters. The ANN model for a completely stirred tank bioreactor had the architecture of 2-5-1 and used concentration and unit flow as units. The ANN models for monolith reactors used concentration, gas–liquid ratio and pressure drop as inputs.

Arnavata et al. (2013) developed an ANN model to predict the volume fraction of CO, CO_2, H_2 and CH_4 in the producer gas obtained from biomass gasification in fluidised bed gasifiers using one ANN model for each of the gas species. The architecture and input for all four models were the same. The model used seven inputs which were ash, carbon, oxygen, hydrogen contents of dry biomass, moisture content of wet biomass, equivalence ratio and gasification temperature. The equivalence ratio is defined as the ratio of the actual air–fuel ratio to the air–fuel ratio for complete combustion of the biomass. The ANN model had only one hidden layer with two nodes.

8.3.2 Process Modelling and Monitoring

ANN-based models have been used to monitor chemical and biochemical process. In these applications, the data were continuously measured and the ANN was used to predict the required change in the process conditions. The use involves online handling of time series. Several other attempts have been limited to model a process only.

Chemical oxidation and coagulation comprise one of the methods to treat wastewater. Syu et al. (2003) studied online control of Fenton treatment of wastewater. A time-delayed back-propagation type ANN with a 7-4-1 structure was used for control of pH by addition of hydrogen peroxide. The seven inputs were the previous three values of COD of influent water at times $t–\Delta t$, $t–2\Delta t$ and $t–3\Delta t$; values of COD of effluent water at times t and $t–3\Delta t$; and amount of hydrogen peroxide added at times $t–\Delta t$ and $t–2\Delta t$. The output was the amount of hydrogen peroxide to be added (i.e. at time t). Here, Δt is the time between two samples collected. The ANN was trained in a dynamic mode (i.e. during the process) with a window size of 15.

Dynamic behaviour of batch and semi-batch operations varies from batch to batch. It is due to variations in the initial conditions, operating conditions of the processing equipment etc. The model, based on using only the initial conditions such as concentration and temperature as input and the process state at the end of the operation, can be fitted by using an ANN model, but this approach does not give information about the optimum trajectory of variables. Rani and Patwardhan (2004) used an ANN to predict the process trajectories. The input and output trajectories were discretised and fitted by ortho-normal polynomials. The ANN accepted the input in the form of coefficients of these polynomials and produced the coefficients for the output polynomials. The input and output trajectories from several batches were used to train the ANN. The developed model was used to optimise the trajectory for batch and semi-batch reactors.

ANN has also been used to predict the thermodynamic properties of mixtures using extended corresponding states (ECSs) (Scalabrina et al. 2006). The input thermodynamic data were only volumetric and coexistence data.

An ANN-based model for cross-flow nanofiltration of NaCl and $MgCl_2$ salts was developed (Darwish et al. 2007). The model predicted the membrane rejection as a function of pressure and permeate flux for three nanofiltration (NF) membranes. The effects of the training method, sigmoid function and training algorithm on the mean square error were studied. Algorithms based on the conjugate gradient, regularisation and quasi-Newton principles were better than those based on the gradient descent and variable momentum gradient descent.

Lipase production using *Bacillus sphaericus* in submerged batch cultures was studied by Rajendran and Thangavelu (2007). The concentration of glucose, sesame oil, peptone, NaCl and $MnSO_4 \cdot H_2O$ was taken as input to an ANN-based model with architecture of 5-3-1 to predict lipase activity. The ANN predicted much better values than that by response surface methodology.

A temperature gradient reactor is used for the synthesis of dimethyl ether from CO using a Cu–Zn catalyst supported at γ-alumina (Omata et al. 2009). In the temperature gradient reactor, the catalyst bed was divided into five zones of temperatures, with the highest temperature at the inlet. To achieve a high one-pass CO conversion, it was desired to find an optimum temperature in each zone. The ANN with a radial basis function with a 5-19-1 architecture was used. The inputs to the ANN were the temperatures of all five zones. The ANN predicted the CO conversion. After designing the experiments, 19 sets of data were used to train the ANNs. For four values of temperature in each zone, a total of $4 \times 4 \times 4 \times 4 = 1024$ temperature gradients are possible. The optimum temperature gradient was searched using the grid search optimisation method. Based on the results, some more experiments were designed and included to train the ANN two more times. The architectures of ANN were 5-23-1 and 5-16-1 in the second and third steps, respectively. In brief, the training and obtaining of experimental data were taken alternatively to reduce the number of experiments.

8.3.3 Pattern Recognition

The term 'pattern recognition' means finding out a particular state of the process. The flow regimes, the failure of a particular sensor in a plant, different kinetic regimes, changes in the operating conditions etc. may be treated as different states of a process. These quantities may not have a definite numerical value, but they will be either present or absent.

The ANN provides numerical outputs; therefore, the presence of a state should be assigned a value. It is done by assigning 1 if the state is present and 0 or −1 if it is absent. It can easily be done by applying a hard-limit sigmoid function (Table 8.1).

Sharma et al. (2006) used a feedforward back-propagation network and radial basis function network to identify the flow patterns and the transition during air–water two-phase flow through circular pipes. The four inputs to the ANN were easily measurable parameters: air superficial velocity, water superficial velocity, pipe diameter and pipe inclination. The ANN was trained using the data reported in the literature. The output of the ANN classified the flow pattern into bubble, slug, churn and annular flow.

A procedure for the online identification of horizontal gas–liquid flow regimes from pressure gradient signals in a pipe was developed (Selli and Seleghim 2007). Using Gabor transform, the constants (called atoms) were obtained and used as inputs to the ANN model. The architecture of the ANN was 380-60-5. The outputs were different flow patterns: annular, bubbly, intermittent, stratified wavy and stratified smooth. Though the intermittent and stratified wavy regimes could be identified with greater certainty, other flows were identified with low certainties.

Development of sensors involves establishing a relation between the measured signal and the type and concentration of the analyte. Meier et al. (2007) used ANN-based feature extraction techniques for the detection of chemical warfare agents, and five weapon-related toxic industrial chemicals from the conductometric signals from microsensors based on metal oxide films. Seven interfering systems were also successfully analysed.

8.3.4 Fault Diagnosis

Otawara et al. (2002) used an ANN-based model to predict chaotic behaviour in three-phase fluidised beds. Similar studies were also carried out in bubble columns (Lin et al. 2003). The time interval between successive arrivals of bubbles in a bubble column is chaotic in nature. The experimental values of the time interval were extracted from the experimental data obtained from an optical probe. An ANN with 25 input nodes and 50 nodes in the hidden layer was trained. The inputs were the gas velocity and successive data points in the time series, and the output was the next value of the time interval. The predicted values were again used as inputs. The ANN predicted next 200 points. The predicted time series showed similar mean, standard deviation, power spectra and Lyapunov exponent.

Detection and diagnosis of process faults in the absorption of CO_2 gas in monoethanolamine (MEA) in an absorption and stripping pilot plant were investigated using ANN (Behbaharsi et al. 2009). The process consisted of an absorption column in which solvent is pumped, and a mixture of air and gas is fed to the column. The best ANN structure was 5-9-8. The five inputs to the ANN were two flow rates of gas and rich solvent streams leaving the column, the temperature of both leaving streams and the pressure drop. Eight types of faults were considered. These were low and high values of the gas and liquid flow rates, a concentration of CO_2 in feed and a concentration of MEA in the column. The ANN was trained using the data with only one fault at a time. However, it predicted not only single faults but also a combination of two faults. The prediction of three faults at a time could not be predicted with certainty.

In a model of predictive control, many variables affect each other. As a consequence, the detection of fault and its magnitude is a major problem. Zumoffen and Basualdo (2008) illustrated the use of ANN to predict offset while designing a fault detection, isolation and estimation (FDIE) system.

8.3.5 Process Control

Probably, the role of ANN in process control cannot be ignored due to the simplicity of application of ANN and the manner in which it can combine with the first principle model. The hybrid models will be discussed in Section 8.6. Here, a few examples are presented to give an idea of various ways in which the ANN can be used for process control purposes.

Loquasto and Seborg (2003) discussed different ANNs used to identify abnormal and normal plant operation, the presence or absence of model–plant mismatch and normal and abnormal disturbances. A pattern classification approach for monitoring a model predictive control (MPC) system was presented. Closed-loop system responses for a variety of disturbances were used to train the ANN. Features extracted from the database were used as inputs to the neural network.

An online ANN-based model for control of the biodiesel trans-esterification reactor followed by control system design was developed (Mjalli and Hussain 2009). The ANN model was used to obtain a linear model of the reactor at each time step.

One of the applications of ANNs in process control is the development of soft sensors, which are nothing but ANNs trained to predict a variable quickly using easily measurable variables as inputs. The fast computations have reduced the measurement time observed when sophisticated equipment such as a chromatograph is used. The soft sensors can substitute measuring devices and even reduce the need for measuring devices, and they can detect faults and failures.

8.4 Advantages of ANN-Based Models

Models based on ANNs have been applied in several chemical engineering applications. At times, they have been used as fashionable methodology. Although the ANN is only a black-box approach, it can be used effectively in several cases due to the features it offers. Some of the advantages of using ANN-based models are as follows:

1. ANNs with suitable architecture can approximate highly non-linear functions with any degree of complexity. Therefore, these are called 'universal approximators'. These can be used to correlate the data when it is difficult to find a suitable function with which to correlate.

2. ANN models can easily be developed and require less time to develop than other models. The availability of software (e.g. the neural network toolbox of MATLAB®) has reduced the time and effort needed to develop ANN-based models.

3. Due to the short time taken for training, the ANN can be trained online. The self-learning feature of the ANN is used for this purpose. The architecture of ANN used for online training, and the activation function (sigmoid or radial basis function) should be decided before online applications. It is done from the previous knowledge (i.e. the data taken earlier, used to develop an ANN model offline). Thus, online application is the readjustment of weight matrices and values of bias.

4. If experimental data are available, the ANN models can be developed even in those cases in which there is a lack of theoretical knowledge of the process and interdependence of parameters of the system. The parameters can be correlated in terms of easily measureable quantities even though an exact relationship is difficult to conceive.

5. The ANNs possess fast computational capabilities. Therefore, they can be used in real-time applications (e.g. process control). The ANN can be used as soft sensors to measure difficult-to-measure variables and as a controller. They can also be used to differentiate between normal and abnormal behaviour of a process.

6. A combination of feature extraction and fast computational technique, the ANN model can be used to take decisions. They can be used to

extract features from video images and use them to classify events. It gives them the capability of human-like behaviour. ANN models thus can be a part of expert systems.

7. ANNs can handle even noisy, incomplete or inconsistent data if over-fitting is avoided.

8. ANNs are capable of predicting cause-and-effect relationships.

8.5 Limitations of ANN-Based Models

There are several applications in which ANNs have been used successfully. However, it has certain shortcomings which require a second thought before using ANNs. Some of the limitations of ANNs are as follows:

1. ANN models are known as black-box models. Therefore, when extrapolated, they may provide poor estimates of the output. Though the ANN models have the ability to predict the output, it is also difficult to understand the interrelationship among the various inputs involved. It is difficult to interpret and analyse the results. Due to this inability, it is not advisable to use ANN for applications such as equipment design or plant design. However, in a goal-oriented approach, engineers are often interested in final results.

2. Training is the most essential phase in a network's life. The success or failure of the ANN is based on it. Since the ANNs mimic human learning procedures (i.e. step-by-step learning), they are trained with a well-mannered algorithm, which generally takes several thousand iterations. To accomplish the training of an ANN, computers take much time, hours or even days.

3. The ANNs are trained using experimental data. Therefore, a large set of experimental data are required. The data must cover wide operating conditions so that the predictions can be reliable. It is difficult to get sufficient data for training of ANNs. Therefore, an ANN model has been attempted to correlate the experimental data in cases where in spite of data available in abundance but clear understanding of the interrelation between various variables is missing.

4. The architecture of an ANN cannot be predicted. If an ANN has been trained for a process, the variation of conditions for which the ANN has been trained can change the architecture.

Data sparsity, over-fitting and poor generalisation are other problems faced by researchers when utilising the basic neural network alone. These weaknesses have recently encouraged researchers to focus on hybrid systems incorporating ANNs.

8.6 Hybrid Neural Networks

The 'black-box' nature of ANN models and unsatisfactory extrapolation by ANN models have led to the development of hybrid neural network models which combine ANNs with simple models. These are expected to perform better than ANNs in process identification tasks, since generalisation and extrapolation are confined only to the uncertain parts of the process while the basic model is always consistent with first principles (Psichogios and Ungar 1992).

ANNs can be used as a sub-model for a complex model. The remaining part of the model may be developed using laws of physics (e.g. laws of conservation, kinetics and thermodynamics). The latter has been referred to in the literature as the first principle model (FPM). These models may have analytical solutions. The governing equations may be solved numerically also. Some of the parameters may have to be determined by using either an empirical correlation or a sub-model. If an ANN is used to estimate any of the parameters and used in the FPM, then this approach has been called a 'series approach' (Ng and Hussain 2004). This approach is followed when either the time required for the estimation of the parameters using the sub-model is large or the sub-model is not available.

Another approach to develop a hybrid neural network is the parallel approach (Ng and Hussain 2004). In this approach, a simple FPM is used. In this model, the difference between the model prediction and the experimental value, called the residue, is correlated by the ANN model. The trained ANN is then used to predict the mismatch between the model values and the experimental value. The residue from the ANN model is added to the values predicted using the FPM.

The ANN has combined with the FPM in three ways depending upon the method of obtaining the solution of the model. The FPM can be solved analytically or numerically. If the model equations are solved analytically, then the ANN can be used to estimate some of the parameters appearing in the model. It is a series-type neural network. Such models are easily developed as the role of the ANN model does no more than provide a kind of correlation between the input and the output.

If the model equations are solved numerically, then the ANN model may be combined either with the time derivative or with the spatial derivative. Combining the ANN model with both derivatives in engineering applications is probably not reported in the literature.

If the ANN model is combined with a time derivative, then the numerical solution procedure is chosen in such a way that the ANN uses the values of the estimated variables at previous time intervals as inputs, and provides outputs a variable which appears in the FPM. The model equations thus get all the necessary variables so that they can be solved to get the values at the next time interval.

8.6.1 Application of Hybrid Neural Networks

Several chemical and biochemical systems involve variations of pH and complex reaction kinetics which make it difficult to develop a model based on first principles alone. Therefore, the development of dynamic models for the purpose of controlling the processes is also difficult. A hybrid neural network model makes use of a simple model and a neural network model in such a manner that some of the process parameters which cannot easily be correlated with state and control variables are described by an ANN model. These values are then used in the simple model (Psichogios and Ungar 1992). In general, a hybrid model can be written in the following form:

$$\frac{dx}{dt} = f(x,u,p) \tag{8.14}$$

and

$$p = g_{ANN}(x,u) \tag{8.15}$$

where g_{ANN} is the function described in the form of an ANN. The subscript to function g has been added to clarify the interaction of the ANN with the FPM. The model equations have to be solved numerically. Both sets of equations can be solved sequentially (a series approach) or simultaneously (a parallel approach) (Figure 8.5). While solving these equations

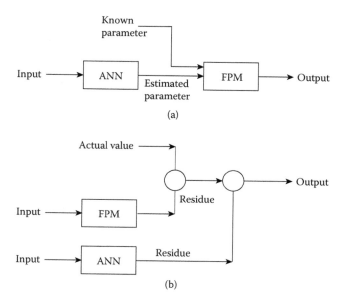

FIGURE 8.5
Structure of hybrid models: (a) series and (b) parallel.

numerically, the variable can be discretised. In a parallel mode, a suitable algorithm should be used to not only solve the FPM model but also train the ANN. Training of the ANN and solution of the model equation take place simultaneously. If the data to train the ANN are available, the solution is in series mode. It is called the 'use of prior knowledge' (i.e. data) to train ANN.

Let us understand the type of hybrid ANN models and their application through a few examples available in the literature.

8.6.1.1 A Hybrid ANN Model for a Bioreactor

Psichogios and Ungar (1992) developed a hybrid model for a fed-batch stirred bioreactor. The FPM was the material balance over the reactor.

$$\frac{dX}{dt} = \mu X - \frac{F}{V} X \tag{8.16}$$

$$\frac{dS}{dt} = -k_1 \mu X - \frac{F}{V}(S_i - S) \tag{8.17}$$

where $F = dV/dt$, X and S are the concentrations of biomass and substrate, respectively. The specific growth rate, μ, is a lumped kinetic parameter and is completely unknown in the hybrid model. To solve Equations 8.16 and 8.17, the value of μ is required. Therefore, the structure of the hybrid neural network consisted of an ANN which produces the specific growth rate as output with the values of X and S as inputs. The hybrid model was trained by minimising the summation of the square of error of dimensionless biomass and substrate concentration for all data points.

Since the specific growth rate was not known, the learning scheme obtained the error between the model prediction and the current experimental value. It was assumed that between two successive measured values, the specific growth rate remains constant. Since the specific growth rate is not known, additional relationships to estimate the specific growth rate are required. The training can start with a guess value of μ. The search for the optimum value of μ is done by minimising the error using a search strategy. It was obtained from sensitivity equations by differentiating Equations 8.16 and 8.17. The detailed description is given in Chapter 9. Following, sensitivity relationships were used.

$$\frac{dG_X}{dt} = \left(\mu - \frac{F}{V} \right) G_X + X \tag{8.18}$$

$$\frac{dG_S}{dt} = -k_1 \mu\, G_X - \frac{F}{V} G_S - k_1 X \tag{8.19}$$

where

$$G_X = \frac{dX}{d\mu}, \; G_S = \frac{dS}{d\mu} \tag{8.20}$$

The terms G_X and G_S are functions of time and can be evaluated from the experimental data as the derivatives can be obtained from two successive values of biomass and substrate concentrations. These are the gradients of the output of the ANN part of the hybrid neural network. The actual specific growth rate is not required. The role of the sensitivity equation is to change the error of the variable estimated from the FPM (X and S in the present case) to estimate the correction in the non-linear parameter (i.e. μ) predicted by the ANN sub-model (Ng and Hussain 2004) so that the error of the variable is minimised. The following initial conditions are used:

$$G_X = G_S = 0 \tag{8.21}$$

The error signal to correct the weight matrix of the ANN was the dimensionless form of the following equation:

$$e_s = \frac{\left(X - X_{exp}\right)G_{Xi} + \left(S - S_{exp}\right)G_{Si}}{\sqrt{\left[G_{Xi}^2 + G_{Si}^2\right]}} \tag{8.22}$$

The error signal is minimised by adjusting the weight matrix. The role of the ANN was to predict the specific growth rate. However, the training scheme does not require its value to be known for various sets of input–output data i.e. the set of X, S for the present and next time intervals. For detailed discussion, the reader is referred to the original work of Psichogios and Ungar (1992).

8.6.1.2 Other Applications

A hybrid model for the performance of a styrene monomer reactor system consisting of an adiabatic radial reactor was developed Lin et al. (2004). The ethylbenzene dehydrogenation reaction has been modelled by six different reactions (Scheel and Crow 1969). An FPM to estimate the compositions, temperature and pressure profiles from the reactor geometry and reaction mechanism was proposed. The dynamic model equation, including the catalyst deactivation factor, φ, is as follows:

$$F\frac{dx_i}{dr} = (2\pi r L)\sum_j \left(r_{ij} M_i \varphi\right) \tag{8.23}$$

$$FC_p \frac{dT}{dt} = (2\pi rL) \sum_j \left(-\Delta H_j \sum_i r_{ij}\phi \right) \tag{8.24}$$

$$\frac{dP}{dr} = \frac{P_o - P_i}{R_o - R_i} \tag{8.25}$$

$$\sum_i x_i = 1 \tag{8.26}$$

The pressure drop along the radial direction is due to the assumption of a linear pressure profile. The term r_{ij} is the rate of reaction for the ith component in the jth reaction.

The catalyst deactivation factor was estimated using an ANN model with a 5-3-1 architecture. The inputs to the ANN model were the temperature, total pressure, partial pressures of steam and ethylbenzene and previous value of ϕ estimated at the previous time interval. The model equations were then solved to get pressure, mole fractions, partial pressures and the temperature at the next time interval. This procedure was repeated to get dynamic modelling of the reactor.

For chemical reactor modelling, an accurate knowledge of kinetics is necessary. A shift of a rate-determining step and lack of basic knowledge of reaction mechanisms may result in an unsuccessful attempt of a model based on first principles. Zahedi et al. (2005) used an ANN model to correlate composition of the product as a function of temperature, pressure and feed composition for CO_2 hydrogenation to methanol (Figure 8.6). For the packed bed, the FPM was obtained by writing a 1D model. To solve the model equations, the bed is divided into several layers. The ANN was adopted to estimate the outlet composition only, and the FPM was used to calculate the outlet pressure and temperature from the reactor. The outlet temperature of

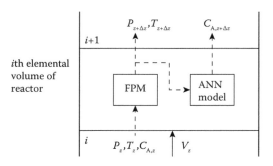

FIGURE 8.6
Hybrid ANNs for packed bed reactor based on Zahedi et al. (2005).

each layer was calculated by simultaneous solution of governing equations (heat, mass and pressure drop). The ANN model is a sub-model of the hybrid model.

The use of ANN for phenol degradation in biological wastewater was given by Jose et al. (1999). The methodology used a hybrid model. The algorithm to train the ANN was based on the Lyapunov stability criteria. The ANN used was a single-layer neural network with a "Mexican hat" wavelet, given as

$$\phi_i(x) = \left(1 - \frac{|\varepsilon_i - x|^2}{\sigma^2}\right) \exp\left(-\frac{|\varepsilon_i - x|^2}{2\sigma^2}\right) \tag{8.27}$$

Here, ε_i is the centre of the node, and σ is called dilation of the wavelet. The wavelets are a local basis function as they quickly approach zero as the distance from the centre increases.

A control strategy for non-linear pH control in wet limestone flue gas desulphurisation (WLFGD) plants based on neural networks was proposed by Perales et al. (2010). One ANN was used for identification of the pH in the oxidation tank. Another ANN and a linear model of the plant were used together to correct the modelling error obtained from the linear model. It was, therefore, a parallel-type hybrid ANN model.

Process control of non-linear processes, such as the polymerisation process, is a challenging task. Every batch may proceed in a different manner, and hence obtaining a rigorous FPM may be a difficult task. ANN models are trained using known data. The training data were generated using a rigorous model. In other words, prior knowledge of the data was used to train the ANN. Ng and Hussain (2004) used mass and energy balances as simple FPMs. The ANN was used to estimate two non-linear parameters: the overall heat transfer coefficient and the rate of polymerisation. The outputs were fed to the FPM based on energy balances. The sensitivity equations were used for adjustment weights. The inverse model for the problem was also given for the hybrid neural network model. The residue between the estimated and experimental values of jacket inlet temperature was used to train the ANN. The inverse model was approximate only as it was obtained by using linearised energy balances. The ANN part predicted the jacket temperature. The inverse model was used as a controller.

The hybrid neural networks keep the process variables consistent with physical models. Therefore, the hybrid neural network models are more reliable than neural networks alone. They show better generalisation and extrapolation capabilities in comparison to the ANN.

The hybrid neural network model requires less data to get trained. The hybrid network model uses a simple but accurate model. In cases of hybrid models having many uncertain parameters, the model predictions may not be good.

While using a hybrid model, it may be argued that the FPM is not accurate and that the ANN is not trained. However, it is the qualitative prediction of gradients of the parameters which helps in controlling the process (Tsen et al. 1996). Therefore, a hybrid model for predictive control is expected to be effective.

8.7 Summary

The methodology of developing ANNs of feedforward was presented. It involves collection of data, choice of input and optimum architecture of ANNs through learning rules. ANNs have been successfully applied in developing a quick model for several chemical and biochemical processes. The ANN can be combined to FPMs also in the form of a hybrid ANN model. Such models have shown their importance in processes with no very clear understanding (e.g. biological treatment of wastewater).

References

Arnavata, M.P., Hernandez, J.A., Bruno, J.C., and Coronas, A. 2013. Artificial neural network models for biomass gasification in fluidized bed gasifiers. *Biomass Bioenergy*. 49: 279–289.

Behbaharsi, R.M., Jazayeri-Rad, H., and Hajmirzaee, S. 2009. Fault detection and diagnosis in a sour gas absorption column using neural networks. *Chem. Eng. Technol.* 32(5): 840–845.

Darwish, N.A., Hilal, N., Al-Zoubi, H., and Mohammad, A.W. 2007. Neural networks simulation of the filtration of sodium chloride and magnesium chloride solutions using nanofiltration membranes. *Chem. Eng. Res. Des.* 85(A4): 417–430.

Jamialahmadi, M., Zehtaban, M.R., Steinhagen, H.M., Sarrafi, A., and Smith, J.M. 2001. Study of bubble formation under constant flow conditions. *Chem. Eng. Res. Des.* 79(A): 523–532.

Jha, S.K., and Madras, G. 2005. Neural network modeling of adsorption equilibria of mixtures in supercritical fluids. *Ind. Eng. Chem. Res.* 45: 7038–7041.

Jose, L., Sanchez, G., Robinson, W.C., and Budman, H. 1999. Development studies of an adaptive on-line softsensor for biological wastewater treatments. *Can. J. Chem. Eng.* 77: 707–717.

Kasabov, N.K. 1998. *Foundations of Neural Networks, Fuzzy Systems and Knowledge Engineering*. Cambridge, MA: MIT Press.

Lim, H., Kang, M., Park, S., and Lee, J. 2004. Modeling and optimization of a styrene monomer reactor system using a hybrid neural network model. *Ind. Eng. Chem. Res.* 43: 6441–6445.

Lin, H.Y., Chen, W., and Tsutsumi, A. 2003. Long-term prediction of nonlinear hydro-dynamics in bubble columns by using artificial neural networks. *Chem. Eng. Process.* 42: 611–620.

Loquasto, F., III., and Seborg, D.E. 2003. Monitoring model predictive control systems using pattern classification and neural networks. *Ind. Eng. Chem. Res.* 42: 4689–4701.

Meier, D.C., Evju, J.K., Boger, Z., Raman, B., Benkstein, K.D., Martinez, C.J., Montgomery, C.B., and Semantik, S. 2007. The potential for and challenges of detecting chemical hazards with temperature-programmed microsensors. *Sens. Actuat. B.* 121: 282–294.

Mjalli, F.S., and Hussain, M.A. 2009. Approximate predictive versus self-tuning adaptive control strategies of biodiesel reactors. *Ind. Eng. Chem. Res.* 48: 11034–11047.

Mohammadi, A.H., and Richon, D. 2006. Estimating the hydrate safety margin in the presence of salt or organic inhibitor using refractive index data of aqueous solution. *Ind. Eng. Chem. Res.* 45: 8207–8212.

Molga, E., Cherbanski, R., and Szpyrkowicz, L. 2006. Modeling of an industrial full-scale plant for biological treatment of textile wastewaters: Application of neural networks. *Ind. Eng. Chem. Res.* 45: 1039–1046.

Nezhad, B.Z., and Aminian, A. 2010. A multi-layer feed forward neural network model for accurate prediction of flue gas sulfuric acid dew points in process industries. *App. Thermal Eng.* 30: 692–696.

Ng, C.W., and Hussain, M.A. 2004. Hybrid neural network—Prior knowledge model in temperature control of a semi-batch polymerization process. *Chem. Eng. Process.* 43: 559–570.

Omata, K., Hashimoto, M., Sutarto, and Yamada, M. 2009. Artificial neural network and grid search aided optimization of temperature profile of temperature gradient reactor for dimethyl ether synthesis from syngas. *Ind. Eng. Chem. Res.* 48: 844–849.

Otawara, K., Fan, L.T., Tsutsumi, A., Yano, T., Kuramoto, K., and Yoshida, K. 2002. An artificial neural network as a model for chaotic behavior of a three-phase fluidized bed. *Chaos Solitons Fractals.* 13: 353–362.

Perales, A.L.V., Ortiz, F.J.G., Barrero, F.V., and Ollero, P. 2010. Using neural networks to address nonlinear pH control in wet limestone flue gas desulfurization plants. *Ind. Eng. Chem. Res.* 49: 2263–2272.

Pollock, G.S., and Eldridge, R. 2000. Neural network modeling of structured packing height equivalent to a theoretical plate. *Ind. Eng. Chem.* 39: 1520–1525.

Psichogios, D.C., and Ungar, L.H. 1992. A hybrid neural network-first principles approach to process modeling. *AIChE J.* 38(10): 1499–1511.

Rajendran, A., and Thangavelu, V. 2007. Sequential optimization of culture medium composition for extracellular lipase production by *Bacillus sphaericus* using statistical methods. *J. Chem. Technol. Biotechnol.* 82: 460–470.

Rani, K.Y., and Patwardhan, S.C. 2004. Data-driven modeling and optimization of semibatch reactors using artificial neural networks. *Ind. Eng. Chem. Res.* 43: 7539–7551.

Rene, E.R., Lopez, M.E., Veiga, M.C., and Kennes, C. 2011. Neural network models for biological waste-gas treatment systems. *N. Biotechnol.* 29(1): 56–73.

Scalabrina, G., Marchf, P., Bettioa, L., and Richonb, D. 2006. Enhancement of the extended corresponding states techniques for thermodynamic modeling. II. Mixtures. *Int. J. Refrig.* 29: 1195–1207.

Scheel, J.G.P., and Crowe, C.M. 1969. Simulation and optimization of an existing ethylbenzene dehydrogenation reactor. *Can. J. Chem. Eng.* 47: 183–187.

Selli, M.F., and Seleghim, P., Jr. 2007. Online identification of horizontal two–phase flow regimes through Gabor transform and neural network. *Heat Transf. Eng.* 28(6): 541–548.

Shaikh, A., and Al-Dahhan, M. 2003. Development of an artificial neural network correlation for prediction of overall gas holdup in bubble column reactors. *Chem. Eng. Proc.* 42: 599–610.

Sharma, H., Das, G., and Samanta, A.N. 2006. ANN-based prediction of two-phase gas-liquid flow patterns in a circular conduit. *AIChE J.* 52(9): 3018–3028.

Syu, M., Chen, B., and Chou, S. 2003. A study on the sedimentation model and neural network online adaptive control of a benzoic acid imitated wastewater oxidation process. *Ind. Eng. Chem. Res.* 42: 6862–6871.

Tsen, A.Y.D.T., Jang, S.S., and Wong, D.S.H. 1996. Predictive control of quality in batch polymerization using hybrid ANN models. *AIChE J.* 42(2): 455–465.

Tsutsumi, A., Chen, W., and Hasegawa, T. 2001. Neural networks for prediction of the dynamic heat-transfer rate in bubble columns. *Ind. Eng. Chem. Res.* 40: 5358–5361.

Xie, T., Ghiaasiaan, S.M., and Karrila, S. 2003. Flow regime identification in gas/liquid/pulp fiber slurry flows based on pressure fluctuations using artificial neural networks. *Ind. Eng. Chem. Res.* 42: 7017–7024.

Yang, H., Fang, B.S., and Reuss, M. 1999. k_L a correlation established on the basis of a neural network model. *Can. J. Chem. Eng.* 77: 838–843.

Yang, M., and Wei, H. 2006. Application of a neural network for the prediction of crystallization kinetics. *Ind. Eng. Chem. Res.* 45: 7070–7075.

Zahedi, G., Elkamel, A., Lohi, A., Jahanmiri, A., and Rahimpor, M.R. 2005. Hybrid artificial neural network: First principle model formulation for the unsteady state simulation and analysis of a packed bed reactor for CO_2 hydrogenation to methanol. *Chem. Eng. J.* 115: 113–120.

Zamankhan, P., Malinen, P., and Lepomaki, H. 1997. Application of neural networks to mass transfer predictions in a fast fluidized bed of fine solids. *AIChE J.* 43(7): 1684–1690.

Zhang, C.L. 2005. Generalized correlation of refrigerant mass flow rate through adiabatic capillary tubes using artificial neural network. *Int. J. Refrig.* 28: 506–514.

Zumoffen, D., and Basualdo, M. 2008. Improvements in fault tolerance characteristics for large chemical plants: 1. Waste water treatment plant with decentralized control. *Ind. Eng. Chem. Res.* 47: 5464–5481.

9

Model Validation and Sensitivity Analysis

The methodology for developing various types of models was presented in Chapters 4 through 8 of this book. Before using a model to simulate a process, it is necessary to ensure that the model predictions represent the process behaviour. A procedure to evaluate the model with this aim is known as a model validation.

A process model can be simple or rigorous depending upon the assumptions involved. Both types of models have few advantages and limitations. The simple models require less computational effort but have certain limited scope. As compared to simple models, the rigorous models require large computational effort and may also be time-consuming. The model predictions may be used in a variety of situations as they may be more accurate. The simple models are easy to interpret; in contrast, the rigorous or elaborate models are not so easy to interpret, nor is easy to study the interrelations between the various parameters involved. The simple model can be used as a sub-model with ease, whereas the use of a rigorous model as a sub-model may require certain modifications in the solution methodology.

The choice between simple and rigorous models should not be made heuristically. In some cases, it may not be possible to develop a simple model. Still, out of several models, some of them may be considered simple models. In several cases, even the black-box approach of artificial neural networks may be favoured. A model should be able to predict the required behaviour of the process for which it has been developed. The model predictions should exhibit the following characteristics:

1. The predicted values should be close to the experimental values. The difference between the predicted values and experimental values should be within acceptable limits.

2. The model should explain the expected dependence of output on different input variables.

3. Several parameters are used in a model. The expected dependence of output on these parameters should also be predicted by the model.

4. The model parameter may have a meaningful interpretation. In that case, it helps in increasing the knowledge of the process.

5. The model may be used to predict the process behaviour within a known range of parameters. It is not always possible to use a model for a wide range of parameters. The range of applicability depends upon the type of assumptions.

A proposed model is to be checked to see that it fulfils the expectations. If more than one model fulfils these criteria, then a systematic procedure is to be followed to evaluate the model by comparing the predicted values with the experimental values. This step is called model verification. When the model represents the actual process to a reasonable extent, it is said to be validated. The validated model reproduces system behaviour satisfactorily and serves the purpose for which it was proposed.

The procedure for model validation depends upon the type of the model and the purpose for which the model is proposed. When a model consists of several sub-models, each of these may be validated in a different manner.

9.1 Model Validation: Objective

A model is to be validated before it is used to study the behaviour of a given process through simulation. Before simulation, confidence in a model's prediction should be established. The validations methodology depends upon the purpose of the model. The following aspects are related to model behaviour and are to be checked during model validation.

9.1.1 Model Output

The main objective of a model is to provide outputs which are similar to those of the actual process. The model output may be the values of a few dependent variables in the case of static models. In the case of dynamic models, the dynamic behaviour of a process should also be similar to that exhibited by the actual process.

A model is not only a way of representation of the experimental or process data. It should predict process behaviour like an actual process. Only after ensuring this fact the model may be simulate the process and to explore the situations in which the data for the actual process are not available. This exploration may involve searching for new optimum operating conditions, a better controller design, conversion in a reactor with a new catalyst, a design of a new reactor etc. However, it requires that the model must be able to predict the dependence of model output on the parameters used in the model.

By changing the model parameters, the model output changes. The variation of model output with variation in parameter is called 'sensitivity' in qualitative terms. The sensitivity of the model should also be reasonably similar to that of the actual process. The model validation step thus involves comparing the model output and sometimes its sensitivity towards various parameters.

If the model output and the parametric sensitivity of the model are satisfactory, it may be accepted for simulation purposes. A detailed model validation may be required in several cases.

9.1.2 Assumptions

The model equations were obtained from first principles. This involved simplification of the process by making the assumptions. For example, let us consider that one of the assumptions is that the properties of fluids such as density, surface tension and viscosity are independent of the temperature. These assumptions are quite common. A model for the process such as the heat or mass transfer coefficient in equipment under this assumption is developed, and model equations are solved.

The model may accept these properties as inputs which are independent of other variables. These inputs do not influence any other parameter. In other words, they do not interact with other parameters. But the model equations consist of differential equations; therefore, the solution may introduce interactions with other parameters. This interaction should be similar to the interaction exhibited by the actual process.

Thus, a method for studying the effect of a parameter on the model output due to interaction effects with other parameters is required. If the effect of a parameter on model output through interaction with other parameters (as predicted by the model) is reasonably similar to that observed experimentally, then the assumption is correct. Otherwise, the model may be refined by modifying the assumptions or may be used in limited situations.

Unfortunately, there is no unique way in which the assumptions are made in the first stage of the model development. The type of assumption depends upon the degree of clear understanding of the process. However, validity of the assumptions can be checked after a model is available.

9.1.3 Model Inputs

In an actual process, the input parameters may change in an uncertain manner. For example, the flow rate, operating pressure and temperature, properties of the process streams, composition of the raw material and ambient conditions may change with time. Due to repeated changes in the parameter, it is essential to know how the optimum operating conditions have to be changed to get the same model output or a different desired output. In process control, it is similar to regulator and servo problems, respectively. Some of the parameters may not affect the model predictions to a significant level, but a small change in other parameters may change the model output to a significant level (i.e. the model may be less sensitive to some of the parameters and may be very sensitive to other parameters).

The random variation in the input parameters may be described by a probability density function, mostly by normal distribution function. The study of the effect of this uncertainty on the uncertainty of the model output (i.e. the propagation of the uncertainty) is known as 'uncertainty analysis'. The strength of the relation between a given uncertain input and the output is known as 'sensitivity analysis' (Saltelli et al. 2005).

9.2 Model Validation Methodology

Earlier attempts to verify that the model is acceptable were based on the graphical presentation of the variation of model output with input parameters. Keeping all of the parameters constant, only one of the parameters is changed, and its effect on the model output is plotted (Figure 9.1a). From the plot, the trend of the line is observed. It may be linear or non-linear, monotonically

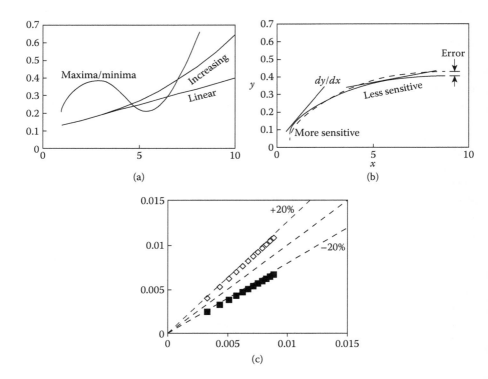

FIGURE 9.1
(a) Variation of model output as a function of a variable and parameter, (b) graphical interpretation of direct differentiation sensitivity and (c) parity plot and error.

increasing or decreasing, with an occurrence of maximum or minimum value. The linearity on the log-log scale or semi-log scale was also checked. The behaviour of the data and the model were compared. Such plots were also used to demonstrate the error in model predictions. The modelling error could be identified as systematic error (Figure 9.1b). These plots allowed the comparison of the model predictions with empirical correlations also.

The acceptability of the model was also checked from a scatter plot between the experimental and predicted values (Figure 9.1c). It is also known as a parity plot. The errors could easily be shown by straight lines passing through the origin. The points lying on $y = x$ represent zero error. The deviation from the line is taken as an error or percentage error. The average error is computed from the error for all of the points.

These methods have the following drawbacks:

1. While varying one parameter, other parameters are to be kept constant. It requires a large set of experimental data.

2. The methodology is based on the fact that the parameters do not have an interacting effect on the model output.

3. The average error depends upon the number of data points and the distribution of data over the range of the study. An uneven distribution may give a misleading value of average error.

4. The methodology considers that there is no uncertainty in specifying the parameters.

5. It does not allow us to compare the relative importance of different parameters affecting the output of a model.

The model validation techniques should address these issues also. The model may be used for several values of the parameter. The error may be analysed statistically. One of the statistical methods in evaluating the model is known as 'goodness of fit'. Some of the commonly used criteria for goodness of fit are presented in Table 9.1. Here y and $\hat{y}$ are experimental and estimated values respectively.

For a linear model, the model validation can easily be achieved by regression. For example, a linear regression model is defined as follows:

$$y = a_0 + \sum_{i=1}^{n} a_i x_i \tag{9.1}$$

The coefficients $a_0 \ldots a_n$ can be determined by making use of correlation coefficients. The amplitude of these coefficients is the measure of relative importance. For large values of the coefficient, a_i, the importance of parameter, x_i, is also large. However, the model should be linear with respect to all the parameters. A small value of a_i indicates that either the model output is less sensitive to that parameter or there exists non-linear dependence of the

TABLE 9.1

Measures of Correlation (Goodness of Fit)

Term	Formula
Coefficient of determination	$$R^2 = \frac{\sum_{i=1}^{N}(y_i - \hat{y}_i)^2}{\sum_{i=1}^{N}(y_i - \bar{y}_i)^2}$$
Percent standard error of the prediction	$$\%SEP = \frac{100}{\bar{y}_i}\sqrt{\frac{\sum_{i=1}^{N}(y_i - \hat{y}_i)}{N}}$$
Coefficient of the efficiency	$$E = \frac{\sum_{i=1}^{N}(y_i - \bar{y}_i) - \sum_{i=1}^{N}(y_i - \hat{y}_i)}{\sum_{i=1}^{N}(y_i - \bar{y}_i)}$$
Average relative variance	$$ARV = \frac{\sum_{i=1}^{N}(y_i - \hat{y}_i)}{\sum_{i=1}^{N}(y_i - \bar{y}_i)}$$

model output on that parameter. Non-linearity can be handled by replacing the parameters by transformed parameters (e.g. the power law–type functions can be converted to linear form by taking logarithms).

$$y = a_0 x_1^{a_1} x_2^{a_2} \ldots x_n^{a_n} \tag{9.2}$$

This equation can be converted into the following linear model:

$$y = \log a_0 + \sum_{i=1}^{n} a_i \left(\log x_i\right) \tag{9.3}$$

The estimators to determine these coefficients become biased since the regression is based on the minimisation of the summation of square of log of errors and not the summation of square of errors.

The method still lacks the interacting effects of various parameters on the model output.

A model having m parameters, $\bar{x} = \{x_1, x_2, \ldots x_m\}$, can be written as

$$y = f(x_1, x_2, \ldots, x_m) \tag{9.4}$$

The model can be expressed as a truncated Taylor series. It is a polynomial representation of the model output and is also called a response function (Akhnazarova and Kafarov 1982).

$$y = a_0 + \sum_{i=1}^{m} a_i x_i + \sum_{\substack{i=j=1 \\ i \neq j}}^{m} a_{ij} x_i x_j + \sum_{i=1}^{m} a_{ii} x_i^2 + \dots \tag{9.5}$$

The response surface methodology is the graphical representation of the response function. More terms in Equation 9.5 can be considered depending upon the number of parameters involved. This methodology includes interactions between parameters. However, it requires the generation of large values of predicted values from the model, and then fitting the response surface to all these data. Since the sensitivity analysis can be performed through simulation studies, the response surface methodology is not very promising.

In the case of dynamic models, the dynamic behaviour of the model output as a function of model parameters is also studied. These models are mainly used in process monitoring and control applications. A vast literature on the description of process behaviour is available. The process has been defined as first order, second order or higher order, with delay time, integrating process etc. The dynamic behaviour has also been studied in the frequency domain. A dynamic model for the process is acceptable if it describes the dynamic process in well-defined types and helps in estimation of the parameters associated with the type of process.

9.2.1 Validating Dynamic Models

In an unsteady process, the parametric sensitivity may change with time. It is possible that during the initial period of the process, the process exhibits high sensitivity to one particular parameter. After some time, another parameter becomes important and has the largest effect on the process output. It is commonly observed in cases of complex reactions or biochemical processes. A dynamic model must also predict the dynamic behaviour of the process.

When a process is simulated to predict the dynamic behaviour and is compared with the experimental dynamic response, the error changes with time. An average error may be used for comparison purposes. The process control engineers frequently use the following averaging methods (Seborg et al. 2004).

If the error between the model prediction and the experimental value at any time t is $e(t)$, then the integral of the absolute value of the error (IAE) is given by

$$\text{IAE} = \int_0^{\infty} |e(t)| dt \tag{9.6}$$

The integral of the squared error (ISE) is defined as

$$ISE = \int_0^\infty e(t)^2 \, dt \tag{9.7}$$

The integral of the time-weighted absolute error (ITAE) is given by

$$ITAE = \int_0^\infty t|e(t)| \, dt \tag{9.8}$$

The dynamic behaviour of the model predictions will be close to the experimental dynamic behaviour if the chosen integrated error measure is less.

9.2.2 Statistical Analysis of the Model

It is quite common to think that a model will predict the same output for a given set of parameters, but this is true with deterministic models only. For stochastic models or population balance models, the model output will be different in different runs. To study the effect of uncertainty in a parameter, the model parameters may be assumed to follow a probability density function. The model output will be different in each run. Due to uncertainty in more than one parameter, it becomes essential to determine the parameter that is significantly affecting the model output. The analysis will also give the insignificant factors.

The uncertainty is defined as the sum of a fixed bias error, β, and random error, ε_μ.

$$u_\mu = \beta + \varepsilon_\mu \tag{9.9}$$

In cases of more than one source of bias error or random error, a root mean square value can be taken.

$$\bar{\beta} = \sqrt{\sum_i \beta_i^2} \; ; \; \bar{\varepsilon}_\mu = \sqrt{\sum_i \varepsilon_{i_\mu}^2} \tag{9.10}$$

For a model having m parameters, $\bar{x} = \{x_1, x_2, \ldots x_m\}$, with uncertainty and model output as $y = f(x_1, x_2, \ldots, x_m)$, the combined uncertainty, u_c, in model output can be defined as

$$u_c^2 = \sum_{i=1}^m \left(\frac{\partial f}{\partial x_i} \right)^2 u^2(x_i) \tag{9.11}$$

The random error can be Gaussian or non-Gaussian. For non-Gaussian random error, a method for estimating the confidence interval was given by Vásquez et al. (2010). It is based on the observation that most of the distribution of probability tails follows either exponential or Pareto distribution irrespective of the type of probability density function. Therefore, the confidence interval could be determined for an arbitrary probability density function by knowing the parameters from the probability tail.

9.2.3 Analysis of Variance

The popular method of studying the effect of a parameter on the model output requires that for each simulation run, only one parameter is varied and all other parameters are kept constant. The same is repeated for all other parameters. This not only requires a large number of simulations but also does not give any information effect due to the interactions between the parameters if more than one parameter is varied simultaneously. In the absence of no interaction among the parameters (i.e. the linear processes), this method is sufficient. In the presence of interacting parameters, an analysis of variance (ANOVA) is used. The variance of the sample runs is separated into various components showing the effect of individual parameters. One of such method is the *F*-test. The details may be found in books on experimental design or statistics.

To conduct ANOVA, simulations may be carried out by changing the parameters. Due to uncertainties in parameters, different values of model output will be obtained. The parameters are varied in a well-defined manner so that the variance of model predictions can be analysed. Various schemes for designs of experiments are discussed in the literature. The aim of the design of the experiments is to decide the set of parameters within which the experiments are to be performed. In the present context, the simulation is a numerical experiment.

9.3 Sensitivity Analysis

The methodology for model validation discussed earlier in this chapter can be considered qualitative as no attempt is made to quantify the sensitivity of the model output to model parameters. A number of quantitative methods were developed and are reported in the literature (Saltelli et al. 2005). A few of them are described here.

9.3.1 Direct Differential Method

The parameter sensitivity can be quantitatively described by the first derivative of the output to a parameter. If the function is analytically differentiable, it can be differentiated directly. Since the models are generally based on first principles, it looks obvious to differentiate the analytical solution in the case of several simple models for which the analytical solutions are possible. This method is called the 'direct differential method'.

Let us consider that there are ϕ_i parameters affecting the output, y, of a model. The aim is to study the effect of variation of ϕ_i on y. The absolute sensitivity, $s^y_{\phi_i}$, is defined as

$$s^y_{\phi_i} = \frac{\partial y}{\partial \phi_i} \qquad (9.12)$$

It is also called a first-order local sensitivity. It can be estimated numerically by changing the parameter by a small amount and finding out the change in the model prediction. If local sensitivities for various parameters are to be compared, then it is appropriate to define a normalised sensitivity, $S^y_{\phi_i}$, as

$$S^y_{\phi_i} = \left(\frac{\phi_i}{y}\right)\left(\frac{\partial y}{\partial \phi_i}\right) \qquad (9.13)$$

This is also known as relative sensitivity.

If the relative sensitivity for a particular model is known, then for a small change in the parameter, the change in the output for the model can be determined by

$$\frac{\Delta y}{y} = S^y_{\phi_i}\left(\frac{\Delta \phi_i}{\phi_i}\right) \qquad (9.14)$$

Equation 9.14 can also be used to determine the required change in the model input so that the model output changes by a desired value:

$$\Delta \phi_i = \left(S^y_{\phi_i}\frac{\Delta y}{y}\right)\phi_i \qquad (9.15)$$

The output of the process predicted by the model should be within the acceptable limits of process output. For example, let the mole fraction of the more volatile component in distillate from the distillation column be 90%. Due to disturbances, the actual distillate will be around this value. Let the acceptable value of the distillate composition be 89%–91%. Since the output has to be controlled by varying parameters, it is necessary to know how much change in the parameter can be tolerated to keep the distillation within a limit. Let the deviation in the input parameter be δ_{ϕ_i}. If the acceptable variation of the process output (model output) is δ_y^{acp}, the normalised sensitivity threshold is defined as

$$S^y_{\phi_i,\text{TH}} = \frac{\delta_y^{acp}}{\delta_{\phi_i}} \qquad (9.16)$$

If the normalised sensitivity is less than the normalised sensitivity threshold, then the plant operation is acceptable (Zanfir and Gavriilidis 2002):

$$S_{\phi_i}^y < S_{\phi_i, \text{TH}}^y \tag{9.17}$$

Thus, the sensitivity analysis provides an operational range for each parameter. This information is important for the design of a process controller.

The first principle model generally contains differential equations. It is also possible to differentiate the differential equation itself. These equations are called 'sensitivity equations'. The model equations and sensitivity equations are solved simultaneously to get both the model output and sensitivity. The solution can be obtained analytically or even numerically.

Let the system be defined by

$$\frac{d\bar{y}}{dx} = f\left(\bar{x}, \bar{y}, \bar{\phi}\right) \tag{9.18}$$

On differentiation, the sensitivity equations are obtained:

$$\frac{ds_{\phi_i}^y}{dx} = \bar{J}s_{\phi_i}^y + \frac{\partial f\left(\bar{x}, \bar{y}, \bar{\phi}\right)}{\partial \phi_i} \tag{9.19}$$

The $\bar{J}$ is the Jacobian matrix for the system and is given by

$$\bar{J} = \begin{pmatrix} \dfrac{\partial f_1}{\partial x_1} & \dfrac{\partial f_1}{\partial x_2} & \cdots & \dfrac{\partial f_1}{\partial x_n} \\[2mm] \dfrac{\partial f_2}{\partial x_1} & \dfrac{\partial f_2}{\partial x_2} & \cdots & \dfrac{\partial f_2}{\partial x_n} \\[2mm] \cdots & \cdots & \cdots & \cdots \\[2mm] \dfrac{\partial f_n}{\partial x_1} & \dfrac{\partial f_n}{\partial x_2} & \cdots & \dfrac{\partial f_n}{\partial x_n} \end{pmatrix} \tag{9.20}$$

For example, let us consider the Laplace equation which is used to describe the two-dimensional (2D) temperature distribution in solids.

$$\frac{\partial^2 T}{\partial X^2} + \sigma \frac{\partial^2 T}{\partial Y^2} = 0 \tag{9.21}$$

Here the temperature, T, X and Y are dimensionless. The width of the solid is L_1 and L_2 in the x and y directions. The parameter $\sigma = \left(\dfrac{L_1}{L_2}\right)^2$ was due to conversion of the Laplace equation into a dimensionless form.

Let us consider the following boundary conditions:

At $X = 0$, at all Y, $\dfrac{\partial T}{\partial X} = 0$

At $X = 1$, at all Y, $\dfrac{\partial T}{\partial X} = 0$

At $Y = 0$, at all X, $T = T_1$

At $Y = 1$, at all X, $T = T_2$

Sensitivity equations can be obtained by differentiating the model equations with respect to parameter φ_i.

$$\frac{\partial^2 s_{\phi_i}^T}{\partial X^2} + \left(\sigma \frac{\partial^2 s_{\phi_i}^T}{\partial Y^2} + \frac{\partial \sigma}{\partial \phi_i} \frac{\partial^2 T}{\partial Y^2} \right) = 0 \tag{9.22}$$

The sensitivity equations are differential equations which require the boundary conditions to be specified. The required boundary conditions can be obtained by differentiating the boundary conditions for the model equations.

At $X = 0$, at all Y, $\dfrac{\partial s_{\phi_i}^T}{\partial X} = 0$

At $X = 1$, at all Y, $\dfrac{\partial s_{\phi_i}^T}{\partial X} = 0$

At $Y = 0$, at all X, $s_{\phi_i}^T = s_{\phi_i}^{T_1}$

At $Y = 1$, at all X, $s_{\phi_i}^T = s_{\phi_i}^{T_2}$

The direct differential method is easily implemented. A major class of optimisation technique involves estimation of the first derivative of the parameter to determine the direction of the search. The process optimisation is based on the use of parameter sensitivity without mentioning it.

The system of equations, consisting of model equations and sensitivity equations, may consist of differential equations and algebraic equations. Such equations are observed in chemical and biochemical processes, too. The equations can be solved numerically. A method for performing sensitivity equations for a zero-order catalytic reaction in a catalytic-walled tube and for affinity membrane separation of protein was described by Caracotsios and Stewart (1995). The sensitivity analysis is based on the direct differential method.

9.4 Global Sensitivity Measures

The sensitivities so far discussed in this chapter are local only (i.e. they depend upon the value of the input parameters). It is required to define global sensitivity measures which will represent sensitivity over the entire range of parameters and should be able to account for the effect on the model output due to interaction with other parameters also. Of several global sensitivity measures, only three of them will be discussed.

9.4.1 Gradient-Based Global Sensitivity Measures

The first-order local sensitivity, calculated as the gradient, depends upon the value of input parameters. A few of these measures have been described by Kiparissides et al. (2009). A global sensitivity measure can be obtained by averaging the entire range of an input parameter, x_i. The value is for a particular parameter.

$$\bar{M}_i = \int_{x_{min}}^{x_{max}} \left| S_{x_i}^y \right| dx_i \tag{9.23}$$

Another measure of global sensitivity is the variance of $\bar{M}_i$:

$$\bar{\Sigma}_i^2 = \int_{x_{min}}^{x_{max}} \left(S_{x_i}^y \right)^2 dx_i - \left(\bar{M}_i \right)^2 \tag{9.24}$$

The integral in Equation 9.24 can also be treated as a measure of global sensitivity:

$$\bar{G}_i = \bar{\Sigma}_i^2 + \left(\bar{M}_i \right)^2 = \int_{x_{min}}^{x_{max}} \left(S_{x_i}^y \right)^2 dx_i \tag{9.25}$$

Another measure of global sensitivity is

$$\frac{\bar{M}_i}{\bar{\Sigma}_i} = \left[\frac{1}{\bar{\Sigma}_i^2} \int_{x_{min}}^{x_{max}} \left(S_{x_i}^y \right)^2 dx_i - 1 \right]^{1/2} \tag{9.26}$$

9.4.2 Variance-Based Global Sensitivity Measures

The variance method of studying sensitivity is based on the definition of sensitivity index in terms of variance (i.e. the second moment of output). An input, x_i, may affect the output, y, either directly or by interacting with other input parameter, x_j. For independent input parameters, x_i, the variance of the output is decomposed as

$$\text{Var}(y) = \sum_i V_i + \sum_i \sum_{j>i} V_{ij} + \sum_i \sum_{j>i} \sum_{k>j} V_{ijk} + \dots + V_{12\dots n} \tag{9.27}$$

where

$$V_i = \text{Var}\left[E\left[y / x_i = a_i\right]\right] \tag{9.28}$$

$$V_{i,j} = \text{Var}\left[E\left[y / x_i = a_i, x_j = a_j\right]\right] - \text{Var}\left[E\left[y / x_i = a_i\right]\right] - \text{Var}\left[E\left[y / x_j = a_j\right]\right] \tag{9.29}$$

where V_i is the conditional variance of the expectation (i.e. mean value) of y when $x_i = a_i$. The first term, V_i, is the variance in the absence of interaction of a parameter with parameters, and all other terms are contributions to variance due to interaction. Other interaction terms can be written in a similar manner. The total number of terms in Equation 9.27 is $2^n - 1$. However, all the terms, if evaluated, are generally not required. Only two sensitivity indices are defined. Sensitivity indices are defined as

$$S_{i,\dots,j} = \frac{V_{i,\dots,j}}{\text{Var}(y)} \tag{9.30}$$

The sum of all sensitivities is equal to 1, that is,

$$S = \sum_{i=1}^{m} S_i + \sum_{1 \le i < j \le m} S_{i,j} + \dots + S_{1,2,\dots,m} \tag{9.31}$$

The first-order effect is given by the sensitivity index, S_i. It does not include sensitivity due to interaction. The right-hand side of Equation 9.31 contains several terms involving a particular parameter, i, but a few terms do not involve the ith parameter. The sum of all the terms involving i the parameter is known as the total sensitivity index, S_{Ti}, for that parameter. It includes the first-order sensitivity index for the ith parameter as well as its interaction with all other parameters. Parametric sensitivity due to interaction only is the difference between the two.

The first-order effect index is written as follows:

$$S_i = \frac{\text{Var}\left[E_{x_{-i}}\left[y / x_i\right]\right]}{\text{Var}\left[y\right]} \tag{9.32}$$

The numerator is the conditional mean value of y is estimated for all parameters except x_i and is denoted by index -1. This index is the effect of x_i on y without considering the interaction of x_i with other input parameters.

The total effect index is defined as

$$S_{Ti} = 1 - \frac{\text{Var}\left[E_{x_{-i}}\left[y / x_{-i}\right]\right]}{\text{Var}\left[y\right]} \tag{9.33}$$

In the absence of interaction, $\sum S_i = 1$ and $S_i = S_{Ti}$. In the presence of interaction, $S_i = S_{Ti}$. A small value of a first-order index does not mean that the parameter is not important. It may affect the output y due to interactions with other parameters.

The first-order sensitivities and total sensitivity index for all the parameters are compared to identify the more important factors and parameters which should be fixed to reduce the uncertainty in the model output. A detailed discussion on sensitivity is given in the reviews by Saltelli et al. (2005, 2012).

The relative importance of parameters is decided by the values of the first-order local sensitivity, S_i. For an important parameter, the value of S_i is high. To have an idea of the magnitude, let us consider that for $S_i > 0.01$ for a model output, the parameter will be considered important (Cosenza et al. 2014). It is called 'factor prioritisation'. Some of the parameters may not influence the model output. The identification should include the linear and non-linear influence. Therefore, the value of S_{Ti} was used to identify non-influential factors. It should be checked for the parameters which are non-important. This procedure is called 'factor fixing'. Such parameters may be treated as constants in the model. The parameters for which $S_i < 0.01$ and $S_{Ti} < 0.1$ may be considered non-influential.

Due to a wide variation in the nature of the model, it is difficult to prescribe a single strategy to conduct the sensitivity analysis. Few attempts have also been made to fit the model by another function and conduct the sensitivity analysis using the fitted function. Androulakis et al. (2006) conducted sensitivity analysis to study the effect of thermochemical parameters on the reaction rate constants for the reactions occurring during the oxidation of fuel. The model predictions were fitted using a response surface methodology. From the coefficient of the response surface, the sensitivity to parameters was discussed. The standard deviation was considered a measure of uncertainty. Thus, the variations of standard deviation in the estimated kinetic rate constants were determined as a function of temperature.

The variance-based method is a statistical method. These methods are meaningful if a linear correlation between the parameters and the model output is expected.

9.4.3 Determination of Variance-Based Sensitivity Indices

The aim of the sensitivity analysis is to test the performance of the model and to determine the sensitivity of the model input to the model parameters. These parameters may be either input to the model or the parameters used in the model. To determine the variances, these parameters are to be varied, and the variance of the model output is to be determined. A scheme to vary the model parameters is required so that the conditional variances can be estimated.

9.4.3.1 Sobol's Method

Sobol's method to choose the points is as follows (Saltelli et al. 2012):

1. Two sets of data matrices $- \bar{x}_j = \{x_1, x_2, \dots x_m\}_j, j = 1 \dots n$ – are generated using random numbers. Among several such methods, the most common is the Monte Carlo method.

2. If one of the sets of data matrices is used to determine the variances, then it is the variance of expectation of y. To determine $\text{Var}\left[E_{x_{-i}}\left[y / x_i\right]\right]$, the new set of data is the second matrix in which all the elements for x_i are replaced by the elements from the first matrix. Similarly, when the new set of data from the second matrix with all the elements for x_i and x_j are replaced by the elements from the first matrix, is used, then $\text{Var}\left[E_{x_{-i}}\left[y / x_i, x_j\right]\right]$ can be determined. The remaining terms are also determined in a similar manner.

3. The total sensitivity indices can now be determined using Equations 9.32 and 9.33, and similar equations for other terms.

9.4.3.2 Fourier Amplitude Sensitivity Test

Sobol's method requires a large number of simulation runs. This is due to the fact that one parameter is changed at a time. The method is good when sensitivity analysis for only a few parameters is to be conducted. But for a large number of parameters, one can use the Fourier amplitude sensitivity test (FAST). The simulations are conducted by varying all the parameters with different periodic variations. Using the Fourier transform, the amplitude of the power gives an estimate of the sensitivity indices. Details regarding Sobol's method and FAST are given in Saltelli et al. (2012).

The first-order sensitivity is sufficient to determine the factor which should be given priority. Total sensitivity is to be considered if the non-significant factors are to be fixed.

9.5 Role of Sensitivity Analysis

The sensitivity analysis reveals important information about the effect of various parameters on the output of the process. This information may be used for a better process design, process monitoring and control. Sensitivity analysis is carried out experimentally or through simulation using the developed model. The sensitivity analysis through experiments is not an aim of the present discussion. A few applications of the sensitivity analysis using simulation studies are presented in this section.

9.5.1 Process Design

The role of sensitivity in optimal process design was discussed by Chen et al. (1970). The method was based on optimisation using a Lagrange multiplier. Two types of optimisation were studied. In one case, the expected value (mean) of objective function or, in the other case, maximum relative sensitivity was minimised. The disturbance following uniform distribution correctly described the large uncertainty in comparison to that following normal distribution. It was shown that the mean system performance may deviate from the deterministic system performance. For uncorrelated parameters, the deviation is due to the second-order derivative of output with respect to the parameter, $\partial^2 y/\partial \phi_i^2$, and the variance of the parameter.

The issue of optimal design of a large system was addressed by Takamatsu et al. (1970). A large system is a set of several types of process equipment. Mathematically, a model for such a large system is composed of several sub-models, each representing process equipment. The model for the large system consists of all equations from all of the sub-models. The range of model output defines the constraints to the optimisation problem. Using the adjoint parameters (Lagrange multiplier), the constrained problem is converted to an unconstrained optimisation problem. The range of the model output is expressed as a sensitivity coefficient. The new objective function using the Lagrange multiplier is formed. Differentiating the new objective function with respect to various parameters, the first-order local sensitivity coefficient is obtained.

Sensitivity analysis of the design of a packed tower reactor for catalytic hydro-chlorination of acetylene was conducted by Priestley and Agnew (1975). The sensitivity of the reactor temperature towards various kinetic, transport parameters was conducted. A 2D model for the packed tower considering inter-phase heat and mass transfer was used. The first- and second-order sensitivity coefficients were evaluated for four temperature profiles. The first- and second-order sensitivity coefficients were defined as

$$s_{\phi_i}^T = \frac{\partial T}{\partial \phi_i} \tag{9.34}$$

$$\sigma^2(T) = \sum_{i=1}^{N} \left(\frac{\partial T}{\partial \phi_i} \right)^2 \sigma^2(\phi_i) \qquad (9.35)$$

Here, $\sigma^2(T)$ is the second moment about the mean or expected value. The second derivatives were also estimated. From the values of the sensitivity coefficients, it was concluded that the hot spot or the reaction temperature was highly influenced by the activation energy.

The sensitivity analysis was carried out using a trace of optimum solution (Seferlis and Hrymak 1996). The method is based on the shift of the optimum solution by changing various parameters. The objective function was the percentage return on the investment. The constraints were handled by using Lagrange multipliers to convert the constrained optimisation problem into an unconstrained optimisation problem. The path of the optimal solution over the range of the disturbance in various parameters such as flow rate, cooling water flow rate, reactor temperature and reactant flow rate was studied. The ratio of the change in the objective function to the change in the parameter was considered a measure of sensitivity.

The effect of the change in relative volatility of a component in a multicomponent mixture on the reflux ratio in the distillation column was investigated by Nelson et al. (1983). The ratio of the percentage change of a reflux ratio to the percentage change in the average relative volatility of one of the components in the system was designated as sensitivity. For a small change in the relative volatility, it is equivalent to the gradient-based first-order local sensitivity. This study shows the importance of sensitivity analysis in process design of the distillation column.

9.5.2 Process Operations

Sensitivity analysis has been used to operate process plants more efficiently. The Fischer–Tropsch (F–T) synthesis exhibits steady-state multiplicity. Song et al. (2003) carried out a sensitivity analysis to study the multiplicity of F-T synthesis in bubble column slurry reactors. The effect of the cooling water flow rate and deactivation of the catalyst on the multiplicity was investigated. A model for the reactor was considered for the analysis. Plug-flow for gas and complete mixing of slurry were assumed. The catalyst particles were assumed to be uniformly distributed, although the non-uniform axial concentration profile of solids is known to exist in the bubble column. The slurry–bubble column is assumed to operate at constant conditions so that the gas-holdup, density, and heat capacity can be taken as constant. The temperature in the reactor was uniform. The model equations were differentiated to obtain the sensitivity equations. The model and sensitivity equations were solved simultaneously to obtain the parametric sensitivity. The sensitivity of the reaction temperature to the coolant temperature showed a sharp increase at the critical temperature.

A direct differential method was used to perform a sensitive analysis in the case of a catalytic plate reactor by Zanfir and Gavriilidis (2002). The catalytic plate reactor consists of plates coated with catalysts and spaced closely to form channels. The endothermic and exothermic reactions took place in alternate channels. A 2D reactor model was differentiated with respect to each parameter – namely, the channel length, wall thickness, thermal conductivity, inlet temperature, composition, fluid velocity and kinetic parameters – to give sensitivity equations. The model and sensitivity equations were solved simultaneously to compute local sensitivity. The sensitivity analysis was used to predict the acceptable limit for variation of catalyst loading to maintain the reactor temperature within acceptable limits.

Fluid-bed agglomerators are generally used in the pharmaceutical industry to prepare granulated material. The sensitivity of various process parameters on the predicted particle size distribution was studied using a discretised population balance model for the fluid-bed granulation (Cryer and Scherer 2003). Numerical experiments were conducted by varying the parameters, which followed either normal distribution or uniform distribution. The variance (fraction of the variation explained) of the output for each factor was compared, and important parameters were selected. For example, particle size was most affected by the size of the binder droplet entering the bed.

A sensitivity analysis for a dynamic model for a continuous stirred tank reactor (CSTR) for acid-catalysed hydrolysis of acetic anhydride was carried out by Jayakumar et al. (2011) using a direct differential method. The dynamic model equations for the CSTR were differentiated to obtain the sensitivity equations. The parametric sensitivity changed with time. The effect of the cooling water flow rate and wall capacitance (J/C) on the transient sensitivity was studied. The analysis successfully predicted that at a high value of wall capacitance, the reactor temperature exhibited multiplicity (i.e. as the wall capacitance increased beyond a certain level, the steady-state temperature changed with time). The parametric sensitivity by the model compared well with the experimental values. The CSTR exhibited parametric sensitivity behaviour for small changes in the cooling water flow rate. The wall capacitance affected the parametric sensitivity for reactor concentration, reactor temperature and cooling water temperature. Also, in the presence of wall capacitance, the time to attain the steady state increased.

A first-order local sensitivity was used to determine the granulating operating parameters (Fung et al. 2006). The sensitivities were determined by using a model based on first principle and empirical correlations.

A sensitive analysis of a phenol degradation reactor was carried out using the direct differential method (Dutta et al. 2001). The governing model equations in dimensionless form were differentiated to obtain sensitivity equations. All the equations were solved numerically using the Runga–Kutta method. The parametric sensitivity was carried out in dimensionless form. The parameters studied were biomass concentration, inhibition coefficient and substrate concentration. The region of parameters in which the system was sensitive was determined.

9.5.3 Model Development

The sensitivity analysis is helpful in model development. In Chapter 8, the role of sensitivity in the development of the hybrid artificial neural network model was mentioned. The sensitivity equations were obtained by differentiating the model equations and were solved simultaneously with the model equations.

Since the sensitivity analysis helps in identifying the significant and insignificant factors, the model may be defined by fixing some of the parameters. The assumptions regarding such parameters may be simplified, resulting in simplification of the model. If a parameter is highly sensitive to the model output, it should be described accurately. The role of the sensitivity analysis in model development may be clear from the examples reported in the literature.

The ultraviolet–hydrogen peroxide process is an effective method for wastewater treatment since ultraviolet radiation can effectively deactivate several waterborne cysts. The most important model parameters were determined by conducting local sensitivity analyses (Audenaerta et al. 2011).

The sensitivity analysis was used to propose a model to predict the activity of glucose-6-oxidases (Moon et al. 2012). One of the enzymes was less active, and the other was more active. Temporal variation of the sensitivity estimated using the direct differential method revealed that the reaction was more sensitive to kinetic parameters for one of the reactions, while at the later stages the kinetic parameters for another reaction affected the reaction rate.

Biological processes such as biological wastewater treatment are very complex, and a model for such a process may involve several parameters. Cosenza et al. (2014) carried out a variance-based sensitivity analysis of a model for a wastewater treatment plant. The plant consisted of an anaerobic reactor, an anoxic reactor, an aerobic reactor, a membrane separator and a tank to collect permeate. All units are connected in series. An extended-FAST method was used to determine the first-order and higher order sensitivity indices S_i and S_{Ti}. The non-influential and important parameters were identified. Significant differences in the values of S_i and S_{Ti} indicated significant interaction among the model parameters. The relationship between the parameters was non-linear. The evaluation of only first-order sensitivity indices could not have revealed this fact.

From the above discussion, it can be said that gradient-based and variance-based sensitivity analysis has been applied in various processes. Most of the models developed for chemical or biochemical processes are based on the first principles. The types of equations are differential equations, algebraic equations or a combination of both. The gradient-based methods can easily be applied since the sensitivity equations can be obtained by differentiating these equations.

The variance-based methods are easily conducted where the experimental data has to be applied. For validation of the models, their use is limited to those applications only in cases in which (1) there is uncertainty of the parameters (Yu and Harrism 2009), or (2) the model accepts random inputs. Depending upon the nature of the model and its application, a suitable method of sensitivity analysis and model validation may be chosen.

Main effects can be effectively used to rank the factors in the case of linear models. But in the case of non-linear models, higher order sensitivity indices and the total sensitivity index should also be considered.

9.6 Summary

Every model has to be validated before it can be used for simulation of a process under new conditions. The model is validated by comparing the model prediction with the experimental data. The sensitivity of the model output towards various model parameters is studied. The sensitivity analysis allows determination of the important and non-significant parameters in the process. This information is useful for better process control and even in model development.

Two broad categories of the sensitivity indices: the gradient-based and variance-based sensitivity indices are popular. The gradient-based sensitivity indices are suitable for model validation. In the case of random nature or uncertainty of the parameters, the variance-based methods can also be used.

References

Akhnazarova, S., and Kafarov, V. 1982. *Experimental Optimization in Chemistry and Chemical Engineering*. Moscow: MIR Publishers.

Androulakis, I.P., Grenda, J.M., Barckholtz, T.A., and Bozzelli, J.W. 2006. Propagation of uncertainty in chemically activated systems. *AIChE J.* 52(9): 3246–3256.

Audenaerta, W.T.M., Vermeersch, Y., Van Hulle, S.W.H., Dejans, P., Dumoulin, A., and Nopens, I. 2011. Application of a mechanistic UV/hydrogen peroxide model at full-scale: Sensitivity analysis, calibration and performance evaluation. *Chem. Eng. J.* 171: 113–126.

Caracotsios, M., and Stewart, W.E. 1995. Sensitivity analysis of initial-boundary-value problems with mixed PDEs and algebraic equations applications to chemical and biochemical systems. *Comput. Chem. Eng.* 19(9): 1019–1030.

Chen, M.S.K., Erickson, L.E., and Fan, L.T. 1970. Consideration of sensitivity and parameter uncertainty in optimal process design. *Ind. Eng. Chem. Process Des. Dev.* 9(4): 514–521.

Cosenza, A., Giorgio, M., Vanrolleghem, P.A., and Neumann, M.B. 2014. Variance-based sensitivity analysis for wastewater treatment plant modelling. *Sci. Total Environ.* 470–471: 1068–1077.

Cryer, S.A., and Scherer, P.N. 2003. Observations and process parameter sensitivities in fluid-bed granulation. *AIChE J.* 49(11): 2802–2809.

Dutta, S., Chowdhury, R., and Bhattacharya, P. 2001. Parametric sensitivity in bioreactor: An analysis with reference to phenol degradation system. *Chem. Eng. Sci.* 56: 5103–5110.

Fung, K.Y., Ng, K.M., Nakajima, S., and Wibowo, C. 2006. A systematic iterative procedure for determining granulator operating parameters. *AIChE J.* 52(9): 3189–3202.

Jayakumar, N.S., Agrawal, A., Hashim, M.A., and Sahu, J.N. 2011. Experimental and theoretical investigation of parametric sensitivity and dynamics of a continuous stirred tank reactor for acid catalyzed hydrolysis of acetic anhydride. *Comput. Chem. Eng.* 35: 298–305.

Kiparissides, A., Kucherenko, S.S., Mantalaris, A., and Pistikopoulos, E.N. 2009. Global sensitivity analysis challenges in biological systems modeling. *Ind. Eng. Chem. Res.* 48: 7168–7180.

Moon, T.S., Nielsen, D.R., and Prather, K.L.J. 2012. Sensitivity analysis of a proposed model mechanism for newly created glucose-6-oxidases. *AIChE J.* 58(8): 2303–2308.

Nelson, A.R., Olson, J.H., and Sandler, S.I. 1983. Sensitivity of distillation process design and operation to VLE data. *Ind. Eng. Chem. Proc. Des. Dev.* 22(3): 547–552.

Priestley, A.J., and Agnew, J.B. 1975. Sensitivity analysis in the design of a packed bed reactor. *Ind. Eng. Chem. Proc. Des. Dev.* 14(2): 171–174.

Saltelli, A., Ratto, M., Tarantola, S., and Campolongo, F. 2005. Sensitivity analysis for chemical models. *Chem. Rev.* 105: 2811–2827.

Saltelli, A., Ratto, M., Tarantola, S., and Campolongo, F. 2012. Sensitivity analysis for chemical models. *Chem. Rev.* 112(5): PR1–PR21.

Seborg, D.E., Edgar, T.E., and Mellichamp, D.A. 2004. *Process Dynamics and Control*, 2nd ed. Hoboken, NJ: John Wiley.

Seferlis, P., and Hrymak, A.N. 1996. Sensitivity analysis for chemical process optimization. *Comput. Chem. Eng.* 20(10): 1177–1200.

Song, H.S., Ramkrishna, D., Trinh, S., Espinoza, R.L., and Wright, H. 2003. Multiplicity and sensitivity analysis of Fischer–Tropsch bubble column slurry reactors: Plug-flow gas and well-mixed slurry model. *Chem. Eng. Sci.* 58: 2759–2766.

Takamatsu, T., Hashimoto, I., and Ohno, H. 1970. Optimal design of a large complex system from the viewpoint of sensitivity analysis. *Ind. Eng. Chem. Proc. Des. Dev.* 9(3): 368–379.

Vásquez, V.R., Whiting, W.B., and Meerschaert, M.M. 2010. Confidence interval estimation under the presence of non-Gaussian random errors: Applications to uncertainty analysis of chemical processes and simulation. *Comput. Chem. Eng.* 34: 298–305.

Yu, W., and Harrism, T.J. 2009. Parameter uncertainty effects on variance-based sensitivity analysis. *Reliab. Eng. Syst. Saf.* 94: 596–603.

Zanfir, M., and Gavriilidis, A. 2002. An investigation of catalytic plate reactors by means of parametric sensitivity analysis. *Chem. Eng. Sci.* 57: 1653–1659.

10

Case Studies

In Chapters 4 through 8, the development of various types of models was discussed. A large number of models used in chemical and biochemical processes are based on the application of laws of conservation of momentum, mass and energy. The models may be of a deterministic or stochastic nature. Simple models have analytical solutions. Rigorous models require numerical solutions of the model equations. Artificial neural network models are not based on any law but are developed using experimental data. The models developed are used for simulation after model validation. Simulation helps us in understanding the behaviour of the process. If the model predictions are similar to behaviour observed experimentally, then the predicted process behaviour can be accepted with more confidence. It is also possible to analyse the model and study the assumptions which are responsible for good predictions. This aspect is related to parametric sensitivity analysis.

The published literature discusses a brief description of the process, the assumptions in the final form of the model, the model equations, the solution methodology and the model validation. It usually does not discuss failures. The model development and simulation studies are carried out almost simultaneously. First, a preliminary model is obtained, and then simulation studies are carried out. From model predictions, it is decided to modify the model equations. Again, the simulation studies are carried out. This process is repeated until a satisfactory model is obtained. A large model may be broken into various sub-models. Each sub-model is developed to the modeller's satisfaction, and then combined. Because these details are generally not readily available, it becomes difficult to appreciate the kind of effort needed to develop a new model and carry out simulation studies. It is more difficult if a new modeller decides not to use any specific-purpose commercial software.

In this chapter, a few case studies of model development of each of the major classes of models are presented. It includes a deterministic model with an analytical solution, a deterministic model solved numerically, a stochastic model and a neural network model.

10.1 Axial Distribution of Solids in Slurry Bubble Columns: Analytical Deterministic Models

The axial distribution of solids in a slurry bubble column has been discussed in Section 5.1. The model required two parameters: a hindered settling velocity which is different than that used in sedimentation and the dispersion coefficient of solids. In fact, the model contains only one term, the ratio of these two. The hindered settling velocity used in the model is not the same as that used in sedimentation and fluidisation. Sometimes, this velocity is more than the terminal velocity of the particles. A new model to predict the axial dispersion of solids in slurry bubble columns is developed by modifying the sedimentation–dispersion model given by Equation 5.2. Analytical solution of the model is given by Equation 5.6.

The model accepts hindered settling velocity that is used in the case of sedimentation. The model involves only one parameter, the solid dispersion coefficient.

The hindered settling velocity was computed from the axial solid distribution in slurry bubble columns and was correlated in terms of terminal velocity of the particles and gas holdup. A few correlations are presented in Table 5.1. Various investigators have used free settling velocity (Cova 1966; Imafuku et al. 1968; Farkas and Leblond 1969; Kojima et al. 1986). Although two constants – hindered settling velocity, U_c, and the solid dispersion coefficient, E_s – appear in the model, in fact there is only one constant, the ratio $U_c{:}E_s$ (Imafuku et al. 1968; Murray and Fan 1989; Zhang 2002). This ratio can be determined by plotting $\ln(C_s)$ versus z. The slope of the line is $U_c{-}E_s$, and the intercept is C_s^0. Reilly et al. (1982) performed leaching studies and did not obtain a straight line.

Smith and Ruether (1985) mentioned that the hindered settling velocity, U_c, was higher than the terminal velocity of the particle, $U_{t,\infty}$. It was attributed to liquid circulation velocity in the column; hence, the gravitational field increased due to the presence of centripetal force. However, in the case of sedimentation, the hindered settling velocity is always lower than the terminal velocity.

The solid dispersion coefficient, E_s, is a measure of solid back-mixing in the column. It is different from the liquid dispersion coefficient, E_l, which is a measure of the liquid dispersion coefficient. Several investigators have taken the solid dispersion coefficient to be equal to liquid dispersion coefficient (Cova 1966; Imafuku et al. 1968; Farkas and Leblond 1969). Friedlander (1957) has shown that particles smaller than 10 μm are carried away along eddies. Other investigators found the solid dispersion coefficient to be different than the liquid dispersion coefficient. Some of the correlations for the solid dispersion coefficient in slurry bubble columns are presented in Table 5.1.

Pe and Fr are the Peclet number and Froude number, respectively. The Peclet number, Pe_{p1}, is based on U_g^c, the critical velocity (i.e. the minimum gas velocity to suspend the solids). The Peclet number, Pe_{p2}, is based on the superficial gas velocity.

The assumption of $E_s = E_1$ may be valid for small particles. However, it does not mean that the values of E_1 for an empty column should be used. The presence of solids itself decreases E_1 (Smith and Ruether 1985).

In all models, the hindered settling velocity has been assumed to be constant throughout the column. Because the concentration of solids varies in the axial direction, the hindered settling velocity is expected to depend upon the axial position. The proposed model uses the same hindered settling velocity as is used in the case of sedimentation by suitably modifying the existing model.

The one-dimensional (1D) sedimentation–dispersion model for batch operation is given by Equation 5.4. Let U_c be the hindered settling velocity used for the sedimentation process. Using the Richardson and Zaki (1954) relationship, we get

$$U_c = U_{t,\infty}\varepsilon^n = U_{t,\infty}\left(1 - \frac{C_s}{\rho_s}\right)^n \tag{10.1}$$

where ε is the fraction of liquid in slurry.

Substituting Equation 10.1 into Equation 5.4 and after rearrangement, we get

$$\frac{E_s}{U_{t,\infty}}\frac{d^2C_s}{dz^2} = -\left(1 - \frac{C_s}{\rho_s}\right)^n\frac{dC_s}{dz} = \frac{\rho_s}{(n+1)}\frac{d}{dz}\left[\left(1 - \frac{C_s}{\rho_s}\right)^{n+1}\right] \tag{10.2}$$

Upon integration, we get

$$\frac{(n+1)E_s}{\rho_s U_{t,\infty}}\frac{dC_s}{dz} = \left(1 - \frac{C_s}{\rho_s}\right)^{n+1} + A_1 \tag{10.3}$$

Substituting the following boundary conditions into Equation 10.3:

As $z \to \infty$, $\dfrac{dC_s}{dz} \to 0$ and $C_s \to 0$ we get $A_1 = -1$.

Substituting the value of A_1 in Equation 10.3 and after rearrangement, the following equation is obtained:

$$\frac{(n+1)E_s}{U_{t,\infty}}\frac{d}{dz}\left(1 - \frac{C_s}{\rho_s}\right) = 1 - \left(1 - \frac{C_s}{\rho_s}\right)^{n+1} \tag{10.4}$$

Integration using the separation of variables method and rearranging:

$$\frac{U_{t,\infty}}{(n+1)E_s}\int_0^z dz = \int_{1-C_s^0/\rho_s}^{1-C_s/\rho_s} \frac{d\left(1-\dfrac{C_s}{\rho_s}\right)}{1-\left(1-\dfrac{C_s}{\rho_s}\right)^{n+1}} \tag{10.5}$$

Because an analytical solution of Equation 10.5 is not easy to obtain, let us try for an approximate solution. Let

$$y = \left(1-\frac{C_s}{\rho_s}\right) \tag{10.6}$$

On substitution of Equation 10.6 in Equation 10.5, we get

$$\frac{U_{t,\infty}z}{(n+1)E_s} = \int_{y_0}^y \frac{dy}{1-y^{n+1}} \tag{10.7}$$

Multiplying the denominator and numerator by y, followed by integration by parts, provides

$$\int_0^y \frac{dy}{1-y^{n+1}} = \int_0^y \frac{ydy}{y\left(1-y^{n+1}\right)} = \left[y\int_0^y \frac{dy}{y\left(1-y^{n+1}\right)}\right]_0^y - \int_0^y \frac{dy}{y\left(1-y^{n+1}\right)} \tag{10.8}$$

The solution of the integral given in the last term is taken from the handbook of integrals (Dwight 1972):

$$\int \frac{dy}{y\left(a+by^m\right)} = \frac{1}{am}\log\left|\frac{y^m}{a+by^m}\right| \tag{10.9}$$

Substituting Equation 10.9 into Equation 10.8 and since $y < 1$:

$$\int \frac{dy}{1-y^{n+1}} = \frac{1}{(n+1)}\log\left|\frac{y^{n+1}}{1-y^{n+1}}\right| = \frac{1}{(n+1)}\log\left(\frac{y^{n+1}}{1-y^{n+1}}\right) \tag{10.10}$$

or

$$\int_0^y \frac{dy}{1-y^{n+1}} = \frac{y}{(n+1)}\left[\log y^{n+1} - \log\left(1-y^{n+1}\right)\right]$$

$$- \frac{1}{n+1}\left[\int \log y^{n+1}\,dy - \int \log\left(1-y^{n+1}\right)dy\right] \tag{10.11}$$

$$= y\left[1-\frac{1}{n+1}\log\left(1-y^{n+1}\right)\right] + \frac{1}{n+1}\int_0^y \log\left(1-y^{n+1}\right)dy$$

Expanding the terms in series and integrating Equation 10.11 gives

$$\int_0^y \frac{dy}{1-y^{n+1}} = y\left[1 - \frac{1}{n+1}\log\left(1-y^{n+1}\right)\right] + \frac{1}{n+1}\int_0^y\left[y^{n+1} + \frac{y^{2(n+1)}}{2} + \frac{y^{3(n+1)}}{3} + \cdots\right]$$

$$= y\left[1 - \frac{1}{n+1}\log\left(1-y^{n+1}\right)\right] + \frac{1}{n+1}\left[\frac{y^{n+2}}{n+2} + \frac{y^{2n+3}}{2n+3} + \frac{y^{3n+4}}{3n+4} + \cdots\right]$$

$$(10.12)$$

Since $y < 1$, higher terms in the series are neglected. The approximate solution is

$$\int_0^y \frac{dy}{1-y^{n+1}} \approx y\left[1 - \frac{1}{n+1}\log\left(1-y^{n+1}\right)\right] \tag{10.13}$$

Substituting the value of y in Equation 10.12, we get

$$\int_{1-C_s^0/\rho_s}^{1-C_s/\rho_s} \frac{d\left(1-\dfrac{C_s}{\rho_s}\right)}{1-\left(1-\dfrac{C_s}{\rho_s}\right)^{n+1}} \approx \left(1-\frac{C_s}{\rho_s}\right)\left[1 - \frac{1}{n+1}\log\left(1-\left(1-\frac{C_s}{\rho_s}\right)^{n+1}\right)\right] \tag{10.14}$$

Using the approximate value of the integral from Equation 10.14 in Equation 10.5:

$$\frac{U_{t,\infty}z}{E_s} = \frac{(n+1)\left(C_s^0 - C_s\right)}{\rho_s} + \left(1-\frac{C_s^0}{\rho_s}\right)\log\left\{1-\left(1-\frac{C_s^0}{\rho_s}\right)^{n+1}\right\}$$

$$-\left(1-\frac{C_s}{\rho_s}\right)\log\left\{1-\left(1-\frac{C_s}{\rho_s}\right)^{n+1}\right\} \tag{10.15}$$

For dilute slurries $\dfrac{C_s}{\rho_s} \gg 1$; hence, $\left(1-\dfrac{C_s}{\rho_s}\right) \approx 1$ and $\left(1-\dfrac{C_s}{\rho_s}\right)^{n+1} \approx 1-(n+1)\dfrac{C_s}{\rho_s}$.

Substitution of these approximations into Equation 10.15 gives

$$\frac{U_{t,\infty}z}{(n+1)E_s} = \left[1 - \frac{1}{(n+1)}\log\left\{(n+1)\frac{C_s}{\rho_s}\right\}\right] - \left[1 - \frac{1}{(n+1)}\log\left\{(n+1)\frac{C_s^0}{\rho_s}\right\}\right] \tag{10.16}$$

Upon rearrangement:

$$\log\frac{C_s}{C_s^0} = -\frac{U_{t,\infty}}{E_s}z \text{ or } \frac{C_s}{C_s^0} = \exp\left(-\frac{U_{t,\infty}}{E_s}z\right) \tag{10.17}$$

Equation 10.17 is the same as Equation 5.6. However, the terminal velocity of the particle, $U_{t,\infty}$, is used in Equation 10.17. Equation 5.6 uses a 'hindered settling velocity'. For dilute suspensions (low values of C_s:ρ_s), the concentration profile is exponential. In cases of dilute suspensions, the particles remain far away from each other; hence, the velocity of the particles is not affected by the neighbouring particles. Therefore, free terminal velocity should be used in the model given by Equation 10.16. If, in any case, it fails to fit the data, Equation 10.15 may be used.

10.1.1 Validation of the Model

The first confidence about the model comes from the fact that under certain conditions, the model reduces to a simple form which is available in the literature and has been already validated. For the validation of Equation 10.15, the experimental data for axial distribution of solids in a slurry bubble column operated in batch mode are taken from the literature. The column diameter, particle diameters, density of the solid and superficial gas velocity used by various investigators are given in Table 10.1.

The Reynolds numbers were less than 200 in all of the cases. The exponent n, in correlation of Richardson and Zaki (1954), was evaluated using Equations 6.43 through 6.47.

The hindered settling velocity was estimated using Equation 10.1. The terminal velocity was calculated as

$$U_{t,\infty} = \sqrt{\frac{4g(\rho_s - \rho_1)d_p}{3C_D\rho_1}} \tag{10.18}$$

The drag coefficient, C_D, was estimated from the graph given by McCabe et al. (1993).

The values of E_s and C_0 were determined using non-linear regression by a simplex method for each data set. The predicted concentration is compared

TABLE 10.1

Parameters and Gas Velocity for the Axial Variation of Solids in a Slurry Bubble Column

Investigator	D, m	$d_p \times 10^6$, m	ρ_s, kgm^{-3}	U_g, ms^{-1}
Imafuku et al. (1968)	0.1	111	2550	0.012028
Farkas and Leblond (1969)	0.0381	126, 559	1405, 1170	0.022–0.0733
Kojima and Asano (1981)	0.055	115	2390, 2490	0.034–0.041
Kato et al. (1972)	0.066, 0.122, 0.214	75.5–163	2520	0.057–0.124
Smith and Ruether (1985)	0.108	48.5–193.5	2420	0.031–0.09
Murray and Fan (1989)	0.076	49, 97, 163	2450, 2990	0.0328–0.1725
Zhang (2002)	0.1	107.5, 126, 180	2636	0.096–0.105

with the experimental values for a couple of concentration profiles in Figure 10.1. A parity plot between the estimated and experimental values of solid concentration for all points, as shown in Figure 10.2, illustrates that the new model successfully predicts the solid concentration profile.

The values of E_s and C_0 estimated using the present model were about 5.6% and 4.9% lower than the values predicted using the exponential model. This small difference is well within the experimental error. It could be due to the fact that the particles were small and the concentration of the slurry was low.

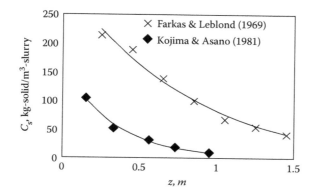

FIGURE 10.1
Comparison of predicted concentration profiles using Equation 10.15 with experimental data of Farkas and Leblond (1969) and Kojima and Asano (1981).

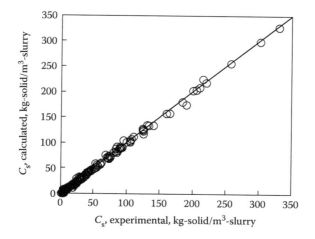

FIGURE 10.2
Comparison between the estimated values of solid concentration using Equation 10.15 and experimental values of Imafuku et al. (1968), Farkas and Leblond (1969), Kojima and Asano (1981), Kato et al. (1972), Smith and Ruether (1985), Murray and Fan (1989) and Zhang (2002).

Under such conditions, the present law is simplified to the exponential law. However, the present model allows use of the same hindered settling velocity as is used for sedimentation. The small error in the estimation of E_s allows the use of correlations for E_s already available in the literature. It also justifies the approximations made while obtaining an analytical solution of the model equation.

The present model does not require any correlation for the setting velocity as compared to the exponential model. The new model requires the values of the terminal velocity of the particle and the Richardson–Zaki exponent. It, however, requires more computational effort. Although the model equations have an analytical solution, estimation of the model parameters is obtained using non-linear regression.

10.1.2 Simulation Studies

The model is now ready for the simulation studies. It can now be used to study the effect of particle size, particle density and the superficial gas velocity on the liquid dispersion coefficient. The values of E_s were recomputed from the experimental data reported in literature; they are presented as a function of the terminal velocity of the particle, $U_{t,\infty}$, and superficial gas velocity, U_g, and are plotted in Figures 10.3 and 10.4, respectively. The solid dispersion coefficient increases linearly with increasing terminal velocity. It increases with increasing U_g at low velocity but becomes constant at high U_g.

The hindered setting velocity of the particle used in the modified sedimentation–dispersion model for axial distribution of solid in a slurry

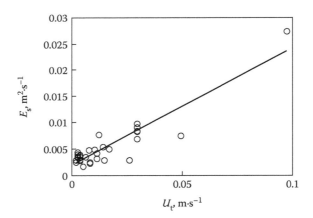

FIGURE 10.3
E_s as a function of $U_{t,\infty}$ in a slurry bubble column using Equation 10.15 for the data of Imafuku et al. (1968), Farkas and Leblond (1969), Kojima and Asano (1981), Kato et al. (1972), Smith and Ruether (1985), Murray and Fan (1989) and Zhang (2002).

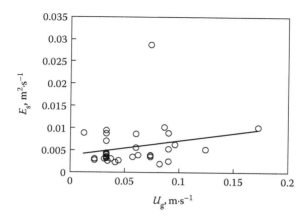

FIGURE 10.4

E_s as a function of U_g in a slurry bubble column. The values of E_s were calculated using Equation 10.15 for the data of Imafuku et al. (1968), Farkas and Leblond (1969), Kojima and Asano (1981), Kato et al. (1972), Smith and Ruether (1985), Murray and Fan (1989) and Zhang (2002).

bubble column can be estimated from sedimentation studies. The model is yet to be validated for large particles or concentrated slurries. Computational effort in the model is more than that in the exponential model, but it can be performed easily with available software such as the optimisation toolbox of MATLAB® or even with the Solver add-in of Microsoft Excel. The present model is valid for only the batch mode of operation of the column.

10.2 Conversion for a Gas–Liquid Reaction in a Shallow Bed: A Numerical Model

Let us consider a case in which a reactant present in gas enters a shallow pool of the second reactant through a sieve plate, as shown in Figure 10.5. An isothermal reaction takes place at the gas–liquid interface. A model to predict conversion in the gas–liquid dispersion is developed and is discussed below.

The model is divided into several sub-models which are combined to form a complete model. The model is based on first principles. Use of empirical correlations is minimal. Differential equations have been replaced by difference equations. The model should consider hydrodynamics, mass transfer and chemical reaction. The reaction is isothermal; therefore, energy balance and heat transfer aspects are not considered. Let us analyse the process from the point of view of making assumptions and breaking the model into several sub-models.

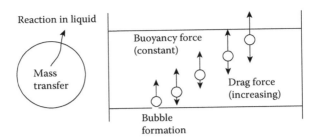

FIGURE 10.5
Gas–liquid reactions in a shallow pool of liquid.

Gas holdup in bubble columns has been extensively studied, and a number of correlations are available. No single correlation fits all the experimental data. The fluid properties, superficial gas and liquid flow rate, column internals, sparger geometry, column diameter and static bed height are the factors which determine the bubble behaviour and hence the gas holdup. These factors are expected to influence the process in shallow beds also. The difference between the behaviour of the shallow bed and bubble column will become clear from the discussion below.

1. As the gas passes through the holes in the sieve plate, it takes the shape of a bubble which grows in size. It remains attached to the hole due to surface tension until it grows to a size so that buoyancy force exceeds the surface tension force and inertial forces. The size of the bubble formed will not be the same as that of the size of the hole. A sub-model to predict the bubble diameter formed at the sparger nozzle may be used.

2. After detachment from the distributor holes, the bubble is under the influence of two forces, the buoyancy force and the drag force. The latter is zero because the initial velocity of the bubble is close to zero. In a conventional column, the bubbles get enough time to attain the terminal velocity, and the region close to the distributor may be neglected in comparison to the remaining column. In a shallow pool of liquid, the bubbles do not get enough time to attain the terminal velocity of the bubble. The velocity of the bubble at any axial position may be estimated using Newton's law of motion if the drag force is known.

   Because the bubble velocity is changing in the axial direction, all properties which depend upon the bubble velocity will also vary in the axial direction. A sub-model for the axial variation of bubble velocity is to be developed.

3. The bubble velocity thus calculated is the velocity of a single bubble in isolation. In a swarm, the bubble behaviour is different. Bubble coalescence and bubble breakup may be present at high gas velocity.

At low gas velocity, bubble coalescence and bubble breakup are absent. Therefore, such an assumption will limit the scope of the model to low gas velocity. Because the bed is shallow, the bubble might leave the gas–liquid dispersion before bubble coalescence and bubble breakup take place. The absence of coalescence may be a reasonable assumption.

Chen et al. (1994) observed two flow regimes in which bubbles move as individuals. A 'dispersed bubble flow regime' was observed up to 0.01 m/s. The bubbles move as individuals even up to 0.021 m/s, but they take spiral paths. This regime was named the 'vortical-spiral flow regime'. The bubble coalescence and breakup are totally absent in both of these flow regimes. Let us assume that the bubbles do not take the spiral path and the flow is in the dispersed bubble flow regime.

The assumption of an absence of bubble breakup is due to another reason. Due to low bubble velocity in the lower region near the distributor plate, the agitation is less and the turbulent forces may not be large enough to result in any breakup of bubbles.

The absence of coalescence and bubble breakup simplifies the model equations. However, the presence of other bubbles will hinder the bubble velocity. There are several correlations available to predict the hindered velocity of particles. They may be used in case of rigid bubbles. Richardson and Zaki (1954) have proposed that the hindered velocity can be expressed in terms of the porosity. In gas–liquid dispersion, the porosity is the fraction of the liquid.

$$\frac{U_b}{U_{t,\infty}} = \varepsilon^n \tag{10.19}$$

A correlation for the exponent, n, was also reported. A simple implicit correlation to estimate the exponent is given by Row (1977):

$$n = 2.35 \frac{\left(2 + 0.175\,Re_t^{3/4}\right)}{\left(1 + 0.175\,Re_t^{3/4}\right)} \tag{10.20}$$

The Reynolds number is evaluated at the terminal velocity. In a shallow bed, the terminal velocity is not achieved. Hence, this approach has to be discarded.

4. If a correlation to estimate the drag force is available, then acceleration of the bubble in each axial section can be estimated. The bubble velocity and the drag force depend upon each other; hence, the equation of motion and the equation for the drag have to be solved simultaneously. A correlation given by Equation 5.90 may be used as it is

applicable in a large range of velocity. Although in a shallow bed and for small bubbles, the shape of the bubbles is likely to be spherical, the use of the Evotos number takes into consideration the shape of the bubble if required.

To account for the effect of a swarm of bubbles, let us make a bold assumption that at any cross-section, as the bubbles move up an equal amount of liquid moves down to replace the empty space created. Because the liquid velocity is in a direction opposite to the velocity of the bubble, the drag force will increase and hence reduce the velocity of the bubble. Thus, the axial distribution of drag force may be estimated in each section in case of a swarm of small bubbles with no coalescence and breakup phenomena.

5. It is assumed that the bubbles move vertically. This assumption is true only for small bubbles having close wakes. At high bubble velocity, the bubbles wobble and take curved and oscillating paths. It is expected that for small holes in the distributor, the size of the bubbles formed will be small and that due to the absence of bubble coalescence, they remain small. Consequently, the radial distribution of bubbles may be assumed to be uniform. It is not true in the case of single holes or very few holes in the sparger.

6. Having known axial variation of the bubble diameter, d_b, and velocity of the bubble, U_b, the axial variation of gas holdup can also be estimated. For a given superficial gas velocity, U_g, the gas holdup, ε, is given by

$$\varepsilon = \frac{U_g}{U_b} \tag{10.21}$$

7. The mass transfer rate in a gas–liquid dispersion is the product of mass transfer coefficient, interfacial area and concentration difference. The mass transfer coefficient can be described by any of the models (e.g. Higbie's theory) or correlations given by Equation 5.97, which is applicable at turbulent conditions. Due to bubble motion, the flow is certainly not laminar. The liquid flow is turbulent but not sufficient to cause bubble breakup. Higbie's theory is applicable for steady-state mass transfer. It is assumed that the concentration profile is fully developed in the liquid (i.e. the time scale for this to happen is smaller than that to attain terminal velocity). Let the gas be pure reactant which diffuses into the liquid. In the absence of any concentration profile in the gas, the concentration gradient exists only in liquid. In the liquid, such an assumption is valid due to a high Schmidt number.

To determine the interfacial area, the number of bubbles should be known. From another sub-model, discussed above, the axial variation of the gas holdup and the bubble size is known. The number of bubbles can be evaluated by dividing the volume of the fraction of the gas in a section of the column by the volume of a single bubble. The former is the product of the gas holdup and total volume of the section.

8. To model the reactor, let us assume that the surface concentration is at the saturation concentration of the reactant present in the gas. It may be evaluated from the solubility data, Henry's law or any other suitable equilibrium data. The liquid is assumed to be in the completely mixed mode (i.e. the concentration of the reactant is uniform in the entire reactor and is also equal to the exit concentration). The material balance gives

$$F_{in}\left[C_A(z)-C_{Ain}\right]=\int_0^Z\left[k_La\{C_{As}-C_A(z)\}-r_A(z)\right]dZ \qquad (10.22)$$

Based on the above discussion, the model can be divided into sub-models. The discussion will also help in making assumptions while developing these models. One of the possibilities is to use the following sub-models:

1. Sub-model for bubble formation to estimate the bubble size
2. Sub-model for estimation of axial distribution of the swarm of bubbles
3. Sub-model for axial distribution of mass transfer coefficient and interfacial area ($k_L a$)
4. A reactor model

These sub-models will be developed separately and will be called from other problems.

10.2.1 Size of the Bubble Formed at the Distributor Plate

The final size of the bubbles formed at the orifice was estimated using a model to predict the size of bubble formed at the sparger. The model was given by Geary and Rice (1991). It was assumed that the liquid motion is inadequate to break bubbles. The bubble formation stage was considered to consist of five different stages: bubble birth, bubble expansion, bubble lift-off, bubble rise with continuous expansion and bubble detachment. The force balance until imbalance occurs at the bubble lift-off was given by (Geary and Rice 1991)

$$\frac{4\pi}{3}\rho_L g r^3+\frac{G^2\rho_G}{\pi r_H^2}-2\pi\sigma\,r_H-5\pi\sqrt{\frac{\rho_L\mu_L}{2}}\left(\frac{G}{4\pi r}\right)^{3/2}=\frac{\alpha\rho_L G^2}{12\pi r^2} \qquad (10.23)$$

where the volume of the bubble at the lift-off, V_d, was given by

$$V = V_d = \frac{4}{3}\pi r_d^3, \text{ at } t = t_d \tag{10.24}$$

The values of α were taken as 11/16. For the detachment stage, the force balance resulted in the following expression:

$$2 = \frac{4gV_d^2}{11r_H G^2}\left[y^2 - 1 - 2\ln y\right] + N_r\left[y - 1 - \ln y\right] - N_\mu\left[2\left(y^{1/2} - 1\right) - \ln y\right]$$

$$-\frac{V_d^{1/3}}{r_H}\left(\frac{1}{36\pi}\right)^{1/3}\left[3\left(y^{1/3} - 1\right) - \ln y\right] \tag{10.25}$$

where y is the ratio of the final volume to the expansion stage volume and is given by

$$y = \frac{V_f}{V_d} \tag{10.26}$$

The other terms, N_r and N_μ, are given by

$$N_\mu = \frac{16}{11}\frac{5\pi\sqrt{V_d}}{r_H G^2}\sqrt{\frac{2\mu_L}{\rho_L}}\left(\frac{G}{4\pi}\right)^{3/2}\left(\frac{4\pi}{3}\right)^{1/2} \tag{10.27}$$

and

$$N_r = \frac{16}{11}\frac{V_d}{\rho_L G^2 r_H}\left(\frac{G^2\rho_L}{\pi r_H^2} - 2\pi\sigma\, r_H\right) \tag{10.28}$$

The smallest positive root ($y > 1$) of Equation 10.25 gives the final volume of the bubble, V_f; hence, the final size of the bubble and from that the final bubble volume were calculated using Equation 10.24 through 10.28. The MATLAB code for the sub-model is given by the file 'bubblediaplate.m' (see Appendix B).

10.2.2 Hydrodynamic Model

After knowing the bubble diameter, other hydrodynamic parameters, the bubble velocity, the residence time in the column and the gas holdup can be evaluated. The column is divided into 1000 vertical sections. Such a large number has been chosen so that any change in any of the properties in the section remains small. Therefore, the properties of the dispersed and continuous phases entering a section can be used in place of the properties in the section. It helps in avoiding the use of a differential equation toolbox. The solution strategy can be termed a 'direct simulation'. Choosing a large number of sections is extremely important. The simulation methodology will not work if the number of sections is small for which no real but complex values of bubble velocities were obtained. The computer programme predicted negative and even complex numbers representing velocity. Therefore, the number of sections should be large. An alternate procedure may be to write the model equations as differential equations and then use a differential equation solver.

The drag force and buoyancy force are estimated at the inlet conditions. The velocity of the bubble leaving the nth section, $U_{b,n}$, is estimated using the following law of motion:

$$U_{b,n+1}^2 = U_{b,n}^2 + 2adz \qquad (10.29)$$

where a is the acceleration of the bubble and can be estimated from the force balance:

$$(F_d - F_b) = \frac{1}{2} C_D \left(\frac{\pi d_b^2 \rho_l}{4} \right) U_{b,n}^2 - \frac{\pi d_b^3 (\rho_l - \rho_g) g}{6} = \frac{\pi}{6} d_b^3 \rho_L a \qquad (10.30)$$

The gas holdup is estimated from the following equation:

$$\varepsilon = \frac{U_g}{U_{b,n+1}} \qquad (10.31)$$

The swarm velocity is modelled by considering that at any cross-section, the volume of liquid flowing downwards is equal to the volume of the bubble moving downwards. The liquid velocity can be written as

$$U_l = U_{b,n} + U_g \left(\frac{\varepsilon}{1-\varepsilon} \right) \qquad (10.32)$$

The corrected velocity of the bubble in a swarm is given by the relative velocity of the bubble.

$$U_{b,n,\text{new}} = U_{b,n,\text{old}} + U_l \qquad (10.33)$$

This, however, again changes the drag force; hence, the axial distribution of the velocity should be corrected again. Fortunately, the bubble velocity converged after two iterations only.

The sub-model is given in the 'hydrodynamics.m' file and uses the previous sub-model ('bubblediaplate.m') and drag force given by Equation 5.88 in the file 'dragfun.m'. The data used in the model are stored in two files: 'material.m' and 'column.m'. The former file stores properties for the liquid and gas. The latter file stores the geometric parameters of the column. These files serve the purpose of data files or files for problem definition. Changes can be made in these files without affecting the files containing equations.

10.2.3 Mass Transfer and Reactor Models

The mass transfer and reactor models are put in one file as both are coupled by Equation 10.22. The mass transfer coefficient is described by Higbie's model, given by Equation 5.79. The interfacial area of mass transfer is written as

$$a(z) = \frac{\left(\dfrac{\pi d_b^2(z)}{4}\right)}{\left(\dfrac{\pi d_b^3(z)}{6}\right)}\left(\frac{\pi D^2}{4}\right)\varepsilon(z)\left(\frac{L}{\Delta z}\right) \tag{10.34}$$

The bubble diameter, d_b, and gas holdup, ε, are functions of the axial position, z.

The following kinetic expression was used in the model:

$$-r_A = k'C_A^n \tag{10.35}$$

For any other kinetic rate expression, an appropriate expression for the rate of reaction should be used. The axial variation of the concentration in the column was described by Equation 10.22.

10.2.4 Model Validation

The MATLAB code an 'm' file for the entire model is given in Appendix B. The model validation is to be done by comparing model predictions with the experimental values reported in the literature. The model for the bubble formation is already validated by Geary and Rice (1991). Therefore, it does not require any further validation. However, the correctness of the programme should be checked, and the same is true for Higbie's model. The material balance and kinetic rate constants also do not require validation. The prediction for gas holdup depends upon several parameters, and the model may have very limited scope for the following reasons:

1. If the number of holes is one or two, the gas velocity may be very high, resulting in jetting. The model for bubble formation by Geary and Rice (1991) will not be applicable.
2. Radial movement of bubbles (spiral path of bubbles) due to liquid circulation is not taken into consideration. This will result in deviation of gas holdup in the case of a single-nozzle sparger.
3. Assumption of no bubble coalescence and breakup does not allow the model to be useful at high gas velocity.

The overall gas holdup can be estimated from the model by taking average of the axial gas holdup profile. The experimental data for overall gas holdup in shallow bubble columns have been reported by Lau et al. (2010). The model parameters required are the static bed height, column diameter, hole diameter and number of holes. The number of holes estimated for an open free area of 9.7% is 211. The nozzle diameter, d_o, and column diameter, D_T, are 0.003 and 0.14 m, respectively. The experimental values of the overall gas holdup for $H_s/D_T = 6$ are compared with the estimated values of average gas holdup using the model 'hydrodynamics', and these are presented in Figure 10.6.

The predicted values were in close agreement with the experimental values for $H_s/D_T = 6$. The model also predicts the dependence of the average gas holdup on the superficial gas velocity. The model does not exhibit enough sensitivity to the H_s/D_T ratio. It is still too early to accept the model, however; some more studies are required. One of the possible reasons for the deviation from the experimental data is the assumption of zero velocity when the bubble leaves the sparger. The model has been tested for a perforated-plate sparger having 3 mm holes. For larger holes, more bubbles will be formed; hence, bubble coalescence and bubble breakup may take place. Also, the model seems to work for a large $H_s:D_T$ ratio; it may be worthwhile to test the applicability of the model in cases of a conventional bubble column (a high $H_s:D_T$ ratio).

The average gas holdup for a single-nozzle sparger and a perforated plate with 100 holes was estimated for an air–water system. The column diameter was 0.1 m, and the hole diameter was 0.004 m. The static bed height is 2 m. The estimated values of gas holdup were compared with correlations of Hughmark (1967), Akita and Yoshida (1973), Kumar et al. (1976) and Hikita et al. (1980) in Figure 10.7. These correlations are presented in Table 10.2. The experimental data of Akita and Yoshida (1973) and Hikita et al. (1980) were obtained for a single-nozzle sparger. The other two were for a perforated-plate sparger. It is difficult to obtain data from the

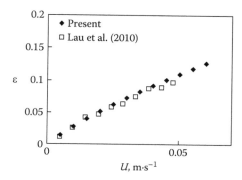

FIGURE 10.6
Comparison of a predicted average gas holdup with an overall gas holdup in a shallow bed reported by Lau et al. (2010).

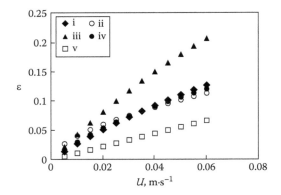

FIGURE 10.7
Comparison of an average gas holdup as a function of superficial gas velocity; correlations of
(i) Hughmark (1967), (ii) Hikita et al. (1980), (iii) Kumar et al. (1976), (iv) Akita and Yoshida (1973)
and (v) predicted by the model.

TABLE 10.2

Correlation for Gas Holdup in Bubble Columns

Investigator	D_T, m	d_o, m	n_h	Correlation
Hughmark (1967)				$\varepsilon = \dfrac{1}{2 + (0.35/U_G)(\rho_L \sigma/72)^{1/3}}$
Akita and Yoshida (1973)	0.152, 0.6	0.005	1	$\dfrac{\varepsilon}{(1-\varepsilon)^4} = c\left(\dfrac{D^2 \rho_L g}{\sigma}\right)^{1/8}\left(\dfrac{D^3 \rho_L^2 g}{\mu_L^2}\right)^{1/12}\left(\dfrac{U_G}{(Dg)^{1/2}}\right)$
				where $c = 0.2$ for pure liquids and non-electrolyte solutions, and $c = 0.25$ for electrolyte solutions
Kumar et al. (1976)	0.05, 0.07, 0.1			$\varepsilon_G = 0.728\, U - 0.485\, U^2 + 0.0975\, U^3$ where $U = U_G\left(\dfrac{\rho_L^2}{\sigma(\rho_L - \rho_G)g}\right)^{1/4}$
Hikita et al. (1980)	0.1	0.011	1	$\varepsilon = 0.672 f\left(\dfrac{U_G \mu_L}{\sigma}\right)^{0.578}\left(\dfrac{\mu_L^4 g}{\rho_L \sigma^3}\right)^{-0.131}\left(\dfrac{\rho_G}{\rho_L}\right)^{0.062}\left(\dfrac{\mu_G}{\mu_L}\right)^{0.107}$ where $f = 1$ for non-electrolytes

literature in which all the required parameters are available. Therefore,
a sparger with 100 holes was treated as a perforated-plate sparger. A further increase of the number of holes is not expected to influence the gas
holdup.

The gas holdups for single-nozzle spargers are consistently lower than the
values of other correlations. This is probably due to the fact that in the case
of a single nozzle, the trailing bubbles move at a faster rate and move in
the radial direction. The drag force on it is more than that estimated using

only the vertical velocity. As a result, the gas holdup is greater. The jetting could also be one of the reasons. Both these are not taken into consideration. The estimated data for $n_h = 100$ are in close agreement with the values estimated using other correlations.

Thus, the model is not validated for a single-nozzle sparger. However, it can be used to predict gas holdup in bubble columns with a perforated-plate sparger. All simulation studies using this model were carried out for perforated-plate spargers.

10.2.5 Simulation Studies

The hydrodynamic model for gas holdup is validated and can now be used to study the effects of various parameters on the gas holdup. The effects of column and sparger geometry and fluid properties for perforated-plate spargers can be investigated through simulation studies. At the same time, the axial distribution of gas holdup and bubble velocity can also be estimated. These are required to estimate axial distribution of the interfacial area for mass transfer.

The simulation studies were carried out to study the effects of liquid viscosity, surface tension, column diameter, number of holes in the gas sparger and nozzle diameter on axial distribution of the gas holdup and mass transfer coefficient. The simulations were limited to $U = 0.06$ m·s^{-1} so that the entire range of applicability is ensured. The flow regime is known to change at $U = 0.01$ m·s^{-1} from a 'bubbly flow regime' to 'churn turbulent regime' for an air–water system.

10.2.5.1 Effects of Column Diameters

The effects of column diameters on gas holdup as a function of superficial gas velocity are presented in Figure 10.8. The nozzle diameter, d_o, was 0.004 m. The gas holdup decreases as the column diameter increases from 0.5 to 5.0 m. Hughmark (1967) observed that the gas holdup depends upon a column diameter smaller than 0.1 m. The present model predicts the dependence even above 0.1 m. It may be attributed to the presence of the bubble coalescence and bubble breakup phenomena.

10.2.5.2 Effect of Nozzle Diameter

The effects of the hole diameter on the gas holdup as a function of superficial gas velocity for $D_T = 0.1$ m, $H_s = 2$ m and $d_o = 0.001, 0.002, 0.003$ and 0.004 mm ($n_h = 100$) are presented in Figure 10.9. For $d_o > 0.001$ m, the gas holdup was independent of d_o. However, for $d_o = 0.001$ m, the gas holdup was higher than that for larger values of d_o. This observation is consistent with the observation reported in the literature (Joshi 1981).

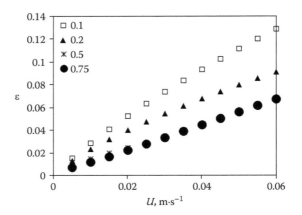

FIGURE 10.8
Effect of the column diameter on gas holdup in a bubble column for $n_h = 100$, $d_o = 0.003$ m and $H_s = 1.5$ m for an air–water system.

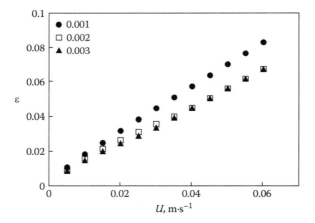

FIGURE 10.9
Effect of the hole diameter on gas holdup in a bubble column for $n_h = 100$, $D_T = 0.5$ m and $H_s = 1.5$ m for an air–water system.

10.2.5.3 Effect of Viscosity on Gas Holdup

Viscosity and surface tension were considered to be the important fluid properties affecting gas holdup. Simulation studies were conducted for a column with a column diameter, $D_T = 0.1$ m; static bed height, $H_s = 2.0$ m; nozzle diameter, $d_o = 0.004$ m; and number of holes, $n_h = 100$. The viscosity was varied in the range of 0.001–0.005 $kg \cdot m^{-1} \cdot s^{-1}$. The effect of viscosity on gas holdup as a function of superficial gas velocity is presented

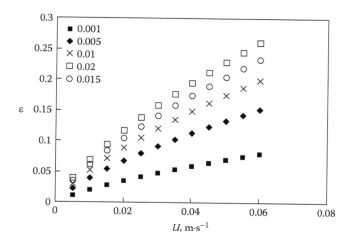

FIGURE 10.10
Effect of viscosity on gas holdup in a bubble column for $n_h = 100$, $D_T = 0.5$ m, $d_o = 0.003$ m and $H_s = 1.5$ m.

in Figure 10.10. The model predicts that the gas holdup increases with increasing viscosity.

10.2.5.4 Surface Tension on Gas Holdup

The effect of surface tension on gas holdup as a function of superficial gas velocity for column with $D_T = 0.1$ m, $H_s = 2$ m, $d_o = 0.003$ m and $n_h = 100$ is presented in Figure 10.11. Surface tension was varied from 0.07 to 0.04 N·m^{-1}. The surface tension of a liquid depends upon the type of the additives such as electrolytes, alcohols and surface active agents. As the surface tension decreased, the gas holdup increased. However, the change in gas holdup is small. It may be due to the fact the presence of foam and its behaviour were not considered in the model.

Thus, the proposed model predicted the effect of sparger, column diameter and viscosity of the liquid on gas holdup. It failed to predict the effect of surface tension. Therefore, the model is applicable for non-foaming systems only. Its use may be extended by multiplying predicted value of gas holdup by a factor dependent upon surface tension. However, its use to correctly predict the mass transfer coefficient and reaction rate is still doubtful. The model should be applied only in homogeneous bubble flow regimes and in cases of perforated-plate distributors.

The main model 'reactor model' given in the file 'masstransfer.m' has little to validate as the equations are based on first principles (i.e. mass balance and kinetic rate expressions, the acceptance of which is beyond doubt). It uses the hydrodynamic model which is developed and validated as above.

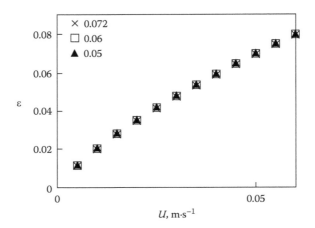

FIGURE 10.11
Effect of surface tension (N·m^{-1}) on gas holdup in a bubble column for $n_h = 100$, $D_T = 0.5$ m, $d_o = 0.003$ m and $H_s = 1.5$ m.

The model predictions for a chemical reaction should be compared with the experimental values, and in case of large deviations a different reactor model may be used.

Even if the model seems rigorous, it misses many important considerations while making assumptions; the model predictions are acceptable in a limited range of parameters. Thus, it is expected that the inclusion of bubble coalescence and bubble breakup phenomena in the hydrodynamic model will improve the large model.

10.3 Stochastic Model to Predict Wall-to-Bed Mass Transfer in Packed and Fluidised Beds

The stochastic models are more complicated than mechanistic models because they involve random variables. Stochastic models are discrete event models and are mostly solved numerically. Analytical solutions are also available in a few cases. The modelling and simulation using a stochastic model to predict the wall-to-bed mass transfer coefficient in packed bed and fluidised beds are developed, and simulation studies based on this model are presented.

The enhancement of wall-to-bed mass transfer coefficient in packed and fluidised beds over the empty column is due to the fluid agitation caused by the particles moving in the vicinity of the transfer surface. Verma and Rao (1988) studied the effect of translation of a single particle and a string of particles on the wall-to-bed mass transfer. The model for fixed bed and fluidised bed using the correlations of Verma and Rao (1988) will be used as a sub-model. The model is based on the addition of enhancement of mass transfer due to a large number of particles translating parallel to the wall. The packed bed model does not consider particle velocity but uses relative velocity between the fluid and particle as the particle velocity.

10.3.1 Improvement in Mass Transfer due to Translation of Particles

A brief introduction to improvement coefficient given by Verma and Rao (1988) is described below.

As a particle sweeps past a transfer surface, the mass transfer rate increases and attains a maximum value, and much after the particle crosses the transfer surface, it decays to zero. The enhancement in mass transfer rate, improvement coefficient, I, is defined as follows (Verma and Rao 1988):

$$I = \int_{-\infty}^{\infty} \left[k(x,t) - k_s \right] dx \qquad (10.36)$$

Here, x is the direction perpendicular to the direction of translation (Figure 10.12). The following equation correlated the experimental data for the improvement coefficient:

$$I = I_m \exp\left[-\frac{c(t_m - t)^2}{t} \right] \qquad (10.37)$$

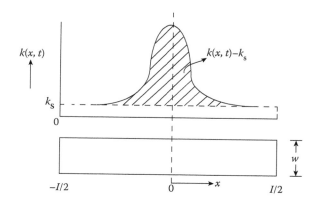

FIGURE 10.12

The improvement coefficient due to translation of a particle parallel to a flat wall (Verma 1984).

The following correlations were proposed for the maximum value of improvement coefficient, I_m, the time at which it occurs, t_m, and the decay constant, c. The additional subscript '1' denotes that it is for the first particle:

$$I_{1,m} = 5.98(d^*)^{-1.11} \text{Re}^{0.56} \tag{10.38}$$

$$t^*_{1,m} = 28.1(d^*)^{1.81} \text{Re}^{0.12} \tag{10.39}$$

$$c_1 = 0.62(d^*)^{0.17} \text{Re}^{-0.38} \tag{10.40}$$

Equations 10.38 through 10.40 are valid for $60 < \text{Re} < 395$ and $0.5 < d_p < 11$

where $d^* = \dfrac{d + \dfrac{d_p}{2}}{d_p}$ and $t^* = \dfrac{t_{1,m} v_p}{d_p}$.

Here, d_p is the particle diameter, d is the shortest distance between the particle and the wall and v_p is the velocity of the particle. The interaction of the flow field induced by the particle and the flow field already present at the time of arrival of the particle was obtained by repeated up-and-down motions of the particle. The improvement coefficient due to jth sweep (i.e. the sweep by the jth particle) was given by

$$I_j = \left[I_{j,m} f \exp\left\{ \frac{c_j t_{j,m}(1 + f - t_{j,m})}{f} t \right\} + I_{j,b} \right] \exp(-c_j t) \tag{10.41}$$

where

$$f = 1 - \frac{I_{j,b}}{I_{j,m}} \exp(-c_j t_{j,m}) \tag{10.42}$$

The parameter $\dfrac{I_{j,b}}{d_p}$ has been identified as the interaction parameter. The parameters, $I_{j,m}$, $t_{1,m}$ and C_j, were correlated as functions of $\dfrac{I_{j,b}}{d_p}$, $t_{1,m}$ and C_1:

$$\ln\left(\frac{I_{j,m} - I_{j,b}}{I_{1,m}} \right) = -3.06 \times 10^{-4}(d^*)^{-1.58} \text{Re}^{0.53} \left(\frac{I_{j,b}}{d_p} \right) \tag{10.43}$$

$$\ln\left(\frac{t_{j,m}}{t_{1,m}} \right) = -3.31 \times 10^{-3}(d^*)^{-1.1} \text{Re}^{-0.75} \left(\frac{I_{j,b}}{d_p} \right) \tag{10.44}$$

$$\ln\left(\frac{C_{j,m}}{C_{1,m}} \right) = 1.94 \times 10^{-3}(d^*)^{0.17} \text{Re}^{-0.86} \left(\frac{I_{j,b}}{d_p} \right) \tag{10.45}$$

for $90 < \text{Re} < 395$ and $0.5 < d_p < 0.84$ and $3.18 \times 10^4 < \left(\dfrac{I_{j,b}}{d_p} \right) < 1.28 \times 10^5$.

The particle diameter, particle velocity, distance from the wall and the time interval between successive sweeps are required to predict the improvement coefficient for several sweeps of a particle.

10.3.2 Model for Wall-to-Bed Mass Transfer in Packed and Fluidised Beds

Let us consider the mass transfer coefficient in a fluidised bed. The fluid agitation at the surface is enhanced by the particles translating parallel to the wall in the case of a fluidised bed. It results in increased mass transfer rates. The effect of a complex flow field around the particle is correlated by the correlations of Verma and Rao (1988). Only the axial component of the velocity was used to describe the effect of a 3D flow field on the mass transfer coefficient. In the case of packed beds, the presence of particles is responsible for enhancement of the mass transfer coefficient. The particle is not moving, but the fluid is. Let us use the relative velocity between the particle and fluid to replace the particle velocity in the correlations given by Equations 10.38 through 10.45.

The improvement in mass transfer rate was found to be negligible if the particle is one diameter away from the surface (Verma and Rao 1988). Therefore, only the first row of the particle in front of the surface was assumed to have any significant influence on the fluid agitation.

To determine the time interval in two successive sweeps, it is assumed that the particles are arranged in an orthorhombic structure, as shown in Figure 10.13. Although there is no direct experimental justification for any particular structure in a liquid fluidised bed, such assumptions have been made by other investigators. The inter-particle distance, L, for orthorhombic arrangement is (Ghanza et al. 1982)

$$\frac{L}{d_p} = \left(\frac{1-\varepsilon_{mf}}{1-\varepsilon}\right)^{1/3}$$

(10.46)

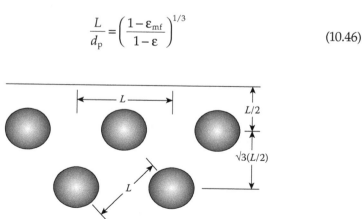

FIGURE 10.13
The average position of particles in a fixed or fluidised bed (Verma 1984).

where ε and ε_{mf} are the porosity of the bed at a superficial fluid velocity and the porosity of the bed at a minimum fluidisation velocity, respectively.

Because only the first row of a particle in close proximity to the transfer surface will affect the mass transfer, the particles can be visualised to form a vertical string. Each of the black particles shown in Figure 10.13 corresponds to a string. If the individual particles are moving down at a velocity, v_p, it corresponds to a string moving across the surface, and the time interval between the arrival of the particle in front of the surface, τ, can be written as

$$\tau = \frac{L}{v_p} \tag{10.47}$$

In case of fixed beds, the fluid velocity will be used in place of particle velocity, v_p. The particles are equally spaced; hence, the inter-particle distance and the time interval are not random.

In a fluidised bed, the particle velocities and the inter-particle distance vary in a random manner. To stimulate the conditions of the fluidised bed as closely as possible, the particle moving across the surface can be assigned velocities according to a probability distribution function. The time interval between the random arrival of the particle and the distance between the particle and the transfer surface can be assumed to follow an appropriate probability distribution function.

The average mass transfer coefficient after several sweeps of the particles can be obtained. However, a model to relate improvement coefficient and mass transfer coefficient has to be obtained.

10.3.3 Relationship between Mass Transfer Coefficient and Improvement Coefficient

Verma (1984) proposed a model which related the improvement coefficient and mass transfer in fixed and fluidised beds. Consider the transfer surface of width, w, and length, L. The width is large enough to ensure uniform mass transfer rate across the surface. The length is equal to the average inter-particle distance. The simplified picture of the fluidised bed is as follows:

1. A string of particles will sweep past the surface; each particle arriving at a random distance from the wall, with random velocity and with random interval between successive particles.

2. The improvement of mass transfer rate due to the motion of a string is confined to the length, L, only. The studies with the repeated up-and-down motion of the particle indicate that the influence beyond one particle diameter is negligible.

3. In absence of the particle, the mass transfer rate from the surface is uniform and corresponds to the mass transfer rate in the empty column at the interstitial velocity, v_I.

The mass transfer rate can be written as

$$m = K(v_I)Lw\Delta C \tag{10.48}$$

where $K(v_I)$ is the mass transfer coefficient in the empty column at v_I.

After several particles sweeping past the transfer surface, the improvement in mass transfer rate attains a steady average value. The mass transfer rate at time, t, after jth particle swept past the surface, $m(T_j + t)$, is

$$m(T_j + t) = m + I_{j,F}(I_{j,b}, t)w\Delta C \tag{10.49}$$

where

T_j = time at which the jth particle arrives at the transfer surface;
t = time elapsed since the arrival of the jth particle; and
$I_{j,F}(I_{j,b}, t)$ = improvement coefficient at time t due to the jth sweep, which represents a quantitative measure of a 3D flow field around the particle.

Now, dividing Equation 10.49 by $(Lw\Delta C)$

$$\frac{m(T_j + t)}{w\Delta C} = \frac{m}{Lw\Delta C} + \frac{I_{j,F}(I_{j,b}, t)}{L} \tag{10.50}$$

or

$$K(T_j + t) = K(v_I) + \frac{I_{j,F}(I_{j,b}, t)}{L} \tag{10.51}$$

The time average mass transfer coefficient, K_{av}, can be written as

$$K_{av} = \lim_{n \to \infty} \frac{1}{T_{n+1}} \sum_{j=1}^{n} \int_0^{(T_{j+1} - T_j)} K(T_j + t)\,dt$$

$$K_{av} = K(V_I) + \lim_{n \to \infty} \frac{1}{LT_{n+1}} \sum_{j=1}^{n} \int_0^{(T_{j+1} - T_j)} I_{j,F}(I_{j,b}, t)\,dt \tag{10.52}$$

The term $I_{j,F}(I_{j,b}, t)$ can be estimated in terms of $I_{j,b}$. Starting from $T = 0$, it is possible to compute $I_{j,F}(I_{j,b}, t)$ and hence $I_{j,b}$. The average mass transfer coefficient can be obtained from Equation 10.52 provided that $I_{j,F}(I_{j,b}, t)$ is known in the fluidised bed.

10.3.4 Evaluation of $I_{j,F}(I_{j,b}, t)$ and K_{av}

The wake structure behind the particle moving close to the transfer surface in the fluidised bed with velocity v_p is assumed to be same as the wake structure of a particle moving with a velocity $V_s = (v_p + V_I)$ in still fluid, provided that the fluid flow field (and, hence, $I_{j,b}$) for both the cases is same. Let us assume that the improvement coefficient in fluidised bed is same as that given by the experimental data for string, although the boundary conditions in both cases are different.

$$I_{j,F}\left(I_{j,b},t\right)=I_{j}\left(I_{j,b},t_{s}\right) \tag{10.53}$$

To an observer moving with the particle, the improvement coefficient at a distance y from the particle would appear to be the same either in the case of still fluid or in the bed. Then, the distance can be expressed in terms of the time as

$$y=tv_{p}=t_{s}V_{s} \tag{10.54}$$

where t and t_{s} are the times taken for the particle to move a distance y from the bed and still fluid, respectively. Since improvement in the still fluid and time in the fluidised bed are known, it can be conveniently written as

$$I_{j,F}\left(I_{j,b},t\right)=I_{j}\left(I_{j,b},tv_{p}/V_{s}\right) \tag{10.55}$$

Substituting Equation 10.55 in Equation 10.54, we get

$$K_{av}=K(V_{I})+\lim_{n\to\infty}\frac{1}{LT_{n+1}}\sum_{j=1}^{n}\int_{0}^{\left(T_{j+1}-T_{j}\right)}I_{j}\left(I_{j,b},tv_{p}/V_{s}\right)dt \tag{10.56}$$

Equation 10.56 relates the mass transfer coefficient in fluidised bed with the correlation of Verma and Rao (1988).

For a packed bed, the variable t can be replaced by y/V_{I}. Equation 10.52 can be written as

$$K_{av}=K(V_{I})+\lim_{n\to\infty}\frac{1}{V_{I}L^{2}}\int_{0}^{L}I_{j}\left(I_{j,b},y/V_{I}\right)dt \tag{10.57}$$

10.3.5 Validation of the Model for Fluidised Beds

The model was validated for local mass transfer coefficient data of Coeuret and Le Goff (1976 a,b) and Storck and Coeuret (1977). The data were obtained for ionic mass transfer coefficient using a cylindrical cathode of 15 mm diameter and 10 mm length in the empty column and packed and fluidised beds. The Schmidt number in their studies was very close to that observed by Verma and Rao (1988). The probability distribution functions for various parameters used in the model are discussed below.

10.3.5.1 Particle Velocities

The particle velocities are known to be predominantly in the upward direction at the centre of the bed. But the axial velocities at which the mass transfer coefficients were measured are not available. However, the average total speeds of the particle at different liquid velocities and densities and for different particle sizes are available in literature. The experimental conditions employed in these studies are given in Table 10.3. Although wide variations in hydrodynamic parameters are employed in these studies, the data are in close agreement with each other except those of Handley et al. (1966). Thus, it may be assumed that mean axial velocities will be insensitive to the hydrodynamic parameters in liquid fluidised beds. To generate axial particle velocities distributed according to Gaussian distribution (Latif and Richardson 1972; Kmiec 1978), the mean axial velocity and standard deviation or mean square velocity as function of U/U_{mf} are required. Carlos and Richardson (1968) have reported the mean square velocity to vary linearly with U/U_{mf}. Later, Latif and Richardson (1972) found the mean axial velocity to vary linearly with U/U_{mf} for a wider range of porosities. The data on extrapolation gave a positive particle velocity at U_{mf}, which is not feasible. Hence, the data of $\overline{V_a^2}$, reported by Carlos and Richardson (1968), were fitted to obtain the following equation:

$$\overline{V_a^2} = 0.0875 \times 10^{-4} \left(\frac{U}{U_{mf}} - 1 \right) \tag{10.58}$$

The particle velocity was assumed to follow the Gaussian distribution. For a given set of U and U_{mf}, the individual particle velocities may be generated.

TABLE 10.3

Experimental Conditions Employed in Various Studies of Particle Speeds in Liquid Fluidised Beds

Investigator	d_p, mm	$\mu_f \times 10^{-3}$, kgm^{-1}s^{-1}	ρ_f, kgm^{-3}	D_σ, m	ε
Handley et al. (1966)	1.003–1.204	2.067	1087	7.5	0.67–0.905
	1.405–1.676				
	2.057–2.812				
Bordet et al. (1968)	1.04	1	1	10	0.46–0.685
Carlos and Richardson (1968)	1.04	10	1.19	10	0.53–0.7
Latif and Richardson (1972)	6.16	10.5	1.05	10	0.55–0.95
Kmiec (1978)	6.17	1.767	1.075	5	0.628–0.78

Normally distributed random numbers, $\widetilde{N}$ with mean = 0 and variance = 1, were generated. They were correlated to obtain Gaussian distribution velocities using the following equation:

$$\widetilde{V_p} = \sqrt{V_a^2}\,\widetilde{N} + \overline{V_a} \qquad (10.59)$$

where $\widetilde{V_p}$ is the random axial velocity.

An absence of data on the average axial velocities of the average particle speeds was used. Carlos and Richardson (1968) found the ratio of the mean to root mean square speeds to be close to 0.92. Hence, $\widetilde{V_p}$ was obtained from V_a^2 by multiplying it by 0.92. Since the particle velocity does not exceed the interstitial fluid velocity, V_I, random velocities greater than V_I were not considered. From the data of Kmiec (1978), it was found that the particle velocities are always higher than 0.005 m·s⁻¹. Hence, random velocities smaller than 0.005 m·s⁻¹ were rejected.

10.3.5.2 Distance of the Particle from the Wall

For an orthorhombic arrangement of the particle in the bed, the first layer (i.e. vertically adjacent to the wall) will be at $\left(\dfrac{L}{2} + \sqrt{3}\dfrac{L}{2}\right)$. It is assumed that in a fluidised bed, the particles can take a path which lies between $\dfrac{d_p}{2}$ and $\left(\dfrac{L}{2} + \sqrt{3}\dfrac{L}{2}\right)$. Because the mass transfer data were measured in the core of the bed, where the porosity is uniform (Kmiec 1978), the dimensionless distance of the particle from the transfer surface is chosen to be uniformly distributed in the following range:

$$0.5 < d^* < \frac{L\left(0.5 + \dfrac{\sqrt{3}}{2}\right)}{d_p} \qquad (10.60)$$

10.3.5.3 Time Interval

The time interval between successive arrivals of the particles at the transfer surface, τ, is another important model parameter for which experimental data are not easily available. Because the number of particles in a small volume is described by the Poisson process, the time interval between the arrivals is assumed to uniformly distributed (Koppel et al. 1970). Therefore, the value of τ was taken as uniformly distributed in the following range:

$$\frac{d_p}{v_p} < \tau < \frac{\left(2L - d_p\right)}{v_p} \qquad (10.61)$$

where $\frac{d_p}{v_p}$ is the minimum time interval due to the finite size of the particle and the upper limit is due to the maximum separation between the particle imposed by the porosity of the bed.

10.3.6 Algorithm for Estimation of Mass Transfer Coefficient

The algorithm for the estimation of a mass transfer coefficient in fixed and fluidised beds is given in Figure 10.14.

The length of the transfer surface for a vertical string of particle was fixed was estimated using the porosity of the bed. At interstitial liquid velocity, V_I, the mass transfer rate per unit width of the transfer surface was computed. Starting from $T = 0$, the velocity of the first particle, d^*, and the time of arrival of the next particle was generated. The values of $I_j\left(I_{j,b}, tv_p/V_s\right)$ was computed considering $I_{1,b}$ to be zero. Likewise, for the second and subsequent particles, $I_j\left(I_{j,b}, tv_p/V_s\right)$ was computed by taking $I_{j,b}$ be equal to $I_j\left(I_{j-1,b}, \tau_j v_p/V_s\right)$, for about 1000 particles. The average mass transfer coefficient, K_{av}, was found using Equation 10.52.

In the case of packed bed, the inter-particle distance in the string was found from the porosity. employing V_I as the slip velocity. The term $I_j\left(I_{j,b}, L/V_I\right)$ was computed starting from the first horizontal layer of the particles adjacent to the distributor to other layers in upward direction until a steady periodic variation was attained. The value of K_{av} was estimated using Equation 10.57.

10.3.7 Model Validation

The mass transfer coefficient was estimated for beads consisting of 1, 2 and 10 mm glass and beads consisting of 10 mm nylon. These were compared by Verma (1984) with the experimental values reported by Coeuret and Le Goff (1976 a,b) and Storck and Coeuret (1977). The mass transfer coefficient in fixed and fluidised beds for 10 mm glass beads is shown in Figure 10.15; the mass transfer coefficients in the empty column at the superficial and interstitial velocities are also shown.

Storck and Coeuret (1977) observed that the motion of the particle has little effect on the mass transfer coefficient, and the bed can be considered to be an expanded packed bed. However, the average mass transfer coefficients estimated considering $\widetilde{V_p}$ to be zero and in the downward direction (labelled 2 and 3, respectively) large difference were observed. A closer look at v_p and V_I values revealed that the difference is less when V_I is several times larger than v_p. It is true for heavier particles (Verma 1984). Their mass transfer data also indicate that the differences between fluidised beds and expanded beds are larger for 10 mm nylon beads compared to glass beads of the same size. The model correctly predicts the trend of the variation of mass transfer

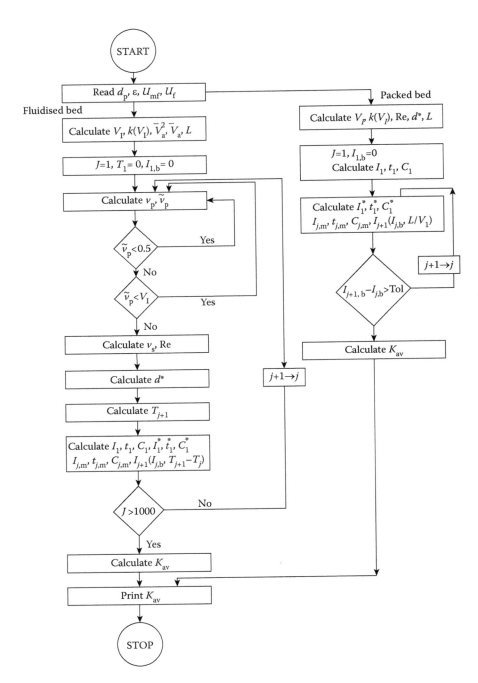

FIGURE 10.14
Algorithm for estimation of the mass transfer coefficient in fluidised and fixed beds (Verma 1984).

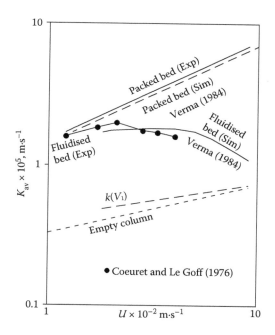

FIGURE 10.15
A comparison of predicted values of the mass transfer coefficient in fixed and fluidised beds (Equations 10.56 and 10.57, respectively) with experimental data of Coeuret and Le Goff (1976 a,b) for 10 mm glass beads (Verma 1984).

coefficient with superficial liquid velocity. The predicted average mass transfer coefficients are in close agreement with the experimental values. The model was not validated for instantaneous values of mass transfer coefficient.

Variations of mass transfer coefficient with y/L for packed beds are presented in Figure 10.16. The mass transfer coefficient increased progressively with the number of sweeps and attained steady periodic variation.

10.4 Artificial Neural Network Model: Heat Transfer in Bubble Columns

The development of an artificial neural network (ANN)–based model to predict the heat transfer coefficient in a bubble column is presented (Verma and Srivastava 2003). There is no single correlation which can fit the heat transfer coefficient in a bubble column studied by various investigators. An ANN-based model may be used under such conditions. The first stage in developing the ANN-based model is to identify the input to the model.

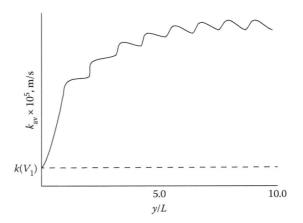

FIGURE 10.16
Axial variation of a mass transfer coefficient in fixed beds as a function of the axial position
(y/L) (Verma 1984).

10.4.1 Selection of Inputs

It is considered that heat transfer is independent of column diameter (Fair
et al. 1962; Kast 1962; Nishikawa et al. 1977). Kolbel et al. (1980) used a sieve
plate, single nozzle and sintered nozzle and observed no effect of sparger
type on the heat transfer coefficient.

A few attempts to model heat transfer in bubble columns were discussed
in Section 5.5. Mass transfer studies in bubble columns show its dependence
on column diameter (Patil and Sharma 1983) and sparger design (Rai 1997).
Heat transfer studies indicated its dependence on gas holdup (Joshi et al.
1980; Verma 1989). The liquid–circulation velocity depends on gas holdup
and contributes to convective heat transfer. The effect of sparger design on
gas holdup in the 'bubbly flow regime' is known (Reilly et al. 1986).

The transfer process depends upon hydrodynamics, which depends upon
the bubble behaviour in the columns. The hydrodynamics varies in the axial
direction. Therefore, the expanded bed height may be taken as model input.
It indirectly relates static bed height and gas holdup. The variables selected
as model input are superficial gas velocity, fluid properties, sparger design,
gas holdup and static bed height. The sparger design can be described at
least two parameters, the number of nozzles and the hole diameter.

10.4.2 Obtaining Experimental Data

The ANN-based models require a large set of data to train the data. Although
there are a number of experimental studies, only a few of them have
all the data required as model input (Table 10.4). The data of Fair et al. (1960),
Hart (1976), Hikita et al. (1981), Lewis et al. (1982) and Verma (1989) have

TABLE 10.4

Geometrical Parameters in Various Heat Transfer Studies in Bubble Columns

Investigator	$d_o \times 10^3$, m	n_h	D_T, m	System
Hikita et al. (1981)	9,13	1	0.1	Air–water, methanol
	13.1, 20.6, 36.2	1	0.19	*n*-Butanol, sucrose solution
Lewis et al. (1982)	1	129	0.292	Air–water, nitrogen–glycol
	1.9	196		
Hart (1976)	6.4	1	0.0991	Air–water, ethylene glycol
Fair et al. (1962)	0.0005	47	0.45	Air–water
Verma (1989)	0.0008	91	0.108	Air–water

reported all the data available that are required as input to the ANN. All data on heat transfer coefficient as a function of superficial velocity are plotted in Figure 10.17, which shows a large scatter of data which could not be described by a single correlation. Geometrical parameters and operating conditions of heat transfer studies reported in the literature are given in Table 10.4.

10.4.3 Architecture of ANNs

The input to the ANN were the superficial gas velocity, U; the Prandtl number, Pr; the number of holes, n_h; the hole diameter, d_o; the column diameter, D_T; the surface tension of the fluid; gas holdup and expanded bed height. Thus, the input layer had eight nodes. The output layer has only one node having heat transfer coefficient as a target. It looks obvious that the superficial gas velocity, fluid properties, sparger geometry and column diameter affect the gas holdup; therefore, it is not necessary to include the gas holdup as input to the model. However, after several attempts, it was observed that inclusion of the gas holdup was necessary; otherwise, even a large architecture could not train the ANN.

An ANN can be developed using several types of available software. The present model was developed using the ANN toolbox in MATLAB. The code is given in Appendix B. Defining the ANN architecture, training and simulation are one-line codes. MATLAB uses default training algorithms. Other algorithms can also be prescribed as per details given in the product manual.

Only one hidden layer was chosen. The number of nodes in the layer was varied between two and six. The ANN was trained, and the performances for 2, 3, 4, 5 and 6 nodes are given in Table 10.5. The performance can be different

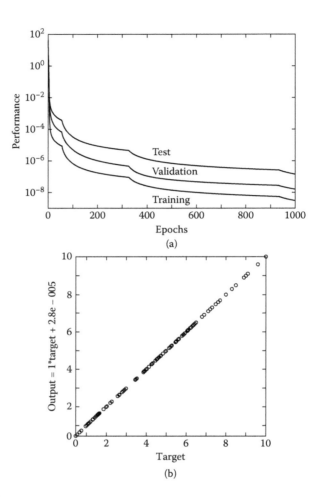

FIGURE 10.17
Developing an ANN model: (a) a performance curve for training of ANN and (b) a parity plot
for an ANN model, with model predictions compared with experimental data of Hikita et al.
(1981), Lewis et al. (1982), Hart (1976), Fair et al. (1962) and Verma (1989).

TABLE 10.5

Performance of ANNs for Different
Numbers of Nodes in the Hidden Layer

Number of nodes	Performance
2	3.3×10^{-8}
3	2.4×10^{-8}
4	1.74×10^{-8}
5	4.2×10^{-8}
6	2.25×10^{-7}

TABLE 10.6

Weights [$w_{ij}(1)$ and $w_{ij}(2)$] and Bias [$b(1)$ and $b(2)$] for Trained ANN

$w_{ij}(1)$ for $j = 1$–8 and $i = 1$–4

−0.3665	1.0652	−1.0863	−0.0256	−0.0860	−0.5173	5.8156	0.5066
1.7008	−5.1564	−2.0771	2.2143	−2.0402	2.4229	0.8952	0.1116
0.0000	−0.0000	0.0000	0.0000	0.0000	−0.0000	0.0505	−0.0000
1.5390	−9.4859	−0.2630	3.9138	5.6892	−3.0500	−4.8404	−0.2529

$w_{ij}(2)$ for $j = 1$–4 and $i = 1$

0.0004	0.0000	19.8176	−0.0002

$b_i(1)$, $i = 1$ to 4

−3.6199	3.9079	0.0103	−1.3354

$b(2)$

−0.2043

in different trainings, as the initial guesses for the weights and bios are random numbers. From Table 10.5, it can be seen that the best performance measured by the sum of the square of errors is for four nodes in the hidden layer.

During the training of the ANN, the data were divided into three sets: 70% for training, 15% for validation and 15% for testing. The performances for each of these are estimated, and the training is stopped when the maximum number of epochs has been reached or the validation failed (i.e. the performance of the validation set started increasing with further training). This step is required to avoid 'over-fitting'. The performance for all three data sets is shown in Figure 10.17a. The final weights and bias are given in Table 10.6. These can be used to predict the heat transfer coefficient for unknown data. When more data are available, the ANN model can be retrained to get a new set of constants.

The parity plot for the heat transfer coefficient is given in Figure 10.17b. It can be seen that the ANN predictions are quite impressive.

10.5 Summary

Four case studies to demonstrate various modelling techniques were presented. These include an approximate analytical model, a numerical model based on laws of conservation, a stochastic model and an ANN-based model. Emphasis is on giving detailed reasoning behind the assumptions made to transform the physics of the process into mathematical representation. The solution methodology has been avoided because of the vast literature and codes that are available. The use of correlations was avoided while developing the models.

References

Akita, K., and Yoshida, F. 1973. Gas holdup and volumetric mass transfer coefficient in bubble columns. _Ind. Eng. Chem. Process Des. Dev._ 12(1): 76–80.

Bordet, J., Borlai, O., Vergnes, F., and Le Goff, P. 1968. Direct measurement of their frequency of collision against a wall in a liquid-solids fluidised bed. _Inst. Chem. Eng. Symp. Ser._ 30: 165–173.

Carlos, C.R., and Richardson, J.F. 1968. Solids movement in liquid fluidised beds: I. Particle velocity distribution. _Chem. Eng. Sci._ 23(8): 813–824.

Chen, R.C., Reese, J., and Fan, L.S. 1994. Flow structure in a three-dimensional bubble column and three-phase fluidized bed. _AIChE J._ 40(7): 1093–1104.

Coeuret, F., and Le Goff, P. 1976a. L'electrode poreuse percolante (epp): I. Transfert de matiere en lit fixe. _Electrochim. Acta._ 21: 185–193.

Coeuret, F., and Le Goff, P. 1976b. L'electrode poreuse percolante (EPP): II. Transfert de matiere en lit fixe ou fluidise de grains non-conducteurs. _Electrochim. Acta._ 21: 195–202.

Cova, D.R. 1966. Catalyst suspension in gas-agitated tubular reactors. _Ind. Eng. Chem. Process Des. Dev._ 5(1): 20–25.

Dwight, H.B. 1972. _Tables of Integrals and Other Mathematical Data_, 4th ed. New York: MacMillan.

Fair, J.R., Lambright, A.J., and Andersen, J.W. 1962. Heat transfer and gas holdup in a sparged gas contactors. _Ind. Eng. Chem. Process Des. Dev._ 1(1): 33–36.

Farkas, E.J., and Leblond, P.F. 1969. Solids concentration profile in the bubble column slurry reactor. _Can. J. Chem. Eng._ 47: 215–218.

Friedlander, S.K. 1957. Behavior of suspended particles in a turbulent fluid. _AIChE J._ 3(3): 381–385.

Geary, N.W., and Rice, R.G. 1991. Circulation in bubble columns: Correction for distorted bubble shape. _AIChE J._ 37(10): 1593–1594.

Ghanza, V.L., Upadhyay, S.N., and Saxena, S.C. 1982. A mechanistic theory for heat transfer between fluidized beds of large particles and immersed surfaces. _Int. J. Heat Mass Trans._ 25(10): 1531–1540.

Handley, B., Doraiswamy, S., Butcher, K.L., and Franklin, N.L. 1966. A study of the fluid and particle mechanics in liquid-fluidized beds. _Trans. Inst. Chem. Eng._ 44: T260.

Hart, W.F. 1976. Heat transfer in bubble-agitated systems: A general correlation. _Ind. Eng. Chem. Process Des. Dev._ 15(1): 109–114.

Hikita, H., Asai, S., Kikukawa, H., Zaike, T., and Ohue, M. 1981. Heat transfer coefficient in bubble columns. _Ind. Eng. Chem. Process Des. Dev._ 20: 540–545.

Hikita, H., Asai, S., Tanigawa, K., Segawa, K., and Kitao, M. 1980. Gas holdup in bubble columns. _Chem. Eng. J._ 20: 59–67.

Hughmark, G.A. 1967. Holdup and mass transfer in bubble columns. _Ind. Eng. Chem. Process Des. Dev._ 6(2): 218–220.

Imafuku, K., Wang, T.Y., Koide, K., and Kubota, H. 1968. The behaviour of suspended solid particles in the bubble column. _J. Chem. Eng. Jpn._ 1(2): 153–158.

Joshi, J.B. 1981. Correspondence: Axial mixing in multiphase contactors: A unified correlation. _Trans. I. ChemE._ 59: 138–143.

Joshi, J.B., Sharma, M.M., Shah, Y.T., Singh, C.P.P., Ally, M., and Klinzing, G.E. 1980. Heat transfer in multiphase contactors. _Chem. Eng. Commun._ 6: 157–271.

Kast, W. 1962. Analyse des wärmeübergangs in blasensäulen. *Int. J. Heat Mass Tranfer.* 5(3–4): 329–336.

Kato, Y., Nishiwaki, A., Fukuda, T., and Tanaka, S. 1972. The behavior of suspended solid particles and liquid in bubble columns. *J. Chem. Eng. Jpn.* 5(2): 112–118.

Kmiec, A. 1978. Particle distribution and dynamics of particle movement in solid-liquid fluidized beds. *Chem. Eng. J.* 15: 1–12.

Kojima, H., Anjyo, H., and Mochizaki, Y. 1986. Axial mixing in bubble column with suspended particles. *J. Chem. Eng. Jpn.* 19(3): 232–236.

Kojima, H., and Asano, K. 1981. Hydrodynamic characteristics of a suspension-bubble column. *Int. Chem. Eng.* 21(3): 473–481.

Kolbel, H., Siemes, W., Maas, R., and Muller, K. 1958. Warmeubergang an Blasensaulen. *Chemie. Ing. Techn.* 20(6): 400–404. English translation: Heat transfer in bubble columns. Retrieved from www.fischer-tropsch.org/DOE/DOE_reports/.../de89012412_toc.htm

Koppel, L.B., Patel, R.B., and Holmes, J.T. 1970. Statistical models for surface renewal in heat and mass transfer: Part III: Residence times and age distributions at wall surface of a fluidized bed: Application of spectral density. *AIChE J.* 16(3): 456–464.

Kumar, A., Degaleesan, T.E., Laddha, G.S., and Hoelscher, H.E. 1976. Bubble swarm characteristics in bubble column. *Can. J. Chem. Eng.* 54: 503–508.

Latif, B.A.J., and Richardson, J.F. 1972. Circulation pattern and velocity distribution for particles in a liquid fluidised bed. *Chem. Eng. Sci.* 27(11): 1933–1949.

Lau, R., Mo, R., and Sim, W.S.B. 2010. Bubble characteristics in shallow bubble column reactors. *Chem. Eng. Res. Des.* 88: 197–203.

Lewis, D.A., Field, R.W., Xavier, A.M., and Edwards, D. 1982. Heat transfer in bubble columns. *Trans. Inst. Chem. Eng.* 60: 40–47.

McCabe, W.L., Smith, J.C., and Harriot, P. (1993). *Unit Operations of Chemical Engineering,* 5th ed. New York: McGraw-Hill.

Murray, P., and Fan, L.S. 1989. Axial solid distribution in slurry bubble column. *Ind. Eng. Chem.* 28: 1697–1703.

Nishikawa, M., Kato, H., and Hashimoto, K. 1977. Heat transfer in aerated tower filled with non-Newtonian liquids. *Ind. Eng. Chem. Process Des. Dev.* 16(1): 133–137.

Patil, V.K., and Sharma, M.M. 1983. Solid-liquid mass transfer coefficient in bubble columns up to 1 m diameter. *Chem. Eng. Res. Des.* 61: 23–61.

Rai, S. 1997. *Studies on Mass Transfer with Immersed Surface in Bubble Columns Using Electrochemical Technique.* PhD dissertation, Institute of Technology, Banaras Hindu University, Varanasi, India.

Reilly, I.G., Scott, D.S., and Abou-El-Hassan, M. 1982. Leaching in bubble column slurry reactor. *Can. J. Chem. Eng.* 60: 399–406.

Reilly, I.G., Scott, D.S., Bruijn, T.De., Jain, A., and Piskorz, J. 1986. A correlation for gas holdup in turbulent coalescing bubble columns. *Can. J. Chem. Eng.* 64(5): 705–717.

Richardson, J.F., and Zaki, W.N. 1954. Sedimentation and fluidization: Part 1. *Trans. Inst. Chem. Eng.* 32: 35–53.

Row, P.N. 1977. A convenient empirical equation for estimation of the Richardson-Zaki exponent. *Chem. Eng. Sci.* 42(11): 2795–2796.

Smith, D.N., and Ruether, J.A. 1985. Dispersed dynamics in a slurry bubble column. *Chem. Eng. Sci.* 40(5): 741–754.

Storck, A., and Coeuret, F. 1977. Wall to liquid mass transfer in fluidized beds. *Can. J. Chem. Eng.* 55(4): 427–431.

Verma, A.K. 1984. *Studies on the Mechanism of Wall to Bed Mass Transfer in Liquid Fluidized Beds: A Single Particle Approach*. PhD dissertation. Indian Institute of Technology Kanpur, India.

Verma, A.K. 1989. Heat transfer mechanism in bubble columns. *Chem. Eng. J.* 42: 205–208.

Verma, A.K., and Rao, D.P. 1988. Enhancement in mass transfer due to motion of a particle: An experimental study. *AIChE J.* 34(7): 1157–1163.

Verma, A.K., and Srivastava, A. 2003. ANN based model for heat transfer from immersed tubes in a bubble column: Effects of immersed surface and sparger geometry. In *Proceedings of Fourth National Seminar on Thermal Systems*, February 22–23, Varanasi, India: Institute of Technology, Banaras Hindu University.

Zhang, K. 2002. Axial solid concentration distribution in tapered and cylindrical bubble columns. *Chem. Eng. J.* 86: 229–307.

11

Simulation of Large Plants

In Chapters 4 through 8 of this book, a methodology to develop various types of models was discussed. These methodologies focussed upon a single process (e.g. a reactor separator and separator). A chemical or biochemical process consists of several such interconnected unit operations. Simulation of an entire process requires the integration of the models for individual processes into a single model for the entire process or plant. The steady-state simulation of large plants based on this kind of methodology is known as flowsheeting. A few brands of flowsheeting software used in the industry are Aspen Plus, Hysys, PRO/II, ProSim, WinSim and ChemCAD. A couple of open-source flowsheeting programmes, COCO and DWSIM, are also available, although they lack the rich library required to perform simulations. The current state of the art for process simulation has been reviewed by Stephanopoulos and Reklaitis (2011).

For the development of new processes, user-written codes using high-level languages are preferred over the commercial packages by many chemical engineers. A comparison among the commercial flowsheeting programme and user-written code reveals the following advantages and disadvantages.

Commercial flowsheeting programmes generally have large libraries for property estimation and unit operations. They also have functions to present the results in graphical formats. These facilities reduce the time for developing simulation programmes, but the cost of such software is high. The commercial flowsheeting programmes lack flexibility. Although they provide ways to write user-defined model equations for a new unit operation, the user has to learn the language before using it to write user-defined functions, sub-routines or classes. Due to various built-in functions, it becomes necessary to be familiar with these functions. Generally, it is also possible to develop such sub-routines using popular computer languages (e.g. C++ and Visual-C) for commercial software.

On the other hand, code developed by the modeller for the model used for simulation consists of only the required component. It may not have a thermodynamic package for hundreds of compounds. Similarly, it may not have a large number of sub-routines for several unit operations. Writing the code by the modeller himself is time consuming. However, it does not involve fixed investment on the flowsheeting software. At the same time, since the modeller is completely familiar with the existing code, modification of the code to make changes in the model equations is an easy job. Having command over the code gives satisfaction to many modellers.

For a modeller, a large plant or a large model means an interconnected network of small models (Figure 11.1) which may also be called 'sub-models'. The sub-models are developed based on the principles discussed in Chapters 4 through 8. The sub-models may be classified as follows.

1. *Equations having analytical solutions*: Most of the model equations are sets of several equations. Many of these equations are differential equations. The model equations for a simple sub-model may have analytical solutions. The model output may be expressed in terms of model inputs either explicitly or implicitly. In the former, the sub-model may be considered to behave as a function. In case of implicit relation, an iterative procedure may be followed.

2. *Equations with no analytical solutions*: Most of the models are described by a set of partial differential equations. These equations are solved numerically. When this kind of model is used as a sub-model representing a unit operation in a flowsheet, then it is not an easy task to interconnect various unit operations.

3. *Algorithm-based models*: A class of models is based on algorithms. These models do not use equations in a traditional sense to describe the process (i.e. the model equations are not a set of only algebraic, transcendental or differential equations). The stochastic models, population balance equations, discrete event models and artificial neural network–based models are some examples. If such models are used as a sub-model, then the large model with interconnections among such models requires a methodology different than that used for the other two types listed here.

The large models consisting of more than one of these types may be called 'hybrid models' and need special methods to interconnect them. A large model may consist of only one type of the above models. In such cases, these can be combined in a much simpler manner.

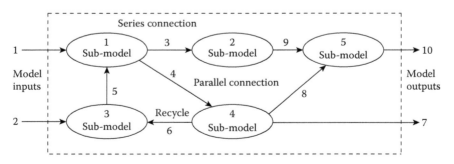

FIGURE 11.1
Interconnected sub-models representing a model for a large plant.

The simulation strategy depends upon the choice of available software. The computer languages can be classified as general-purpose languages and special-purpose languages. At the model development stage, it is required that the time should not be lost for writing codes for every task. Special-purpose languages provides necessary feature to reduce this effort so that the modeller may concentrate on the modelling of the process only. A special-purpose simulation language provides a large number of features required for programming a simulation model. It reduces the effort and time taken to write the computer programme. The features generally available with special-purpose programming are methods to solve simultaneous linear and non-linear equations, differential equation solvers, optimisation tools and generation of random variables following a specific probability distribution. These also have various features for data input and data presentation in various graphic formats, error-handling capabilities. As a consequence, the simulation languages require less code to write the programme. Due to the small size of the code, the changes at later stages can be made easily. The special-purpose languages also have debugging facilities. From this point of view, MATLAB® can be considered a special-purpose language. It can be used for the purpose of simulation, as shown in Chapter 10.

On the other hand, writing the computer programme using a general-purpose language is a time-consuming task. It has to be debugged. The general-purpose languages are available easily and involve less cost. A general-purpose programme may require less execution time if written efficiently.

11.1 Interconnecting Sub-Models

A sub-model may be looked on as a computational sequence which produces output if all the inputs are specified. The complete model will be a network of these sub-models. An example of such a network is shown in Figure 11.1. The direction of general-purpose arrow indicates the direction of the flow of information processing. The information from model input flows in the direction towards model output. Some of the sub-models are connected in series, and some sub-models are connected in a parallel manner. Sometimes, it will be difficult to identify whether a sub-model is connected in series or in parallel arrangement. Few interconnections can be considered a recycling of the information.

The direction of flow of the information depends upon the algorithm used to solve the problem. Therefore, the connectivity of the unit operations may be understood by the process engineer. However, the direction of the interconnected flow information sub-models may be different than the direction of the flow of the material in a process flowsheet. By changing the algorithm to solve the problem, the direction of the interconnection of the sub-model may change. When two sub-models are solved simultaneously, no direction can be assigned to the interconnection. It is better to disregard the interconnectivity in such cases. The information flow diagram for the large plant remains undefined if the methodology to solve the model equations is not decided because it affects the direction of the interconnection.

11.2 Simulation Study

If a model output is estimated for a given set of model inputs, then it can be considered a single simulation run. The simulation study involves several repetitions of such simulation runs and studying the changes in the model outputs. The variance-based sensitivity analysis discussed in Chapter 9 is also a kind of simulation study. The model inputs are varied with a certain purpose. The applications of simulation studies are discussed in Chapter 2. The model inputs have to be chosen in an appropriate way to achieve the aims of the simulation.

A large model has several equations and parameters. Some of the parameters are associated with the feed or input streams. Few parameters can be called design variables (i.e. the values of these parameters should be adjusted to meet design requirements, which is called 'constraints'). To solve the model equations, all the parameters should be specified. If model equations are solved, then a run is called a 'simulation'. Several simulation runs may be conducted by varying parameters, and the effect of the parameters may be studied. These runs may be conducted to study the sensitivity analysis or for the determination of design parameters. The problem of solving a large model or flowsheeting programme has been termed a 'simulation problem' (Shacham 1982).

Design variables are not specified in design problems. For each design variable, equality constraints are specified. The number of unspecified parameters is equal to the number of constraints. Several simulation runs for various values of design variables are conducted, and an acceptable solution is sought. The design procedure involves repeated simulations.

Shacham (1982) defined optimisation problems as those in which some of the model input and design variables are not specified. It involves a cost function which is minimised. Optimisation problems may consist of equality as well as inequality constraints for the unspecified variables. The number of the unspecified variables and number of constraints can be different.

Let us consider that the interconnections of the sub-models are known, but the direction of the interconnections is not known. The method for a single simulation run is known for each of the sub-models. Since the direction of the interconnections is not known, the method for a single simulation run on a large plant is also not known. The approaches used to carry out a simulation study for a large model made of several sub-models are as following.

The strategy for the simulation of a process depends upon the type of processes. The process can be either a batch process or a continuous process. Several processes are a combination of batch and continuous processes.

11.3 Flowsheeting and Continuous Processes

In continuous processes, the materials flow from one unit to another in a continuous manner. The flow rate remains independent of time. Hence, a steady-state simulation model can be used to describe the process behaviour. The dynamic models are required for the process' control purpose only.

Three types of solution methodologies were mentioned by Shacham (1982):

1. The recycle streams were solved using iterative procedures.
2. The equations were grouped according to the presence of recycle streams.
3. The set of linear equations was solved simultaneously.

The flowsheeting programmes are steady-state simulation software. A flowsheeting programme helps in carrying out all of the essential activities associated with simulation of a process plant. The main components of a flowsheeting programme are as follows. It is being assumed that all the sub-models are available and are already validated so that these can be used with confidence.

11.3.1 Data Input and Verification

In the first stage, all the model inputs are provided to the programme. It includes the details about all the unit operations, interconnections between them and the process fluids flowing from one unit to another. The required data can be categorised as follows:

1. *Material*: Any process requires several materials or chemical species. These are raw material, products formed due to chemical reactions, process utilities (e.g. heat transfer medium) etc. Various process streams can be specified as a mixture of these chemical species. The specifications are used for the estimation of thermodynamic properties.

   The thermodynamic properties should also be specified, but a commercial package has a large library to evaluate these properties.

TABLE 11.1

A Few Commonly Used Types of Equipment in Flowsheeting Programmes

Mass Transfer	Heat Transfer	Reactors	Miscellaneous
Flash drum, distillation column, absorber, adsorption and mixers	Heat exchangers, evaporators and condensers	Batch reactor, plug flow reactor, packed beds and fluidised beds	Pumps, compressors and 'T'

Thermodynamic properties of new compounds can be entered. The user can choose not to use the thermodynamic library to estimate a particular property. Instead, the properties can be entered. The thermodynamic data commonly used are enthalpies, equilibrium data and the like.

2. *Chemical reactions*: The next step is the specification of all the chemical reactions taking place in a process. The rate expression is expressed in terms of concentration, pressure and temperature. The flowsheeting software may allow specification of the rate expression as long expressions. In the case of catalytic reactions, the bulk density of the catalyst and bed porosity is also specified. It is used for the estimation of pressure drop in the reactor and volume change due to a change of pressure within the reactor. The heat of the reaction should also be specified at this stage. The latest software accepts expressions for heat of reaction also.

3. *Process topology*: Various sub-models or units and their interconnection define the flowsheet. The details of the unit operation include the type of the equipment, the reactions taking place in it (if any) and the type of the sub-model and the number of input and output streams. A list of a few common types of equipment used in flowsheeting programmes is given in Table 11.1. The units are so categorised that material and energy balance equations can be written easily.

From the point of material balance, the units can be of four types.

i. *Mixers*: Various streams are added to form a new stream. No reactions take place in a mixer. More than one streams enter the unit, and a single stream leaves the mixer. For n inlet streams and m chemical species, component material balance and overall material balance equations are written as (Figure 11.2a)

$$\sum_{j=1}^{n} z_{in,jk} F_{in,j} + z_{out,k} \left(-F_{out}\right) = 0; \text{ for } k = 1 \ldots m \qquad (11.1)$$

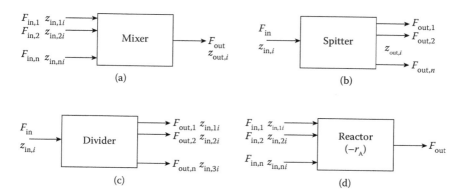

FIGURE 11.2
Types of material-processing equipment: (a) mixer, (b) splitter, (c) separator and (d) reactor.

$$\sum_{j=1}^{n} F_{in,j} - F_{out} = 0 \qquad (11.2)$$

The overall material balance is used to get the composition of the output stream. The consistency equations should also be satisfied:

$$\sum_{k=1}^{m} z_{in,jk} = 1, \text{ for } j = 1 \ldots n \qquad (11.3)$$

$$\sum_{k=1}^{m} z_{out,k} = 1 \qquad (11.4)$$

ii. *Branching or splitting*: To use only a part of a stream, a splitter has to be used. It is really not a unit process. For example, if a stream is branched in a pipeline by using a 'T' connection, then the 'T' is a splitter. A stream is divided into at least two streams. The compositions of the inlet and outlet streams are the same (Figure 11.2b). Hence, only the overall material balance equation is obtained.

$$F_{in} - \sum_{j=1}^{n} F_{out,j} = 0 \qquad (11.5)$$

$$z_{out,jk} - z_{in,jk} = 0, \text{ for } k = 1 \ldots m \text{ and } j = 1 \ldots n \qquad (11.6)$$

iii. *Separator*: The purpose of a separation unit is to separate a stream into at least two streams of different compositions. Distillation accepts at least one stream as input and at least two output streams, distillate and bottom. The separation is carried out by using heat. Extraction, absorption and the like require another stream to carry out separation.

The distillation may have more than one feed streams and may have several side streams also. Thus, a separator has more than one inputs and outputs. Also, the compositions of the input and output streams are different (Figure 11.2c). The material balance equations for n input and l output steams are written as

$$\sum_{j=1}^{n} z_{in,jk} F_{in,j} + \sum_{i=1}^{l} z_{out,ik} \left(-F_{out,k}\right) = 0, \text{ for } k = 1 \dots m \tag{11.7}$$

Equations 11.3, 11.4, 11.8 and 11.9 are the consistency equations.

$$\sum_{k=1}^{m} z_{in,jk} = 1, \text{ for } j = 1 \dots n \tag{11.8}$$

$$\sum_{k=1}^{m} z_{out,ik} = 0, \text{ for } i = 1 \dots l \tag{11.9}$$

Consideration of the equilibrium relationship completes the set of equations for separators. The equilibrium data may be represented in several forms depending upon the type of the separation process.

iv. *Reactor or bioreactor*: This is the most important unit operation. The reactors may have several inputs and outputs. Generally, there is only one output to the reactor, such as in the case of a homogeneous reactor. The compositions of inlet and output streams are different. New chemical species are produced, and the reactants are consumed (Figure 11.2d). The material balance equations consider the rate of reaction:

$$\sum_{j=1}^{n} z_{in,jk} F_{in,jk} + z_{out,k} \left(-F_{out}\right) + (r_k)V = 0, \text{ for } k = 1 \dots m \tag{11.10}$$

The consistency equation for one output is

$$\sum_{k=1}^{m} z_{in,jk} = 1, \text{ for } j = 1 \dots n \tag{11.11}$$

$$\sum_{k=1}^{m} z_{out,k} = 0 \tag{11.12}$$

In the case of more than one output, Equation 11.9 will replace Equation 11.12. The reaction rate term in Equation 11.10 is for a well-mixed reactor. The actual term will depend upon the type of the reactor model. In the case of multiphase reactors, the output streams may be more than one (e.g. the air-lift reactor or trickle bed reactor has two streams, each corresponding to a phase). Equation 11.10 is modified for the multi-outlet streams. The mass transfer between the two phases is described to get the relationship between compositions of different phases. These considerations are required to get an exact material balance equation.

The reactions taking place in the reactor and the operating temperature and pressure should also be specified. A particular flowsheeting programme may also provide a list of options to choose the type of the model (e.g. a lumped-parameter or distributed-parameter model). Some of the model inputs still remain unspecified because these are related to the process streams entering the unit operation. Another property which every stream has is the direction of flow of the material (i.e. to which unit operation it enters or leaves).

There are several unit operations in which the flow rate and composition of the streams do not change. For example, in a heat exchanger, the flow rates of cold and hot fluid inlets are the same as those of cold and hot outlet streams, respectively. The energy balance equations describe the amount of heat or energy added or removed in these units. The heat transferred in a multi-pass shell-and-tube heat exchanger may be estimated as follows:

$$Q = \dot{m}C_p \left(T_{in} - T_{out}\right) = UA_H F(\Delta T)_{LN} \tag{11.13}$$

Heat transfer may also take place in process units in which a material balance is required, except in the case of splitters, in which the temperatures of the inlet and outlet streams are the same.

In the mixer, the inlet streams may be at different temperatures, and hence the energy balance will provide the outlet temperature. The temperature may also change if the heat of mixing is considerable. The energy balance in the mixer gives the following:

$$\sum_{j=1}^{n}\left(\sum_{k=1}^{m} h_{in,jk} F_{in,j}\right) + \sum_{k=1}^{m} h_{out,k}\left(-F_{out}\right) + \left(-\Delta H_{mix}\right)V = 0 \tag{11.14}$$

If the number of reactions is q, then the energy balance in the reactor takes into consideration the heat of the reaction for all q reactions.

$$\sum_{k=1}^{m}\sum_{j=1}^{n} h_{in,jk} F_{in,jk} + \sum_{k=1}^{m} h_{out,k}\left(-F_{out}\right) + \sum_{p=1}^{q}\left(-\Delta H_{rxn}\right)_p V = 0 \tag{11.15}$$

In the separators, the heat of mixing, heat of phase change etc. are taken into account. The energy balance equation can be written as

$$\sum_{k=1}^{m} h_{in,jk} F_{in,j} + \sum_{k=1}^{m} h_{out,k}(-F_{out}) + (-\Delta H_{vap})V + (-\Delta H_{mix})V = 0 \qquad (11.16)$$

4. *Stream specification*: The interconnections in a flowsheet represent a process fluid and are called 'streams', which are specified by temperature, pressure, composition and the flow rate. These are the quantities required to make material and energy balance. If the stream is solid, as in the case of feed to a size reduction unit, then the particle size is also a property of the stream. The property of the stream does not change during transfer from one unit to another. A few streams may be unspecified as these will be known after the simulation.

   In other words, all streams need not be specified. However, to decide the extent of specification, the flowsheeting software has a validation step which should be followed before solving the model.

5. *Convergence criteria*: The tolerances for various parameters are specified.

After entering, the data should be checked for any mistakes. After verification of the data, the programme enters the second stage of the simulation.

If a high-level language is used to write own code, then the initial guess value for a few unspecified parameters may have to be entered. It is required because the solution methodology uses iterative procedures. The commercial software usually handles it in a different way. The object-oriented programming always uses a default value.

Since the simulation studies for a large model cannot be conducted in one session, the intermediate values are stored in temporary files, which are accessed in the next session. For example, after entering the data, the information may be stored or saved.

The graphic facilities in the present software have facilitated the data input stage. A flowsheet can be entered as interconnected graphics. Double clicking on any of the unit operations opens a dialogue box to edit the data related to that unit operation. Double clicking on any of the stream opens a dialogue box to edit the stream properties.

The data input and the verification are pre-processing, and they comprise a preparatory stage before the actual simulation starts.

11.3.2 Thermodynamic Properties and Unit Operations Libraries

There are several thermodynamic properties which are required frequently. For example, the properties of air, water and steam are required in every process plant. Properties of various solvents, heat transfer mediums, and material properties are used repeatedly. Several properties are estimated only once in each run or once in all of the simulation runs for a process. It is a good practice to keep all the data in data files. Commercial packages have thermodynamic libraries for a large number of compounds. In the pre-processing stages, all of the required properties for all relevant compounds are either obtained using the database or estimated using the thermodynamical package. These data and the data entered by the user are put in an intermediate file. All of the chemical species appearing in a flowsheet and the thermodynamic procedures to estimate the properties during the solution of the flowsheet are stored in a data file which is accessed by the executive programme. The modeller has the choice to choose an appropriate thermodynamic method from the various options.

A process flowsheet contains several units. These units are the equipment used in process industries (e.g. reactors, distillation columns, gas absorbers, membrane separators and heat exchangers). All of the process plants use these units. A commercial package has a simulation model for these units. These libraries may have the following features:

1. Different types of equipment may be used for the same purpose. For example, the distillation may be carried out in a packed tower or a plate column. The plate column may have sieve or valve trays. Though both are distillation columns, all such variations are considered different units.

2. Models of different complexities may be available for similar units. The 1D model and 2D model for the packed tower are considered different units. This is due to different solution methodologies used in solving sub-models.

3. The same type of units may be used more than one time. These are different units in a process flowsheet, but they are of the same type. Same-model equations are used for each of the similar units.

The libraries have all the data related to process calculations associated for various unit operations. For example, data for packing factor and correlation for the estimation of flooding point used to perform calculations for packed beds are available in libraries.

All reactions taking place in a unit are defined after specifying the chemical species (i.e. material). The reactions are configured by entering the chemical species, the stoichiometric coefficients, the rate expression and the heat of reaction.

Technology development demands the development of new equipment. The library for unit operation equipment does not have any such equipment. The commercial flowsheeting programme generally allows users to write the code for sub-models in a popular computer language. Sometimes, they also have their own cement and semantics.

11.3.3 Calculation Phase

The next phase of the simulation is the calculation phase (i.e. the start of the simulation run). The following approaches to solve the flowsheeting programme are in use.

11.3.3.1 Sequential Approach

One of the earliest and most natural ways of handling flowsheeting problems is to simulate one sub-model at a time and then use its output as input to another sub-model connected to it. This approach is called the 'sequential approach'. First, one of all the sub-models is solved. Here, 'solution' means for a given model input, a model output is generated. This model output is a quantity which can be used as input to the other sub-model. For example, a distributed-parameter model generating flow field in the process vessel may not be considered as input to another model. The effect of the field on the outgoing stream can, however, be used as input to the next unit. It limits the type of sub-models which can be interconnected easily. For flowsheeting programme material and energy balances, equations are used. These equations use terms which may be obtained using simple or rigorous models.

For this first sub-model, all the inputs should be known. The input to the large model depends upon the kind of the problem. For example, if the simulation is performed for the design of a process plant and the capacity of the plant is known, the sub-model representing the unit operation (last unit in the process flowsheet) providing the final product becomes the first sub-model to be solved. If a water treatment plant is to be designed, the amount of water treated, and hence the first unit operation in the process flowsheet, is known. Thus, it is important to identify the order in which the sub-models are to be used.

The advantages of the sequential approach are as follows:

1. The statements in the computer programme are clearly written.
2. The input to the large model or flowsheet is clearly known. Thus, the problem is well defined.
3. The major advantage of the sequential approach is to use a sub-model of any complexity since only one sub-model has to be used at one time. The sub-model is already developed and validated to provide outputs for the given input.
4. The number of equations is limited; hence, the solution to each sub-model can be obtained easily.
5. During the design stage, it may be required to add one or more units to the process flowsheet. The sequential approach does not pose any problem when, in a flowsheet, a unit is included or deleted.

Due to these characteristics, the sequential approach is suitable for rating problems as well defined parameters.

One of the problems in using the sequential approach is the presence of a recycle stream. For a sub-model having a recycle stream, the model input is not completely known as it also involves model output from another unit. Therefore, the output cannot be determined. The problem can be solved by an iterative procedure, but it increases the computational effort.

For a large process plant, there may be several recycle streams. Here, the recycle stream is referred to a stream for which input to the sub-model is unknown output from another sub-model. In cases of unknown output from a sub-unit, the sub-model itself can be incorporated in the form of model equations. The order of calculation will determine the number of recycle streams. It is possible that by changing the order of calculations, the number of recycle streams in the large model will change.

11.3.3.2 Tearing of Streams

The recycle streams are handled by disassociating the value of the recycle input with the recycle output (Figure 11.3). It is called a tearing of a stream. The unknown recycle input can be assigned a value, and the recycle output (sub-model output) is estimated. The difference or residual is minimised using any optimisation method (e.g. the Newton–Raphson method). The non-linear equations are usually solved by optimisation techniques. These techniques are a class of iterative procedures in which the direction of the search for the solution may be decided by the gradient or any other suitable criteria.

An efficient scheme will reduce the difference between the assumed and estimated values of the recycle stream. Once convergence is achieved, the calculation must be stopped. The criteria for the convergence can be based on the absolute error or relative error, which should be less than the tolerance decided by the modeller.

11.3.3.3 Order of Calculation

More than one sequence in which the sub-models can be evaluated is possible. The sub-model can be entered in the form of a matrix. For example,

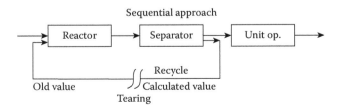

FIGURE 11.3
Sequential approach to the solution methodology and tearing of the recycle stream.

TABLE 11.2

Representation of a Sub-Model Network in Matrix Form

Model Name	1	2	3	4	5	6	7	8	9	10
Sub-model 1	1	0	-1	-1	1	0	0	0	0	0
Sub-model 2	0	0	1	0	0	0	0	0	-1	0
Sub-model 3	0	1	0	0	-1	1	0	0	0	0
Sub-model 4	0	0	0	1	0	-1	-1	-1	0	0
Sub-model 5	0	0	0	0	0	0	0	1	1	-1

Note: Numbers in the first row denotes the stream number in Figure 11.1.

the network shown in Figure 11.1 is presented in matrix form in Table 11.2. The rows from the top are the sequence or order of calculation. The first row is the name or number of the connecting stream. The entries in a row denote the connectivity of a stream to the unit. The entering stream is denoted by '1', and the leaving streams are denoted by '-1'. If a stream is not connected to the sub-model (unit) mentioned in the first column of the same row, then it either is left as blank or may be denoted by '0'. The sequence of calculation in both diagrams is from top to bottom. The following can be observed:

1. If a stream appears only once, then it is raw material if its value is '1' and product if its value is '-1'.
2. All other streams appear twice in the column for that stream. These are intermediate streams.
3. The raw materials are known. In terms of information, it can be said that the known input streams are the raw materials.
4. The product streams or output are to be calculated.
5. For an intermediate stream, looking from the top, if '-1' appears first and then comes '1', then the stream appears as the product first and then it is used as input in the units appearing later in the sequence of calculation. Since it is not a raw material, it is estimated from the unit, and the value is used as input to the unit appearing later. It is a conventional intermediate stream and not a recycle stream.
6. For an intermediate stream, if '1' appears first and then comes '-1', then the stream is being used before it is estimated (i.e. it is a recycle stream).
7. From Table 11.2, it can be seen that due to a different sequence of calculation, the recycle streams are different. The number of recycle streams is also different.

Thus, one can see that the number of recycle streams depends upon the sequence of calculation, which should be decided to reduce the number of recycle streams to the minimum possible value. It can be easily achieved by interchanging the appropriate rows in Table 11.2.

The sequential approach is not suitable for design problems since the solution strategy is based on the relationship between model input and model output. If a few model parameters for a sub-model are not specified, then it is not possible to obtain model output for that sub-model or unit operation. The calculation cannot be carried out further. However, there is a way to handle this situation. Design is considered a repeated simulation. Therefore, several simulation runs with different unspecified design parameters may be conducted. The values of these parameters may be chosen so that they satisfy the equality constraints.

11.3.3.4 Simultaneous or Equation-Solving Approach

In a large process plant, it is very common to have a large number of recycle streams. Under these circumstances, the sequential approach not only is time consuming but is also associated with the problem of convergence. The presence of a large number of recycle streams, which usually is observed in the case of a model for large process plants, makes the sequential approach of simulation inefficient.

The problem of recycle streams can be avoided if all the model equations are solved simultaneously. This approach is also called the 'equation-solving' approach. Using this approach, any equation solver software can be used for the simulation purpose. The optimisation methods can also be used as equation solvers if the residuals of the equations are defined as the objective function. However, with a large set of equations, the solution may not converge; hence, another solution methodology has to be attempted. The simultaneous approach handles the problem in a better way than the sequential approach, but it was difficult to say that the convergence to a solution improved (Umeda and Nishio 1972).

The sub-models to the problems related to chemical, biochemical or environmental processes involve differential equations also. These equations can be converted to finite difference equations and can be solved easily.

Earlier flowsheeting programmes assumed that the model inputs and model outputs are related by linear equations (Shacham 1982). All unit operations were divided into four or five different types of units; these were reactor, separator, mixer and branching. The approach was popular as the solution of linear equations can be obtained with ease. This approach can now be considered a short-cut method. The set of linear equations was solved simultaneously.

If the sub-models in the flowsheet are described by non-linear equations with no differential equations, then the equations can be solved either by linearising the equations or by using equation solvers to solve non-linear equations.

The equation-solving approach can handle design and optimisation problems because the solution methodology to solve a set of non-linear equations is based on methods which are the same as those used for optimisation problems.

Adding a unit or sub-model in the existing model will change the entire set of equations; therefore, it is not easy to add or delete even a single unit. The simulation run for the entire model has to be conducted again.

Due to the problem of having a large number of non-linear equations to represent a model or flowsheet, the equations may be grouped in such a manner that the equations in a group are to be solved simultaneously. Distributing all equations in such groups is called partitioning. It helps in reducing the problem size.

11.3.4 Modular Approach

The existing processes are modified to conserve energy, fuel, water and raw materials. These are also modified to add new technologies that improve plant safety and pollution control measures. These activities require that new units and new streams, including recycle streams, are added. It means that the structure of the model changes with the inclusion of more sub-models and interconnections.

None of the sequential and simultaneous approaches to solve the equations representing a large model is perfect. These approaches are also inefficient to handle the changes in the existing processes. A sequential approach is suitable for a large model with fewer recycle streams, and the simultaneous approach is suitable for small plants with many recycle streams. But the models for a large plant normally have a large number of recycle streams. The modular approach combines the benefits of both approaches. A large model is solved in two stages. A group of the sub-models is combined into one unit called a 'module'. A large model or plant may have several such modules. Each module is solved first, and the relationship between the input and output for the entire module is established. The large model now is represented by interconnected modules. All recycle streams are in the modules, and rarely is any recycle stream outside the module. The number of modules is much less than the total number of sub-models in the large model. The entire simulation is divided into two steps: simulation of the modules and simulation of the large plants. This approach is called the 'modular approach'. Due to conducting simulations in two stages, it is also called a 'two-tier approach'.

11.3.4.1 Choosing Modules

While choosing the sub-models as a module, an attempt should be made to include the recycle streams within the module. This reduces the number of recycle streams outside the modules (i.e. in the structure of the large plant). Fortunately, chemical and biochemical processing plants have a natural combination of units which include recycle streams. Some of these are as follows:

1. *Reactor separator system*: In most reactors, the conversion never reaches 100%. The product is separated, and the unreacted reactants

are recycled to the reactor. In process plants, the reactor is usually followed by a separator. The reactor–separator system is a natural combination as a module.

2. *Distillation columns with reboiler and condenser*: A single distillation column has a reboiler and a condenser. The reboiler recycles the vapour after heating the liquid stream leaving the column. The condenser is used to condense the vapours leaving the column, separate the liquid from it and recycle a part of the liquid to the column.

3. *Series of separators*: Many process plants have a series of separators. They may be of the same type. For example, azeotropic distillation is carried out in more than one column. A series of distillation columns is used when these are designed with the aim of heat integration (e.g. Petlyuk columns).

4. *Multistage reactors with intermediate heat exchangers*: The reactor used in sulphuric acid plants is used in combination heat exchangers in between two stages of the reactor to reduce the temperature of the reactor.

5. *Grinding and screening operations*: The solid particles are grinded to produce powders followed by screening of the product. The oversize is recycled to the grinder.

6. *Separation with recovery of solvent*: In separation processes such as gas absorption, leaching and liquid–liquid extraction, the solvent is recovered in a separator and is recycled back.

Several such examples which can be considered modules are obtained by looking at the flowsheet for various processes. All such examples include at least one recycle stream. Choosing such modules on the basis of the material flow seems to be a natural choice. It is possible that such natural modules may give a much simpler input–output relationship for the module than that for the individual unit operation. Such a relationship may even be linear. However, one should not expect it to always happen.

There are two ways in which the large models can be simulated using the modular approach. A model is chosen so that it includes a recycle. In such cases, the equation-solving approach is better than the sequential approach. Therefore, the modules are solved using the equation-solving approach. The large models as interconnected modules have only a few or no recycle streams. Therefore, in the second stage, any of sequential or simultaneous approaches may be used.

11.3.4.2 Sequential Modular Approach

The model equations for a module are solved simultaneously. If the large model consisting of all these modules is solved sequentially, it is called a sequential modular approach. Such a methodology requires a non-linear equation solver to solve the modules. After the solution, new equations can be obtained to represent a module. To solve the network of modules, a methodology based on a sequential approach is required. Thus, in this approach, two different types of solvers are required for simulation purposes. The executive to solve the equations becomes large. If some more recycles are present, then the approach has the same problems as those associated with the sequential approach.

11.3.4.3 Simultaneous Modular Approach

After simulating the modules by solving the equations simultaneously, the equations to represent the network of modules are solved simultaneously. The recycle streams which are included in the modules are handled without any problem. At the same time, the software has only one type of solver that is a non-linear equation solver.

Earlier commercial packages developed for flowsheeting used the sequential approach. The later packages used simultaneous and modular approaches.

11.3.5 Output Phase

After the simulation is over, the results of the solutions are available. The output contains all the properties of the unspecified streams and size of the various equipment (units). The output contains the operating conditions, the parameters at these conditions. The design data for all the equipment are specified on the required format. The output includes the amount of the raw material and the utility requirement (e.g. the energy requirement for each piece of equipment, the amount of heat transfer medium and the heat load in a heat exchanger).

For example, the output data for a plate column report the number of plates, location of the feed plate, reflux ratio, composition and flow rate of the distillate and the bottom product, heat load in the reboiler and condenser, heat transfer area in these heat exchangers, column diameter, pressure drop etc.

Although the properties of not only the product streams but also all intermediate streams are known, not all of the information is required for the design purpose. Some of the values not required are used to trace the progress of the process.

11.3.6 Simulation Study Using a Flowsheeting Programme

To carry out a simulation study for a large plant using a flowsheeting programme, the first requirement is a model. The model or flowsheet is entered as input. The calculation approach used in the flowsheeting programme is immaterial. The features available are mentioned in the help file associated with the software. The streams and the unit operations are defined. A flowsheet is only a model. It does not optimise the parameters. The flowsheeting programme solves the model equations to determine mainly the flow rates, composition of streams, heat duties and quantities associated with steady-state material and energy balances. For example, it can solve the model equations only if the length of a packed bed reactor is specified or conversion in a continuous stirred-tank reactor is specified. To study the effect of change in the conversion or reactor length, several simulation runs have to be carried out, although it requires changing the values and issuing a command in every run.

The simulation study means solving the model for several values of a parameter. For economic design, the cost of the equipment has to be estimated by varying a parameter and studying its effect on the cost. For example, a plug flow reactor is a 1D model and can be used as a packed-bed catalytic reactor.

11.4 Short-Cut Methods and Rigorous Methods

Short-cut methods use simple models to predict the process behaviour. It is useful at the preliminary stage of analysis and process design. While the detailed design of the process is done by using rigorous methods, a preliminary design using short-cut methods can be used to choose a feasible process among various alternate processes available. Less computational effort, and quick output, while using short-cut methods reduces the design cost.

Short-cut methods have limited scope; therefore, appropriate methods should be used to fulfil the scope of the simulation. However, the use of short-cut methods does not allow accepting arbitrary outputs. The uncertainty in short-cut methods is greater, and it should be evaluated.

The short-cut methods either can be developed by connecting simple submodels or may be a trimmed version of rigorous models. A few of the ways to obtain short-cut methods are given below.

1. Lumped-parameter models may be used in short-cut methods, and the distributed-parameter model may be used in rigorous models.

2. The number of spatial coordinates can be reduced. For example, a 1D model for the packed bed may be used in short-cut methods, and a 2D model can be used in rigorous models.

3. A few unit operations may be eliminated from a process flowsheet to simplify the problem. For example, a simple model may conduct only material balance. The heat exchangers and pumps are eliminated from the flowsheet as these are not required for making material balance.

4. Several simple methods are already in practice. The McCabe–Thiele method for determining number of plates is a short-cut method.

Plate-to-plate calculation involving material and energy balances is a rigorous method.

5. Some of the model inputs may be fixed as constant, thus reducing the number of equations in the sub-model. The model equations are easily solved. For example, the conversion in a reactor may be fixed after getting information from an existing plant.

6. Several empirical correlations and thumb rules are available in the literature (Branan 2011). These may be used as sub-models.

7. If a recycle stream is fixed, the computational loop disappears, resulting in computational effort and cost.

8. The sensitivity analysis of the sub-model may reveal sensitivity of a parameter for the sub-model, but it is possible that the model for a large system may not be sensitive to that parameter. If the parameter remains important, the relation may be approximated as a linear model.

In Chapters 4 through 8, it has been discussed that as the complexity of the model increased, the solution of the model equations changed from analytical solutions to numerical solutions. The simplification of the rigorous models thus makes it possible to obtain a solution quickly as some of the sub-models may have analytical solutions.

11.5 Dynamic Simulation

To study the transient behaviour of a process, a dynamic model for large plants is required. The modes are interconnected dynamic sub-models. These models are used for the following purposes, as discussed in Chapter 2.

1. To study the transient behaviour of the process during plant start-up and shut-down. The effect of failure of a unit such as a pump on the transient behaviour of the process can also be studied.

2. To study the dynamic behaviour due to variation in the feed.

3. To analyse the problems related to process control (i.e. the controller design).

4. To use it for integrated plant design (i.e. the design of the process control system at the plant design stage).

5. To model the batch process.

6. To use as the plant simulator and train the industry personnel and plant operators.

The model equations which connect various sub-models are obtained by writing material and energy balances. The steady-state models do not have temporal derivatives, and hence these equations are algebraic equations. The dynamic simulation includes differential equations with temporal derivatives. The sub-models used in the steady-state simulation may have differential equations, but they have only spatial derivative. These model equations for a sub-model are not global (i.e. their solution relates the input and output relation for the sub-model). The dynamic simulation requires the sub-models to include temporal derivative also.

Similar to steady-state simulation, the dynamic simulation also has the following phases.

11.5.1 Data Input and Verification

Due to the presence of the differential equation, the specifications of the streams and unit operations are different. The initial conditions for all the streams and unit operations are specified (i.e. all the streams, raw material, product and intermediate streams are specified). In steady-state simulation, several streams were not specified. The purpose of the steady-state simulation was to determine these streams at steady state.

The specifications of the process units include the process topology and operating conditions as in the case of steady-state simulation. However, they also include the information about its dynamic nature (e.g. the start-up time and shut-down time).

Several quantities are irrelevant in steady-state simulation. For example, the delay in the pipeline, residence time distribution and holdup in process units do not appear in the material and energy balances. While writing unsteady-state material and energy balances, the terms associated with temporal derivatives appear in dynamic simulation.

11.5.2 Calculation Phase

The executive used in the calculation phase has different requirements. The dynamic simulation requires the solution of differential equations. The solvers generally use the Runga–Kutta method. Since the initial conditions for all streams are known, the recycle streams do not pose any problems. They evolve during the solution of the model equations. The recycle streams do not affect the convergence as they do not have any immediate effect on the units that follow. The gradual variations of these recycle streams are calculated after every time step.

The order of calculation in dynamic simulations is important. Solution of the differential equations is based on finding all the values at the next time step. Therefore, one unit is taken at a time. The order of calculation is therefore is the order of unit operations specified by the topology of the process. There is thus no simultaneous approach.

11.5.3 Data Output

The data are in the form of values of all the parameters at each time step. These may be plotted as per the desired format.

11.6 Batch Processes

Many process plants have unit operations which operate in batch mode. A batch process unit has the following characteristics.

1. The processes in a batch are dynamic in nature (i.e. the process conditions change with time). Therefore, the batch process is described by a dynamic model.

2. The batch operation starts at a particular time and ends after a definite time.

3. During the process, operating conditions may change (e.g. the stirrer speed may change due to a change in viscosity of the contents of the reactor). The change in the temperature or pressure may be required to vary during the operation. These are called events which take place at a definite time and suddenly change the dynamic behaviour of the batch process.

4. One or more reactant streams may be added at a particular time or may be added continuously. The process is called a 'semi-batch process'. It has the same behaviour as the batch process.

11.6.1 Simulation Methodology

The problems for batch processes can be of two types. A single simulation run requires all specification of the resources and the recipe. The design problem is called a 'scheduling problem' and is achieved by repeated simulation until the optimal solution is obtained.

The batch processes are dynamic in nature. It requires a dynamic model to predict the state of the process at any time. For example, a reactor model will include the reaction kinetics, the transfer processes within the reactor, and the operating pressure and temperature. The input to the reactor is the reactants, and output is the conversion for a given volume of the reactor and processing time in a reactor of known volume. The general model for a batch reactor is discussed in Section 4.7.2. However, a detailed model can also be used.

The batch processes are discrete in nature. The simulation methodology for the batch process will be different than that used for the continuous process. The time needed for the addition of reactants and the time consumed while filling and emptying the vessel should be considered while scheduling.

The entire process plant may have several such processes. The batch processes are different than continuous processes in the sense that the same process may be carried out in a different unit (equipment) if that unit is free. It also means that a unit can be used to carry out several processes one by one. The batch processes are multipurpose or multiproduct processes. The simulation and design of batch processes therefore involve scheduling all production activities to maximise the production and hence the profits.

Each operation in a batch process is a series of various operations. The processing is complete after all these operations are completed. The sequence of these operations is called a 'recipe'. Every operation in a recipe is carried out for a fixed amount of time, but the time for each operation in a recipe may be different.

The dynamic models for the processes are known. Various types of units such as reactors and separators are available, and the models for each unit along with their capacity limits are known. After completion of the reaction, the role of the reactor is over. The contents can be taken out immediately (zero wait), or they may be stored for some time using the reactor as an intermediate storage. The available storage capacity for each of the materials is also required. The batch processes can be used in two ways:

1. *Sequential process*: In these processes, the sequences of the unit operations are predetermined. Once a unit is used for one order only, then the next order is processed.

2. *Multipurpose and multitask process*: The sequence of the unit operation is not fixed; instead, a unit can be used for many different tasks. As a result, more than one order can be processed at a time. This requires a scheduling programme for the batch processes. The design problem is now to find an optimal proper production schedule.

11.6.2 Scheduling of Batch Processes

The scheduling problem is solved using one of the following two approaches. The choice of the approach depends upon the process dynamics.

11.6.2.1 Standard Recipe Approach

A recipe is the information about the schedule (e.g. processing time and control strategy) for different batch sizes. It is obtained by optimisation or by employing an empirical approach. The scheduling problem is formulated on the basis of a standard recipe. In this approach, there is no degree of freedom; hence, the solution may be suboptimal only. The scheduling models using this approach use either a mixed-integer non-linear programming (MINLP) or a mixed-integer linear programming (MILP) to obtain the solution.

11.6.2.2 Overall Optimisation Approach

To get the global solution instead of a suboptimal solution, the dynamic models of the processes are directly included in the scheduling problem. The recipes are not fixed but are flexible. This approach is called an 'overall optimisation approach'. Due to more degrees of freedom in the model, this approach yields better solutions. The scheduling model is of the mixed-integer dynamic optimisation type, which can be converted to MINLP problems after discretisation. However, the problem becomes very large in size (Mishra et al. 2005). It requires a large amount of computational effort and time. It becomes difficult to apply this method in cases of large plant and unit operations involving hybrid process dynamics.

11.6.3 Representation of Batch Processes

Due to the discrete nature of the batch processes, it becomes necessary to represent the batch processes in a form suitable for solution. Simulation strategies for simulation of the batch processes can be conducted using any of the following representations of the process.

11.6.3.1 State–Task Network

Kondili et al. (1993) used a graphical depiction of the processes in terms of 'states' and 'tasks'. All of the materials (e.g. raw materials, intermediate products and process utilities) are termed 'states'. These correspond to 'streams' in the flowsheeting programme. The processing steps (e.g. reaction, separation and mixing) are called 'tasks'. In comparison to flowsheeting, where a 'unit' is used, the concept of the task is due to the multipurpose and multitask nature of the batch process. The entire batch process plant is now represented as a combination of 'states' and 'tasks', as shown in Figure 11.4a.

11.6.3.2 Resource–Task Network

This representation is based on interconnected nodes of two types: resources and tasks. The 'resource' includes all entities involved in the process steps, such as materials and processing equipment. It can account for different states of the equipment and locations of material. The 'task' is any operation that transforms a resource into another resource (Figure 11.4b). The graphical representation is analogous to a flowsheet for a continuous process. In Figure 11.4, the task is shown by a rectangular block. Within this block, the name of the task and the duration of the processing are written. The material is attached to this task and is shown by a one-sided arrow. The unit operation as a resource is shown by a circle connected to the task by a double arrow.

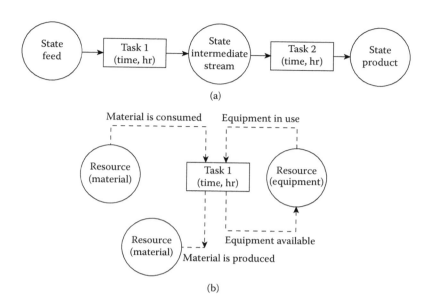

FIGURE 11.4
Representation of batch processes: (a) state-task representation and (b) resource-task representation.

11.6.4 Discrete Time Formulation

This is one of the important issues in deciding the solution strategy. One of the discrete formulations is as follows:

Continuous-time formulation: The time horizon is the time which is required to process all the orders. The time is considered continuous (i.e. the dynamic behaviour of a process unit can be accounted considering time as a continuous variable). The differential equations can be solved using numerical schemes such as the Runga–Kutta method. To account for the starting time and finishing time of various events, the time is divided into intervals, called 'events'.

The time intervals may be equal or unequal. The location of the events on the time horizon for all equipment may be the same. It is called a 'single time-grid formulation'. If the locations of the events on the time horizon for each piece of equipment are different, it is called a 'multiple time-grid formulation'. The discretisation of the time affects the efficiency of the solution strategy. Detailed comparisons are available in the literature (Alle and Pinto 2004; Castro et al. 2004a,b, 2005, 2006; Castro and Novais 2007; Janak et al. 2004, 2006a,b; Mishra et al. 2005; Shaik et al. 2006; Stefansson et al. 2006; Xue and Yuan 2008).

One type of commercial software which can be used for scheduling batch processes is SuperPro Designer. It has a mathematical model to perform material and energy balance calculations for every unit operation. The equipment-sizing calculations are performed using material balance equations. An appropriate equipment size is chosen so that it is just sufficiently large. These types of considerations are written in terms of constraints. Batch-scheduling software has three components. An objective function such as cost and profit is to be minimised or maximised. The constraints for processing are to be defined. Sub-routines for unit operations to estimate size or processing time are available. The type of the constraints reported in the literature can be of the following types.

11.6.5 Types of Constraints

The scheduling problem is a challenging task. The constraints produce feasible values and screen out the possibility of unrealistic numerical values. The types of constraints depend upon the discrete time formulation and the notations to represent the batch processes. Therefore, without going into the detail of the mathematical symbols, some of the commonly used constraints are given below.

1. *Excess resource balance constraint*: An excess resource material at any event may increase or decrease due to production or consumption, respectively. This amount can be estimated by making material balance at the event points on the time grid. Since the excess resource material is to be stored but the storage capacity of the plant is limited, an upper limit of the excess resource should not be exceeded.

2. *Time constraint*: The time of all tasks can be added to give the total time of a cycle of operation because only one task can be carried out in a unit at a time. The total time should always be shorter than the time horizon.

3. *Capacity constraint*: The amount of material processed by the task should be less than the size of the equipment. Processing of the material should be avoided if a task is not executed.

4. *Demand constraint*: No process plant can handle every order. Similarly, a small order will not be economical to accept. Therefore, the demand must have a lower as well as upper limit (Castro et al. 2005).

The objective function, the constraints and the equipment models complete the specification of the problem. It is a constrained optimisation problem. The mathematical techniques of optimisation are beyond the scope of this book. One may look at the examples in the demos given by GAMS and SuperPro Design.

The size of large-size scheduling problems can be reduced by using scheduling rules. Some of these rules are assigning the order on the unit with its (1) earliest start time, (2) shortest change over time, (3) shortest process time, (4) shortest sum of possible start time and process time, (5) shortest sum of change over time and process time and (6) earliest completion time (He and Hui 2007). These rules are heuristics, and there are two approaches to use them. In the automatic rule selection method, there is a database of large rules, and the rules are selected from them automatically as already programmed. A second approach termed 'rule sequence evolution methods' uses candidate heuristic rules to combine and form new rules on the basis of impact factors. He and Hui (2007) used a genetic algorithm to achieve it.

The examples of flowsheeting programmes using dedicated software are available on the web and in manuals of the software. Therefore, these are not discussed.

11.7 Summary

The large plants consist of several pieces of equipment. This equipment is connected to form a flowsheet. The processes in a large plant are of two types: (1) continuous and (2) batch. The batch process plants may have a few continuous units also. The model for continuous processes can be considered a large model consisting of interconnected sub-models representing various units in a flowsheet. The flowsheeting programmes are steady-state simulation programmes. They have a thermodynamic library, a unit operation library and an executive programme to accept input data, communicate with the databases and provide output to the user. Such flowsheeting programmes based on sequential, equation-oriented and modular approaches are available. The object-oriented approach has made the development of such programmes easier.

The steady-state simulation includes algebraic equations and a solver to solve these. The dynamic simulation includes differential equation and an appropriate solver.

The batch processes combine the unit models in a different manner. Due to problems of scheduling the processes, it is a constrained optimisation problem minimising the cost or increasing profits and production. The batch processes are expressed as state-task networks or resource-task networks.

References

Alle, A., and Pinto, J.M. 2004. Global optimization for the cyclic scheduling and operation of multistage continuous plants. *Ind. Eng. Chem. Res.* 43: 1485–1498.

Branan, C.R. 2011. *Rules of Thumb for Chemical Engineers*, 4th ed. Amsterdam: Gulf Profession.

Castro, P.M., Barbosa-Povoa, A., Matos, H.A., and Novais, A.Q. 2004a. Simple continuous-time formulation for short-term scheduling of batch and continuous processes. *Ind. Eng. Chem. Res.* 43: 105–118.

Castro, P.M., Barbosa-Povoa, A., and Novais, A.Q. 2004b. A divide and conquer strategy for the scheduling of process plants subject to changeovers using continuous-time formulations. *Ind. Eng. Chem. Res.* 43: 7939–7950.

Castro, P.M., Barbosa-Povoa, A., and Novais, A.Q. 2005. Simultaneous design and scheduling of multipurpose plants using resource task network based continuous-time formulations. *Ind. Eng. Chem. Res.* 44: 343–357.

Castro, P.M., Grossmann, I.E., and Novais, A.Q. 2006. Two new continuous-time models for the scheduling of multistage batch plants with sequence dependent changeovers. *Ind. Eng. Chem. Res.* 45: 6210–6226.

Castro, P.M., and Novais, A.Q. 2007. Optimal periodic scheduling of multistage continuous plants with single and multiple time grid formulations. *Ind. Eng. Chem. Res.* 46: 3669–3683.

He, Y., and Hui, C.W. 2007. Automatic rule combination approach for single-stage process scheduling problems. *AIChE J.* 53(8): 2026–2046.

Janak, S.L., Floudas, C.A., Kallrath, J., and Vormbrock, N. 2006a. Production scheduling of a large-scale industrial batch plant. I. Short-term and medium-term scheduling. *Ind. Eng. Chem. Res.* 45: 8234–8252.

Janak, S.L., Floudas, C.A., Kallrath, J., and Vormbrock, N. 2006b. Production scheduling of a large-scale industrial batch plant. II. Reactive scheduling. *Ind. Eng. Chem. Res.* 45: 8253–8269.

Janak, S.L., Lin, X., and Floudas, C.A. 2004. Enhanced continuous-time unit-specific event-based formulation for short-term scheduling of multipurpose batch processes: Resource constraints and mixed storage policies. *Ind. Eng. Chem. Res.* 43: 2516–2533.

Kondili, E., Pantelides, C.C., and Sargent, R.W.H. 1993. A general algorithm for short-term scheduling of batch operations-I. MILP formulation. *Comput. Chem. Eng.* 17: 211–227.

Mishra, B.V., Mayer, E., Raisch, J., and Kienle, A. 2005. Short-term scheduling of batch processes. A comparative study of different approaches. *Ind. Eng. Chem.* 44: 4022–4034.

Shacham, M. 1982. Equation oriented approach to process flowsheeting. *Comput. Chem. Eng.* 6(2): 79–95.

Shaik, M.A., Janak, S.L., and Floudas, C.A. 2006. Continuous-time models for short-term scheduling of multipurpose batch plants: A comparative study. *Ind. Eng. Chem. Res.* 45: 6190–6209.

Stefansson, H., Shah, N., and Jensson, P. 2006. Multiscale planning and scheduling in the secondary pharmaceutical industry. *AIChE J.* 52(12): 4133–4149.

Stephanopoulos, G., and Reklaitis, G.V. 2011. Process systems engineering: From Solvay to modern bio- and nanotechnology. A history of development, successes and prospects for the future. *Chem. Eng. Sci.* 66: 4272–4306.

Umeda, T., and Nishio, N. 1972. Comparison between sequential and simultaneous approaches in process simulation. *Ind. Eng. Chern. Process Des. Dev.* 11(2): 153–160.

Xue, Y., and Yuan, J. 2008. Scheduling model for a multi-product batch plant using a pre-ordering approach. *Chem. Eng. Technol.* 37(3): 433–439.

Appendix A

Equations of motion in cylindrical and spherical coordinates are given here. The terms involving shear stress can be simplified for incompressible Newtonian fluids (i.e. constant ρ and μ). These are also given for both coordinate systems.

A.1 Equation of Motion in Cylindrical Coordinates

$$\rho\left(\frac{\partial v_r}{\partial t} + v_r \frac{\partial v_r}{\partial r} + \frac{v_\theta}{r}\frac{\partial v_r}{\partial \theta} - \frac{v_\theta^2}{r} + v_z \frac{\partial v_r}{\partial z}\right)$$

$$= -\frac{\partial p}{\partial r} - \left[\frac{1}{r}\frac{\partial}{\partial r}(r\tau_{rr}) + \frac{1}{r}\frac{\partial \tau_{r\theta}}{\partial \theta} - \frac{\tau_{\theta\theta}}{r} + \frac{\partial \tau_{rz}}{\partial z}\right] + \rho g_r \tag{A.1}$$

$$\rho\left(\frac{\partial v_\theta}{\partial t} + v_r \frac{\partial_r v_\theta}{\partial r} + \frac{v_\theta}{r}\frac{\partial v_\theta}{\partial \theta} + \frac{v_r v_\theta}{r} + v_z \frac{\partial v_\theta}{\partial z}\right)$$

$$= -\frac{1}{r}\frac{\partial p}{\partial \theta} - \left[\frac{1}{r^2}\frac{\partial}{\partial r}(r^2\tau_{r\theta}) + \frac{1}{r}\frac{\partial \tau_{\theta\theta}}{\partial \theta} + \frac{\partial \tau_{\theta z}}{\partial z}\right] + \rho g_\theta \tag{A.2}$$

$$\rho\left(\frac{\partial v_z}{\partial t} + v_r \frac{\partial v_z}{\partial r} + \frac{v_\theta}{r}\frac{\partial v_z}{\partial \theta} + v_z \frac{\partial v_z}{\partial z}\right) = -\frac{\partial p}{\partial z} - \left[\frac{1}{r}\frac{\partial}{\partial r}(r\tau_{rz}) + \frac{1}{r}\frac{\partial \tau_{\theta z}}{\partial \theta} + \frac{\partial \tau_{zz}}{\partial z}\right] + \rho g_z \tag{A.3}$$

For incompressible Newtonian fluids (i.e. constant ρ and μ), the terms involving shear stress on the right-hand side of Equations A.1 through A.3 can be replaced by the terms represented by Equations A.4 through A.6, respectively.

$$-\left[\frac{1}{r}\frac{\partial}{\partial r}(r\tau_{rr}) + \frac{1}{r}\frac{\partial \tau_{r\theta}}{\partial \theta} - \frac{\tau_{\theta\theta}}{r} + \frac{\partial \tau_{rz}}{\partial z}\right] = \mu\left[\frac{\partial}{\partial r}\left\{\frac{1}{r}\frac{\partial}{\partial r}(rv_r)\right\} + \frac{1}{r^2}\frac{\partial^2 v_r}{\partial \theta^2} - \frac{2}{r^2}\frac{\partial v_\theta}{\partial \theta} + \frac{\partial^2 v_r}{\partial z^2}\right]$$

$$\tag{A.4}$$

$$-\left[\frac{1}{r^2}\frac{\partial}{\partial r}\left(r^2\tau_{r\theta}\right)+\frac{1}{r}\frac{\partial\tau_{\theta\theta}}{\partial\theta}+\frac{\partial\tau_{\theta z}}{\partial z}\right]=\mu\left[\frac{\partial}{\partial r}\left(\frac{1}{r}\frac{\partial}{\partial r}(rv_\theta)\right)+\frac{1}{r^2}\frac{\partial^2 v_\theta}{\partial\theta^2}+\frac{2}{r^2}\frac{\partial v_r}{\partial\theta}+\frac{\partial^2 v_\theta}{\partial z^2}\right]$$

$$\tag{A.5}$$

$$-\left[\frac{1}{r}\frac{\partial}{\partial r}\left(r\tau_{rz}\right)+\frac{1}{r}\frac{\partial\tau_{\theta z}}{\partial\theta}+\frac{\partial\tau_{zz}}{\partial z}\right]=\mu\left[\frac{1}{r}\frac{\partial}{\partial r}\left(r\frac{\partial v_z}{\partial r}\right)+\frac{1}{r^2}\frac{\partial^2 v_z}{\partial\theta^2}+\frac{\partial^2 v_z}{\partial z^2}\right] \quad (A.6)$$

A.2 Equation of Motion in Spherical Coordinates

$$\rho\left(\frac{\partial v_r}{\partial t}+v_r\frac{\partial v_r}{\partial r}+\frac{v_\theta}{r}\frac{\partial v_r}{\partial\theta}+\frac{v_\phi}{r\sin\theta}\frac{\partial v_r}{\partial\phi}-\frac{v_\theta^2+v_\phi^2}{r}\right)$$

$$=-\frac{\partial p}{\partial r}-\left(\frac{1}{r^2}\frac{\partial}{\partial r}\left(r^2\tau_{rr}\right)-\frac{1}{r\sin\theta}\frac{\partial}{\partial\theta}(\tau_{r\theta}\sin\theta)+\frac{1}{r\sin\theta}\frac{\partial\tau_{r\phi}}{\partial\phi}-\frac{\tau_{\theta\theta}+\tau_{\phi\phi}}{r}\right)+\rho g_r$$

$$\tag{A.7}$$

$$\rho\left(\frac{\partial v_\theta}{\partial t}+v_r\frac{\partial v_\theta}{\partial r}+\frac{v_\theta}{r}\frac{\partial v_\theta}{\partial\theta}+\frac{v_\phi}{r\sin\theta}\frac{\partial v_\theta}{\partial\phi}+\frac{v_r v_\theta}{r}-\frac{v_\phi^2\cot\theta}{r}\right)$$

$$=-\frac{1}{r}\frac{\partial p}{\partial\theta}-\left(\frac{1}{r^2}\frac{\partial}{\partial r}\left(r^2\tau_{r\theta}\right)-\frac{1}{r\sin\theta}\frac{\partial}{\partial\theta}(\tau_{\theta\theta}\sin\theta)+\frac{1}{r\sin\theta}\frac{\partial\tau_{\theta\phi}}{\partial\phi}+\frac{\tau_{r\theta}}{r}-\frac{\cot\theta}{r}\tau_{\phi\phi}\right)+\rho g_\theta$$

$$\tag{A.8}$$

$$\rho\left(\frac{\partial v_\phi}{\partial t}+v_r\frac{\partial v_\phi}{\partial r}+\frac{v_\theta}{r}\frac{\partial v_\phi}{\partial\theta}+\frac{v_\phi}{r\sin\theta}\frac{\partial v_\phi}{\partial\phi}+\frac{v_r v_\phi}{r}+\frac{v_\theta v_\phi}{r}\cot\theta\right)$$

$$=-\frac{1}{r\sin\theta}\frac{\partial p}{\partial\phi}-\left(\frac{1}{r^2}\frac{\partial}{\partial r}\left(r^2\tau_{r\phi}\right)+\frac{1}{r}\frac{\partial\tau_{\theta\phi}}{\partial\theta}+\frac{1}{r\sin\theta}\frac{\partial\tau_{\phi\phi}}{\partial\phi}+\frac{\tau_{r\phi}}{r}+\frac{2\cot\theta}{r}\tau_{\theta\phi}\right)+\rho g_\phi$$

$$\tag{A.9}$$

For incompressible Newtonian fluids, the terms involving shear stress on the right-hand side of Equations A.7 through A.9 can be replaced by the terms represented by Equations A.10 through A.12, respectively.

$$-\left(\frac{1}{r^2}\frac{\partial}{\partial r}\left(r^2\tau_{rr}\right)-\frac{1}{r\sin\theta}\frac{\partial}{\partial\theta}(\tau_{r\theta}\sin\theta)+\frac{1}{r\sin\theta}\frac{\partial\tau_{r\phi}}{\partial\phi}-\frac{\tau_{\theta\theta}+\tau_{\phi\phi}}{r}\right)$$

$$=\mu\frac{2}{r^2}\left[\frac{r^2}{2}\nabla^2 v_r-v_r-\frac{\partial v_\theta}{\partial\theta}-v_\theta\cot\theta-\frac{1}{\sin\theta}\frac{\partial v_\phi}{\partial\phi}\right]$$

$$\tag{A.10}$$

$$-\left(\frac{1}{r^2}\frac{\partial}{\partial r}\left(r^2\tau_{r\theta}\right)-\frac{1}{r\sin\theta}\frac{\partial}{\partial\theta}\left(\tau_{\theta\theta}\sin\theta\right)+\frac{1}{r\sin\theta}\frac{\partial\tau_{\theta\phi}}{\partial\phi}+\frac{\tau_{r\theta}}{r}-\frac{\cot\theta}{r}\tau_{\phi\phi}\right)$$

$$=\mu\frac{2}{r^2}\left[\frac{r^2}{2}\nabla^2 v_\theta+\frac{\partial v_r}{\partial\theta}-\frac{v_\theta}{2\sin^2\theta}-\frac{\cos\theta}{\sin^2\theta}\frac{\partial v_\phi}{\partial\phi}\right]$$

(A.11)

$$-\left(\frac{1}{r^2}\frac{\partial}{\partial r}\left(r^2\tau_{r\phi}\right)+\frac{1}{r}\frac{\partial\tau_{\theta\phi}}{\partial\theta}+\frac{1}{r\sin\theta}\frac{\partial\tau_{\phi\phi}}{\partial\phi}+\frac{\tau_{r\phi}}{r}+\frac{2\cot\theta}{r}\tau_{\theta\phi}\right)$$

$$=\mu\frac{2}{r^2}\left[\frac{r^2}{2}\nabla^2 v_\phi+\frac{1}{\sin\theta}\frac{\partial v_r}{\partial\phi}-\frac{v_\phi}{2\sin^2\theta}+\frac{\cos\theta}{\sin^2\theta}\frac{\partial v_\theta}{\partial\phi}\right]$$

(A.12)

where $\nabla^2=\dfrac{1}{r^2}\dfrac{\partial}{\partial r}\left(r^2\dfrac{\partial}{\partial r}\right)+\dfrac{1}{r^2\sin\theta}\dfrac{\partial}{\partial\theta}\left(\sin\theta\dfrac{\partial}{\partial\theta}\right)+\dfrac{1}{r^2\sin^2\theta}\left(\dfrac{\partial^2}{\partial\phi^2}\right)$

A.3 Equation of Continuity in Cylindrical Coordinates

The equation of continuity for a chemical species in cylindrical coordinates in the absence of a chemical reaction is as follows:

$$\frac{\partial C_A}{\partial t}=-\left[\frac{1}{r}\frac{\partial}{\partial r}\left(rN_{Ar}\right)+\frac{1}{r}\frac{\partial N_{A\theta}}{\partial\theta}+\frac{\partial N_{Az}}{\partial z}\right]$$

(A.13)

For constant density and diffusivity, Equation A.13 will be in the following form:

$$\frac{\partial C_A}{\partial t}+v_r\frac{\partial C_A}{\partial r}+\frac{v_\theta}{r}\frac{\partial C_A}{\partial\theta}+v_z\frac{\partial C_A}{\partial z}=D_{AB}\left[\frac{1}{r}\frac{\partial}{\partial r}\left(r\frac{\partial C_A}{\partial r}\right)+\frac{1}{r^2}\frac{\partial^2 C_A}{\partial\theta^2}+\frac{\partial^2 C_A}{\partial z^2}\right]$$ (A.14)

A.4 Equation of Continuity in Spherical Coordinates

The equation of continuity for a chemical species in spherical coordinates in the absence of a chemical reaction is as follows:

$$\frac{\partial C_A}{\partial t}=-\left[\frac{1}{r^2}\frac{\partial}{\partial r}\left(r^2 N_{Ar}\right)+\frac{1}{r\sin\theta}\frac{\partial}{\partial\theta}\left(N_{A\theta}\sin\theta\right)+\frac{1}{r\sin\theta}\frac{\partial N_{A\phi}}{\partial\phi}\right]$$

(A.15)

For constant density and diffusivity, Equation A.15 will be in the following form:

$$
\frac{\partial C_A}{\partial t} + v_r \frac{\partial C_A}{\partial r} + \frac{v_\theta}{r} \frac{\partial C_A}{\partial \theta} + v_\phi \frac{1}{r \sin \theta} \frac{\partial C_A}{\partial \phi}
$$

$$
= D_{AB} \left[\frac{1}{r^2} \frac{\partial}{\partial r} \left(r^2 \frac{\partial C_A}{\partial r} \right) + \frac{1}{r^2 \sin \theta} \frac{\partial}{\partial \theta} \left(\sin \theta \frac{\partial C_A}{\partial \theta} \right) + \frac{1}{r^2 \sin^2 \theta} \frac{\partial^2 C_A}{\partial \phi^2} \right]
$$

(A.16)

Appendix B

MATLAB® codes for the examples given in various chapters of this book are presented here. These codes work with the 2009 release. In case of difficulty while using the latest release of MATLAB, a few modifications may be required. Comments have been included to make it self-explanatory.

B.1 Solution of Equation 4.34

The code uses the pdepe tool. Further details can be found in the MATLAB manual. The differential equation is written in a separate file using a well-defined manner. The initial and boundary conditions are also written in separate files. The main programme calls all three files and solves for the temperature profile. The results are plotted in a three-dimensional (3D) format.

```
% File to solve the temperature profile using the pdepe tool
m = 0;                        % Specifies Cartesian coordinates
x = linspace(0,0.05,25);     % Generation of grid
t = linspace(0,0.1,25);
% Solving the differential equations in file
   'chap4ex1equation.m'
% Initial conditions (i.e. at t = 0) in file 'chap4ex1_
   initcond.m'
% Boundary conditions in file 'chap4ex1_boundcond.m'
sol = pdepe(m,@chap4ex1equation,@chap4ex1_initcond,@chap4ex1_
   boundcond,x,t);
u = sol(:,:,1);       % Solution
% Creating plot and labels
mesh(x,t,u,'EdgeColor','black')
xlabel('distance from wall,m'); ylabel('time, t');
   zlabel('Wall-Temperature, C')

function [c,f,s] = chap4ex1equation(x,t,u,DuDx)
% c = x; this file contains differential equations
c = 0.05*((2*x/0.05)-(x/0.05)^2);
f = 0.00000003*DuDx;       s = 0;

function [pl,ql,pr,qr] = chap4ex1_boundcond(xl,ul,xr,ur,t)
% The file contains boundary conditions
pl = ul-100; ql = 0; pr = ur-40; qr = 0;
```

```
function u0 = chap4ex1_initcond(x)
% The file contains initial conditions
u0 = 40;
```

B.2 The MATLAB Code for the Numerical Model to Predict Mass Transfer with a Chemical Reaction

```
File name: masstransfer.m
% Mass transfer with a chemical reaction in a shallow
  gas-liquid dispersion
% Problem is defined in the following files: Material.m and
  Column.m
hydrodynamics; % Estimate axial distribution velocity and
   axial gas goldup
% Model assumptions: Gas in plug flow; liquid well mixed
% Note: Rate expression, -rA = k_rxn * CA^rxn_order; B is in
   excess
rxn_order = 1;                 % order of reaction, exponent of
                                 concentration of A
k_rxn = 0.05;                  % rate constant (units consistent
                                 with concentration)
CA_interface = 100;            % Concentration of A at the surface
CA_in = 0;                     % Inlet concentration
F_in = 1000;                   % Flow rate of inlet
Diffusivity = 1.0e-8;
% Other calculated quantities
n_sect = Hs/n_z;               % Number of sections
A_column = pi*DC^2/4;          % Cross-sectional area of column
A_bubble = pi*bdia^2;          % Surface area of a single bubble
V_bubble = (pi/6)*bdia^3;      % Volume of a single bubble
for sect = 1:n_z
   % kL(sect) = mass transfer coefficient in a section,
      Higbie, Eqn.(5.92)
   kL(sect) = 1.13*sqrt(us_out(sect)*Diffusivity/bdia);
   % a(sect) = Interfacial area in a section
   a(sect) = A_bubble*A_column*n_sect*gas_holdup(sect)/V_bubble;
   rA = k_rxn*CA_in^rxn_order*A_column*n_sect; % Reaction rate
   % Reactor model
   CA_out(sect) = (kL(sect)*a(sect)*(CA_interface-CA_in))/
      F_in+CA_in-rA;
   CA_in = CA_out(sect);
end
plot(CA_out)

File name: hydrodynamics.m
% Estimate the axial distribution of velocity gas goldup
```

```
% Problem is defined in the following files: Material.m and
    Column.m
material;        % Defines the properties of air and water
column;          % Defines the geometrical parameters of a
                   bubble column
bubblediaplate;  % Estimates the bubble diameter at the
                   distributor plate

% Data preparation
par1(1) = bdia;par1(2) = 0.01;
par1(3) = den_water; par1(4) = den_air; par1(5) = surften_
    water; par1(6) = vis_water;

n_z = 1000;              % The shallow bed is divided into n_z
                           axial sections
Column_length = 0.001; % Length of the bubble column (m)
s = Column_length/n_z; % s = Length of a section
u0 = 0;                  % Initial bubble velocity
Force_buoy = pi*(bdia^3)*(den_water-den_air)*9.81/6; %
    Buoyancy force

% Inlet conditions, drag force = 0
% u_in = inlet bubble velocity (= 0), m/s
% u_out = bubble velocity leaving the section, m/s
% Total_time = time of stay by the bubble in the section,
  s u_in = u0; Total_time = 0;
% The loop calculates the axial velocity distribution for
  a single bubble
for sect = 1:n_z
  par(2) = u_in;
  Cdrag = dragfun(par1);
  Force_drag = Cdrag*(pi/8)*den_water*(bdia^2)*(u_in^2);
      % Drag force
  acceleration(sect) = (Force_buoy-Force_drag)/
      ((pi*bdia^3/6)*den_water);
  u_out(sect) = sqrt(u_in^2+2*acceleration(sect)*s);
  u_in = u_out(sect);
end
plot(u_out)
% Swarm velocity
u_in = u_out(1);
for sect = 1:n_z
  % Cross-sectional area occupied by all bubbles
  gas_holdup(sect) = Ug/u_out(sect);
  vel_water = u_in+Ug*(gas_holdup/(1-gas_holdup));
  par1(2) = u_out(sect)+vel_water;
  Cdrag = dragfun(par1);
  Force_drag = Cdrag*(pi/8)*den_water*(bdia^2)*(par1(2)^2);
      % Drag force
```

```
    acceleration_s(sect) = (Force_buoy-Force_drag)/
        ((pi*bdia^3/6)*den_water);
    us_out(sect) = sqrt(u_in^2+2*acceleration_s(sect)*s);
    u_in = us_out(sect);
end
plot(us_out)
plot(gas_holdup)

File name: dragfun.m
function Cd = dragfun(par1)
% Calculates the Cd for a bubble using Equation (5.79)
% par1 contains [bubbledia bubblevelocity den_liq den_gas
  surfacetension vis_liq]
% Bubble velocity is an initial guess only
bdia = par1(1); bvel = par1(2);            % Initialisation
rhol = par1(3); rhog = par1(4); surften = par1(5); visl =
  par1(6);
Eo = 9.81*(rhol-rhog)*bdia^2/surften;    % Evotos number
We = rhol*bvel^2*bdia/surften;           % Weber number
Re = bdia*rhol*bvel/visl;                % Reynolds number
% Equation (5.79) used and the value of Cd returned
Cd = max(min(max(16/Re,13.6/Re^0.8),48/Re),min([Eo/3,0.47*Eo^0
    .25*We^0.5,8/3]));

File name: bubblediaplate.m

% This programme computes bubble size formed from the % of
  a given sparger in a column
denL = den_water;denG = den_air;visL = vis_water;sigma =
  surften_water;g = 9.81;rh = dh/2;
a1 = (4*pi/3)*denL*g; a2 = denG/(pi*rh*rh); a3 =
  2*pi*rh*sigma;
a4 = 5*pi*sqrt(denL*visL/2)/((4*pi)^1.5); alpha = 11/16;
  a5 = alpha*denL/(12*pi);
% rd = radius at departure time, m
rd = rh; G = pi*DC*DC*Ug/(4*nh);
var1 = a1*rd^3+(G*G)*a2-a3-a4*(G/rd)^1.5-a5*(G/rd)^2;
rd = rh+0.00001;
var2 = a1*rd^3+(G*G)*a2-a3-a4*(G/rd)^1.5-a5*(G/rd)^2;
while var1*var2>0
    var1 = a1*rd^3+(G*G)*a2-a3-a4*(G/rd)^1.5-a5*(G/rd)^2;
    rd = rd+0.00001;
    var2 = a1*rd^3+(G*G)*a2-a3-a4*(G/rd)^1.5-a5*(G/rd)^2;
end
% Vd = volume of bubble at the time of departure, m³
% Vf = final volume, m³
% bdia = bubble diameter, m
% Calculation of Vd
Vd = 4*pi*rd*rd*rd/3;
Nr = (16*Vd/(11*denL*G*G*rh))*((G*G*denG/(pi*rh*rh))
    -2*pi*sigma*rh);
```

```
Nmu = (16*5*pi*sqrt(Vd)/(11*G*G*rh))*sqrt(2*visL/denL)
     *((G/(4*pi))^1.5)*sqrt(4*pi/3);
% Calculation of y
y = 1; y1 = log(y);
term1 = (4*g*Vd*Vd/(11*rh*G*G))*(y*y-1-2*y1);
term2 = Nr*(y-1-y1);
term3 = Nmu*(2*(sqrt(y)-1)-y1);
term4 = (Vd^(1/3)/(rh*(36*pi)^(1/3)))*(3*(y^(1/3)-1)-y1);
var1 = 2-term1-term2+term3+term4;
y = y+0.01;
y1 = log(y);
term1 = (4*g*Vd*Vd/(11*rh*G*G))*(y*y-1-2*y1);
term2 = Nr*(y-1-y1);
term3 = Nmu*(2*(sqrt(y)-1)-y1);
term4 = (Vd^(1/3)/(36*pi*rh))*(3*(y^(1/3)-1)-y1);
var2 = 2-term1-term2+term3+term4;
counter = 0;
while var1*var2>0
  if var1< = 0
      break;
  end
  y1 = log(y);
  term1 = (4*g*Vd*Vd/(11*rh*G*G))*(y*y-1-2*y1);
  term2 = Nr*(y-1-y1);
  term3 = Nmu*(2*(sqrt(y)-1)-y1);
  term4 = (Vd^(1/3)/(36*pi*rh))*(3*(y^(1/3)-1)-y1);
  var1 = 2-term1-term2+term3+term4;
  y = y+0.001;
  y1 = log(y);
  term1 = (4*g*Vd*Vd/(11*rh*G*G))*(y*y-1-2*y1);
  term2 = Nr*(y-1-y1);
  term3 = Nmu*(2*(sqrt(y)-1)-y1);
  term4 = (Vd^(1/3)/(36*pi*rh))*(3*(y^(1/3)-1)-y1);
  var2 = 2-term1-term2+term3+term4;
  counter = counter+1;
end
% Calculation of db
Vf = y*Vd;
bdia = 2*(3*Vf/(4*pi))^(1/3);

File name: column.m
% Bubble column dimensions
DC = 0.1;     % Column diameter, m
Hs = 0.05;    % Static bed height, m
nh = 10;      % Number of holes in sparger
dh = 0.0005;  % Hole diameter, m
Ug = 0.0001;  % Superficial gas velocity

File name: material.m
% Properties of air and water at 300 K and 1 atm
den_air = 1.1774;          % kg/m³
```

```
vis_air = 0.000018462;        % kg/m.s
cp_air = 1.0057;              % kJ/kg C
k_air = 0.02624;             % W/m C
den_water = 995.8;           % kg/m³
vis_water = 0.00086;         % kg/m.s
cp_water = 4.179;            % kJ/kg C
k_water = 0.614;             % W/m C
surften_water = 0.072;       % N/m²
```

B.3 MATLAB File to Develop an Artificial Neural Network (ANN) Model for Heat Transfer Coefficient in the Bubble Column

```
% This is the main programme to train the data
clear net; clear s; clear T; % Clear the old data
Heattrbc;                    % Data file (s = input, T = target)
net = newfit(s',T,4);        % Creates the architecture (8-4-1)
[net, tr] = train(net,s',T); % Trains with the default
                               training method
outputs = sim(net,s');       % Predicted values using weights
                               and bias
% Various plots shown on screen and eps files
plotperf(tr)
print -deps -tiff myfigure1
plotregression(T,outputs)
print -deps -tiff myfigure2
% Display the weights
net.IW{1}  % w(i,j) between the input and hidden layer
net.LW{2}  % w(i,j) between the hidden layer and output node
net.b{1}   % bias for the hidden layer
net.b{2}   % bias for the output layer
```

Index

CPSIA information can be obtained
at www.ICGtesting.com
Printed in the USA
LVHW01*1836280218
568196LV00014B/287/P